STUDENT'S SOLUTIONS MANUAL PART TWO

WILLIAM ARDIS
Collin County Community College – Preston Ridge

UNIVERSITY CALCULUS ELEMENTS WITH EARLY TRANSCENDENTALS

Joel Hass
University of California, Davis

Maurice D. Weir
Naval Postgraduate School

George B. Thomas
Massachusetts Institute of Technology

PEARSON
Addison Wesley

Boston San Francisco New York
London Toronto Sydney Tokyo Singapore Madrid
Mexico City Munich Paris Cape Town Hong Kong Montreal

Reproduced by Pearson Addison-Wesley from electronic files supplied by the author.

Publishing as Pearson Addison-Wesley, 75 Arlington Street, Boston, MA 02116.

ISBN-13: 978-0-321-55917-3
ISBN-10: 0-321-55917-7

4 5 6 OPM 11 10 09 08

PEARSON
Addison
Wesley

PREFACE TO THE STUDENT

The Student's Solutions Manual contains the solutions to all of the odd-numbered exercise in UNIVERSITY CALCULUS: ELEMENTS by Joel Hass, Maurice Weir and George Thomas, excluding the Computer Algebra System (CAS) exercises. We have worked each solution to ensure that it

- conforms exactly to the methods, procedures and steps presented in the text
- is mathematically correct
- includes all of the steps necessary so you can follow the logical argument and algebra
- includes a graph or figure whenever called for by the exercise, or if needed to help with the explanation
- is formatted in an appropriate style to aid in its understanding

How to use a solution's manual

- solve the assigned problem yourself
- if you get stuck along the way, refer to the solution in the manual as an aid but continue to solve the problem on your own
- if you cannot continue, reread the textbook section, or work through that section in the Student Study Guide, or consult your instructor
- if your answer is correct by your solution procedure seems to differ from the one in the manual, and you are unsure your method is correct, consult your instructor
- if your answer is incorrect and you cannot find your error, consult your instructor

Acknowledgments

Solutions Writers
William Ardis, Collin County Community College-Preston Ridge Campus
Joseph Borzellino, California Polytechnic State University
Linda Buchanan, Howard College
Duane Kouba, University of California-Davis
Tim Mogill
Patricia Nelson, University of Wisconsin-La Crosse

Accuracy Checkers
Karl Kattchee, University of Wisconsin-La Crosse
Debra McGivney
Marie Vanisko, California State University, Stanislaus
Tom Weigleitner, VISTA Information Technologies

Thanks to Elizabeth Bernardi, Rachel Reeve, Christine O'Brien, Sheila Spinney, Elka Block, and Joe Vetere for all their guidance and help at every step.

TABLE OF CONTENTS

CHAPTER 9 VECTORS AND THE GEOMETRY OF SPACE

9.1 THREE-DIMENSIONAL COORDINATE SYSTEMS

1. The line through the point $(2, 3, 0)$ parallel to the z-axis

3. The x-axis

5. The circle $x^2 + y^2 = 4$ in the xy-plane

7. The circle $x^2 + z^2 = 4$ in the xz-plane

9. The circle $y^2 + z^2 = 1$ in the yz-plane

11. The circle $x^2 + y^2 = 16$ in the xy-plane

13. (a) The first quadrant of the xy-plane (b) The fourth quadrant of the xy-plane

15. (a) The solid ball of radius 1 centered at the origin
(b) The exterior of the sphere of radius 1 centered at the origin

17. (a) The closed upper hemisphere of radius 1 centered at the origin
(b) The solid upper hemisphere of radius 1 centered at the origin

19. (a) $x = 3$ (b) $y = -1$ (c) $z = -2$

21. (a) $z = 1$ (b) $x = 3$ (c) $y = -1$

23. (a) $x^2 + (y-2)^2 = 4,\ z = 0$ (b) $(y-2)^2 + z^2 = 4,\ x = 0$ (c) $x^2 + z^2 = 4,\ y = 2$

25. (a) $y = 3,\ z = -1$ (b) $x = 1,\ z = -1$ (c) $x = 1,\ y = 3$

27. $x^2 + y^2 + z^2 = 25,\ z = 3 \Rightarrow x^2 + y^2 = 16$ in the plane $z = 3$

29. $0 \le z \le 1$

31. $z \le 0$

33. (a) $(x-1)^2 + (y-1)^2 + (z-1)^2 < 1$ (b) $(x-1)^2 + (y-1)^2 + (z-1)^2 > 1$

35. $|P_1P_2| = \sqrt{(3-1)^2 + (3-1)^2 + (0-1)^2} = \sqrt{9} = 3$

37. $|P_1P_2| = \sqrt{(4-1)^2 + (-2-4)^2 + (7-5)^2} = \sqrt{49} = 7$

39. $|P_1P_2| = \sqrt{(2-0)^2 + (-2-0)^2 + (-2-0)^2} = \sqrt{3 \cdot 4} = 2\sqrt{3}$

41. center $(-2, 0, 2)$, radius $2\sqrt{2}$

43. center $\left(\sqrt{2}, \sqrt{2}, -\sqrt{2}\right)$, radius $\sqrt{2}$

45. $(x-1)^2+(y-2)^2+(z-3)^2=14$

47. $(x+2)^2+y^2+z^2=3$

49. $x^2+y^2+z^2+4x-4z=0 \Rightarrow (x^2+4x+4)+y^2+(z^2-4z+4)=4+4$
$\Rightarrow (x+2)^2+(y-0)^2+(z-2)^2=\left(\sqrt{8}\right)^2 \Rightarrow$ the center is at $(-2,0,2)$ and the radius is $\sqrt{8}$

51. $2x^2+2y^2+2z^2+x+y+z=9 \Rightarrow x^2+\frac{1}{2}x+y^2+\frac{1}{2}y+z^2+\frac{1}{2}z=\frac{9}{2}$
$\Rightarrow \left(x^2+\frac{1}{2}x+\frac{1}{16}\right)+\left(y^2+\frac{1}{2}y+\frac{1}{16}\right)+\left(z^2+\frac{1}{2}z+\frac{1}{16}\right)=\frac{9}{2}+\frac{3}{16} \Rightarrow \left(x+\frac{1}{4}\right)^2+\left(y+\frac{1}{4}\right)^2+\left(z+\frac{1}{4}\right)^2=\left(\frac{5\sqrt{3}}{4}\right)^2$
$\Rightarrow$ the center is at $\left(-\frac{1}{4},-\frac{1}{4},-\frac{1}{4}\right)$ and the radius is $\frac{5\sqrt{3}}{4}$

53. (a) the distance between (x,y,z) and $(x,0,0)$ is $\sqrt{y^2+z^2}$
(b) the distance between (x,y,z) and $(0,y,0)$ is $\sqrt{x^2+z^2}$
(c) the distance between (x,y,z) and $(0,0,z)$ is $\sqrt{x^2+y^2}$

55. $|AB|=\sqrt{(1-(-1))^2+(-1-2)^2+(3-1)^2}=\sqrt{4+9+4}=\sqrt{17}$
$|BC|=\sqrt{(3-1)^2+(4-(-1))^2+(5-3)^2}=\sqrt{4+25+4}=\sqrt{33}$
$|CA|=\sqrt{(-1-3)^2+(2-4)^2+(1-5)^2}=\sqrt{16+4+16}=\sqrt{36}=6$
Thus the perimeter of triangle ABC is $\sqrt{17}+\sqrt{33}+6$.

9.2 VECTORS

1. (a) $\langle 3(3),3(-2)\rangle=\langle 9,-6\rangle$
(b) $\sqrt{9^2+(-6)^2}=\sqrt{117}=3\sqrt{13}$

3. (a) $\langle 3+(-2),-2+5\rangle=\langle 1,3\rangle$
(b) $\sqrt{1^2+3^2}=\sqrt{10}$

5. (a) $2\mathbf{u}=\langle 2(3),2(-2)\rangle=\langle 6,-4\rangle$
$3\mathbf{v}=\langle 3(-2),3(5)\rangle=\langle -6,15\rangle$
$2\mathbf{u}-3\mathbf{v}=\langle 6-(-6),-4-15\rangle=\langle 12,-19\rangle$
(b) $\sqrt{12^2+(-19)^2}=\sqrt{505}$

7. (a) $\frac{3}{5}\mathbf{u}=\left\langle \frac{3}{5}(3),\frac{3}{5}(-2)\right\rangle=\left\langle \frac{9}{5},-\frac{6}{5}\right\rangle$
$\frac{4}{5}\mathbf{v}=\left\langle \frac{4}{5}(-2),\frac{4}{5}(5)\right\rangle=\left\langle -\frac{8}{5},4\right\rangle$
$\frac{3}{5}\mathbf{u}+\frac{4}{5}\mathbf{v}=\left\langle \frac{9}{5}+\left(-\frac{8}{5}\right),-\frac{6}{5}+4\right\rangle=\left\langle \frac{1}{5},\frac{14}{5}\right\rangle$
(b) $\sqrt{\left(\frac{1}{5}\right)^2+\left(\frac{14}{5}\right)^2}=\frac{\sqrt{197}}{5}$

9. $\langle 2-1,-1-3\rangle=\langle 1,-4\rangle$

11. $\langle 0-2,0-3\rangle=\langle -2,-3\rangle$

13. $\left\langle \cos\frac{2\pi}{3},\sin\frac{2\pi}{3}\right\rangle=\left\langle -\frac{1}{2},\frac{\sqrt{3}}{2}\right\rangle$

15. This is the unit vector which makes an angle of $120^\circ+90^\circ=210^\circ$ with the positive x-axis;
$\langle \cos 210^\circ,\sin 210^\circ\rangle=\left\langle -\frac{\sqrt{3}}{2},-\frac{1}{2}\right\rangle$

17. $\overrightarrow{P_1P_2}=(2-5)\mathbf{i}+(9-7)\mathbf{j}+(-2-(-1))\mathbf{k}=-3\mathbf{i}+2\mathbf{j}-\mathbf{k}$

19. $\overrightarrow{AB}=(-10-(-7))\mathbf{i}+(8-(-8))\mathbf{j}+(1-1)\mathbf{k}=-3\mathbf{i}+16\mathbf{j}$

21. $5\mathbf{u}-\mathbf{v}=5\langle 1,1,-1\rangle-\langle 2,0,3\rangle=\langle 5,5,-5\rangle-\langle 2,0,3\rangle=\langle 5-2,5-0,-5-3\rangle=\langle 3,5,-8\rangle=3\mathbf{i}+5\mathbf{j}-8\mathbf{k}$

23. The vector **v** is horizontal and 1 in. long. The vectors **u** and **w** are $\frac{11}{16}$ in. long. **w** is vertical and **u** makes a 45° angle with the horizontal. All vectors must be drawn to scale.

(a)

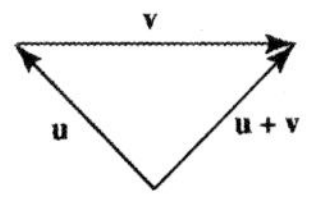

(b)

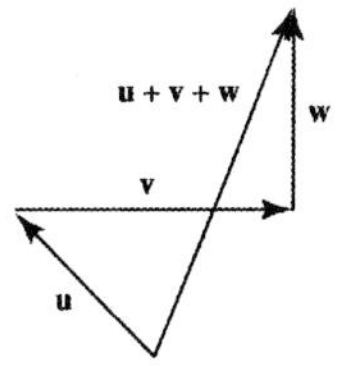

(c)

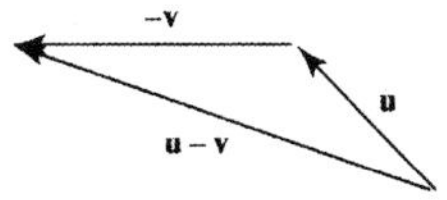

(d)

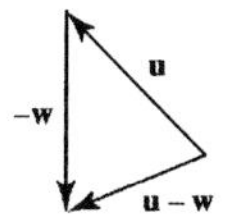

25. length $= |2\mathbf{i}+\mathbf{j}-2\mathbf{k}| = \sqrt{2^2+1^2+(-2)^2} = 3$, the direction is $\frac{2}{3}\mathbf{i}+\frac{1}{3}\mathbf{j}-\frac{2}{3}\mathbf{k} \Rightarrow 2\mathbf{i}+\mathbf{j}-2\mathbf{k} = 3\left(\frac{2}{3}\mathbf{i}+\frac{1}{3}\mathbf{j}-\frac{2}{3}\mathbf{k}\right)$

27. length $= |5\mathbf{k}| = \sqrt{25} = 5$, the direction is $\mathbf{k} \Rightarrow 5\mathbf{k} = 5(\mathbf{k})$

29. length $= \left|\frac{1}{\sqrt{6}}\mathbf{i}-\frac{1}{\sqrt{6}}\mathbf{j}-\frac{1}{\sqrt{6}}\mathbf{k}\right| = \sqrt{3\left(\frac{1}{\sqrt{6}}\right)^2} = \sqrt{\frac{1}{2}}$, the direction is $\frac{1}{\sqrt{3}}\mathbf{i}-\frac{1}{\sqrt{3}}\mathbf{j}-\frac{1}{\sqrt{3}}\mathbf{k}$
$\Rightarrow \frac{1}{\sqrt{6}}\mathbf{i}-\frac{1}{\sqrt{6}}\mathbf{j}-\frac{1}{\sqrt{6}}\mathbf{k} = \sqrt{\frac{1}{2}}\left(\frac{1}{\sqrt{3}}\mathbf{i}-\frac{1}{\sqrt{3}}\mathbf{j}-\frac{1}{\sqrt{3}}\mathbf{k}\right)$

31. (a) $2\mathbf{i}$ (b) $-\sqrt{3}\mathbf{k}$ (c) $\frac{3}{10}\mathbf{j}+\frac{2}{5}\mathbf{k}$ (d) $6\mathbf{i}-2\mathbf{j}+3\mathbf{k}$

33. $|\mathbf{v}| = \sqrt{12^2+5^2} = \sqrt{169} = 13$; $\frac{\mathbf{v}}{|\mathbf{v}|} = \frac{1}{13}\mathbf{v} = \frac{1}{13}(12\mathbf{i}-5\mathbf{k}) \Rightarrow$ the desired vector is $\frac{7}{13}(12\mathbf{i}-5\mathbf{k})$

35. (a) $3\mathbf{i}+4\mathbf{j}-5\mathbf{k} = 5\sqrt{2}\left(\frac{3}{5\sqrt{2}}\mathbf{i}+\frac{4}{5\sqrt{2}}\mathbf{j}-\frac{1}{\sqrt{2}}\mathbf{k}\right) \Rightarrow$ the direction is $\frac{3}{5\sqrt{2}}\mathbf{i}+\frac{4}{5\sqrt{2}}\mathbf{j}-\frac{1}{\sqrt{2}}\mathbf{k}$
(b) the midpoint is $\left(\frac{1}{2}, 3, \frac{5}{2}\right)$

37. (a) $-\mathbf{i}-\mathbf{j}-\mathbf{k} = \sqrt{3}\left(-\frac{1}{\sqrt{3}}\mathbf{i}-\frac{1}{\sqrt{3}}\mathbf{j}-\frac{1}{\sqrt{3}}\mathbf{k}\right) \Rightarrow$ the direction is $-\frac{1}{\sqrt{3}}\mathbf{i}-\frac{1}{\sqrt{3}}\mathbf{j}-\frac{1}{\sqrt{3}}\mathbf{k}$
(b) the midpoint is $\left(\frac{5}{2}, \frac{7}{2}, \frac{9}{2}\right)$

39. $\overrightarrow{AB} = (5-a)\mathbf{i}+(1-b)\mathbf{j}+(3-c)\mathbf{k} = \mathbf{i}+4\mathbf{j}-2\mathbf{k} \Rightarrow 5-a=1, 1-b=4$, and $3-c=-2 \Rightarrow a=4, b=-3$, and $c=5 \Rightarrow$ A is the point $(4,-3,5)$

41. $2\mathbf{i}+\mathbf{j} = a(\mathbf{i}+\mathbf{j})+b(\mathbf{i}-\mathbf{j}) = (a+b)\mathbf{i}+(a-b)\mathbf{j} \Rightarrow a+b=2$ and $a-b=1 \Rightarrow 2a=3 \Rightarrow a=\frac{3}{2}$ and $b=a-1=\frac{1}{2}$

43. 25° west of north is $90°+25° = 115°$ north of east. $800\langle\cos 155°, \sin 115°\rangle \approx \langle -338.095, 725.046\rangle$

45. (a) the midpoint of AB is $M\left(\frac{5}{2},\frac{5}{2},0\right)$ and $\overrightarrow{CM} = \left(\frac{5}{2}-1\right)\mathbf{i}+\left(\frac{5}{2}-1\right)\mathbf{j}+(0-3)\mathbf{k} = \frac{3}{2}\mathbf{i}+\frac{3}{2}\mathbf{j}-3\mathbf{k}$
(b) the desired vector is $\left(\frac{2}{3}\right)\overrightarrow{CM} = \frac{2}{3}\left(\frac{3}{2}\mathbf{i}+\frac{3}{2}\mathbf{j}-3\mathbf{k}\right) = \mathbf{i}+\mathbf{j}-2\mathbf{k}$
(c) the vector whose sum is the vector from the origin to C and the result of part (b) will terminate at the center of mass $\Rightarrow$ the terminal point of $(\mathbf{i}+\mathbf{j}+3\mathbf{k})+(\mathbf{i}+\mathbf{j}-2\mathbf{k}) = 2\mathbf{i}+2\mathbf{j}+\mathbf{k}$ is the point $(2,2,1)$, which is the location of the center of mass

47. Without loss of generality we identify the vertices of the quadrilateral such that $A(0,0,0)$, $B(x_b,0,0)$, $C(x_c,y_c,0)$ and $D(x_d,y_d,z_d)$ $\Rightarrow$ the midpoint of AB is $M_{AB}\left(\frac{x_b}{2},0,0\right)$, the midpoint of BC is $M_{BC}\left(\frac{x_b+x_c}{2},\frac{y_c}{2},0\right)$, the midpoint of CD is $M_{CD}\left(\frac{x_c+x_d}{2},\frac{y_c+y_d}{2},\frac{z_d}{2}\right)$ and the midpoint of AD is $M_{AD}\left(\frac{x_d}{2},\frac{y_d}{2},\frac{z_d}{2}\right)$ $\Rightarrow$ the midpoint of $M_{AB}M_{CD}$ is $\left(\frac{\frac{x_b}{2}+\frac{x_c+x_d}{2}}{2},\frac{y_c+y_d}{4},\frac{z_d}{4}\right)$ which is the same as the midpoint of $M_{AD}M_{BC}=\left(\frac{\frac{x_b+x_c}{2}+\frac{x_d}{2}}{2},\frac{y_c+y_d}{4},\frac{z_d}{4}\right)$.

49. Without loss of generality we can coordinatize the vertices of the triangle such that $A(0,0)$, $B(b,0)$ and $C(x_c,y_c)$ $\Rightarrow$ a is located at $\left(\frac{b+x_c}{2},\frac{y_c}{2}\right)$, b is at $\left(\frac{x_c}{2},\frac{y_c}{2}\right)$ and c is at $\left(\frac{b}{2},0\right)$. Therefore, $\overrightarrow{Aa}=\left(\frac{b}{2}+\frac{x_c}{2}\right)\mathbf{i}+\left(\frac{y_c}{2}\right)\mathbf{j}$, $\overrightarrow{Bb}=\left(\frac{x_c}{2}-b\right)\mathbf{i}+\left(\frac{y_c}{2}\right)\mathbf{j}$, and $\overrightarrow{Cc}=\left(\frac{b}{2}-x_c\right)\mathbf{i}+(-y_c)\mathbf{j}$ $\Rightarrow$ $\overrightarrow{Aa}+\overrightarrow{Bb}+\overrightarrow{Cc}=\mathbf{0}$.

9.3 THE DOT PRODUCT

NOTE: In Exercises 1-8 below we calculate $\text{proj}_\mathbf{v}\,\mathbf{u}$ as the vector $\left(\frac{|\mathbf{u}|\cos\theta}{|\mathbf{v}|}\right)\mathbf{v}$, so the scalar multiplier of $\mathbf{v}$ is the number in column 5 divided by the number in column 2.

	$\mathbf{v}\cdot\mathbf{u}$	$\lvert\mathbf{v}\rvert$	$\lvert\mathbf{u}\rvert$	$\cos\theta$	$\lvert\mathbf{u}\rvert\cos\theta$	$\text{proj}_\mathbf{v}\,\mathbf{u}$
1.	-25	5	5	-1	-5	$-2\mathbf{i}+4\mathbf{j}-\sqrt{5}\mathbf{k}$
3.	25	15	5	$\frac{1}{3}$	$\frac{5}{3}$	$\frac{1}{9}(10\mathbf{i}+11\mathbf{j}-2\mathbf{k})$
5.	2	$\sqrt{34}$	$\sqrt{3}$	$\frac{2}{\sqrt{3}\sqrt{34}}$	$\frac{2}{\sqrt{34}}$	$\frac{1}{17}(5\mathbf{j}-3\mathbf{k})$
7.	$10+\sqrt{17}$	$\sqrt{26}$	$\sqrt{21}$	$\frac{10+\sqrt{17}}{\sqrt{546}}$	$\frac{10+\sqrt{17}}{\sqrt{26}}$	$\frac{10+\sqrt{17}}{\sqrt{26}}(5\mathbf{i}+\mathbf{j})$

9. $\theta=\cos^{-1}\left(\frac{\mathbf{u}\cdot\mathbf{v}}{|\mathbf{u}|\,|\mathbf{v}|}\right)=\cos^{-1}\left(\frac{(2)(1)+(1)(2)+(0)(-1)}{\sqrt{2^2+1^2+0^2}\sqrt{1^2+2^2+(-1)^2}}\right)=\cos^{-1}\left(\frac{4}{\sqrt{5}\sqrt{6}}\right)=\cos^{-1}\left(\frac{4}{\sqrt{30}}\right)\approx 0.75\text{ rad}$

11. $\theta=\cos^{-1}\left(\frac{\mathbf{u}\cdot\mathbf{v}}{|\mathbf{u}|\,|\mathbf{v}|}\right)=\cos^{-1}\left(\frac{\left(\sqrt{3}\right)\left(\sqrt{3}\right)+(-7)(1)+(0)(-2)}{\sqrt{\left(\sqrt{3}\right)^2+(-7)^2+0^2}\sqrt{\left(\sqrt{3}\right)^2+(1)^2+(-2)^2}}\right)=\cos^{-1}\left(\frac{3-7}{\sqrt{52}\sqrt{8}}\right)$
$=\cos^{-1}\left(\frac{-1}{\sqrt{26}}\right)\approx 1.77\text{ rad}$

13. $\overrightarrow{AB}=\langle 3,1\rangle$, $\overrightarrow{BC}=\langle -1,-3\rangle$, and $\overrightarrow{AC}=\langle 2,-2\rangle$. $\overrightarrow{BA}=\langle -3,-1\rangle$, $\overrightarrow{CB}=\langle 1,3\rangle$, $\overrightarrow{CA}=\langle -2,2\rangle$.
$\left|\overrightarrow{AB}\right|=\left|\overrightarrow{BA}\right|=\sqrt{10}$, $\left|\overrightarrow{BC}\right|=\left|\overrightarrow{CB}\right|=\sqrt{10}$, $\left|\overrightarrow{AC}\right|=\left|\overrightarrow{CA}\right|=2\sqrt{2}$,

Angle at A $=\cos^{-1}\left(\frac{\overrightarrow{AB}\cdot\overrightarrow{AC}}{\left|\overrightarrow{AB}\right|\left|\overrightarrow{AC}\right|}\right)=\cos^{-1}\left(\frac{3(2)+1(-2)}{\left(\sqrt{10}\right)\left(2\sqrt{2}\right)}\right)=\cos^{-1}\left(\frac{1}{\sqrt{5}}\right)\approx 63.435°$

Angle at B $=\cos^{-1}\left(\frac{\overrightarrow{BC}\cdot\overrightarrow{BA}}{\left|\overrightarrow{BC}\right|\left|\overrightarrow{BA}\right|}\right)=\cos^{-1}\left(\frac{(-1)(-3)+(-3)(-1)}{\left(\sqrt{10}\right)\left(\sqrt{10}\right)}\right)=\cos^{-1}\left(\frac{3}{5}\right)\approx 53.130°$, and

Angle at C $=\cos^{-1}\left(\frac{\overrightarrow{CB}\cdot\overrightarrow{CA}}{\left|\overrightarrow{CB}\right|\left|\overrightarrow{CA}\right|}\right)=\cos^{-1}\left(\frac{1(-2)+3(2)}{\left(\sqrt{10}\right)\left(2\sqrt{2}\right)}\right)=\cos^{-1}\left(\frac{1}{\sqrt{5}}\right)\approx 63.435°$

15. (a) $\cos\alpha = \frac{\mathbf{i}\cdot\mathbf{v}}{|\mathbf{i}|\,|\mathbf{v}|} = \frac{a}{|\mathbf{v}|}$, $\cos\beta = \frac{\mathbf{j}\cdot\mathbf{v}}{|\mathbf{j}|\,|\mathbf{v}|} = \frac{b}{|\mathbf{v}|}$, $\cos\gamma = \frac{\mathbf{k}\cdot\mathbf{v}}{|\mathbf{k}|\,|\mathbf{v}|} = \frac{c}{|\mathbf{v}|}$ and

$\cos^2\alpha + \cos^2\beta + \cos^2\gamma = \left(\frac{a}{|\mathbf{v}|}\right)^2 + \left(\frac{b}{|\mathbf{v}|}\right)^2 + \left(\frac{c}{|\mathbf{v}|}\right)^2 = \frac{a^2+b^2+c^2}{|\mathbf{v}|\,|\mathbf{v}|} = \frac{|\mathbf{v}|\,|\mathbf{v}|}{|\mathbf{v}|\,|\mathbf{v}|} = 1$

(b) $|\mathbf{v}| = 1 \Rightarrow \cos\alpha = \frac{a}{|\mathbf{v}|} = a$, $\cos\beta = \frac{b}{|\mathbf{v}|} = b$ and $\cos\gamma = \frac{c}{|\mathbf{v}|} = c$ are the direction cosines of $\mathbf{v}$

17. Let $\mathbf{u}$ and $\mathbf{v}$ be the sides of a rhombus $\Rightarrow$ the diagonals are $\mathbf{d}_1 = \mathbf{u} + \mathbf{v}$ and $\mathbf{d}_2 = -\mathbf{u} + \mathbf{v}$ $\Rightarrow \mathbf{d}_1 \cdot \mathbf{d}_2 = (\mathbf{u}+\mathbf{v})\cdot(-\mathbf{u}+\mathbf{v}) = -\mathbf{u}\cdot\mathbf{u} + \mathbf{u}\cdot\mathbf{v} - \mathbf{v}\cdot\mathbf{u} + \mathbf{v}\cdot\mathbf{v} = |\mathbf{v}|^2 - |\mathbf{u}|^2 = 0$ because $|\mathbf{u}| = |\mathbf{v}|$, since a rhombus has equal sides.

19. Clearly the diagonals of a rectangle are equal in length. What is not as obvious is the statement that equal diagonals happen only in a rectangle. We show this is true by letting the adjacent sides of a parallelogram be the vectors $(v_1\mathbf{i} + v_2\mathbf{j})$ and $(u_1\mathbf{i} + u_2\mathbf{j})$. The equal diagonals of the parallelogram are $\mathbf{d}_1 = (v_1\mathbf{i} + v_2\mathbf{j}) + (u_1\mathbf{i} + u_2\mathbf{j})$ and $\mathbf{d}_2 = (v_1\mathbf{i} + v_2\mathbf{j}) - (u_1\mathbf{i} + u_2\mathbf{j})$. Hence $|\mathbf{d}_1| = |\mathbf{d}_2| = |(v_1\mathbf{i} + v_2\mathbf{j}) + (u_1\mathbf{i} + u_2\mathbf{j})| = |(v_1\mathbf{i} + v_2\mathbf{j}) - (u_1\mathbf{i} + u_2\mathbf{j})|$ $\Rightarrow |(v_1+u_1)\mathbf{i} + (v_2+u_2)\mathbf{j}| = |(v_1-u_1)\mathbf{i} + (v_2-u_2)\mathbf{j}| \Rightarrow \sqrt{(v_1+u_1)^2 + (v_2+u_2)^2} = \sqrt{(v_1-u_1)^2 + (v_2-u_2)^2}$ $\Rightarrow v_1^2 + 2v_1u_1 + u_1^2 + v_2^2 + 2v_2u_2 + u_2^2 = v_1^2 - 2v_1u_1 + u_1^2 + v_2^2 - 2v_2u_2 + u_2^2 \Rightarrow 2(v_1u_1 + v_2u_2)$ $= -2(v_1u_1 + v_2u_2) \Rightarrow v_1u_1 + v_2u_2 = 0 \Rightarrow (v_1\mathbf{i} + v_2\mathbf{j})\cdot(u_1\mathbf{i} + u_2\mathbf{j}) = 0 \Rightarrow$ the vectors $(v_1\mathbf{i} + v_2\mathbf{j})$ and $(u_1\mathbf{i} + u_2\mathbf{j})$ are perpendicular and the parallelogram must be a rectangle.

21. horizontal component: $1200\cos(8^\circ) \approx 1188$ ft/s; vertical component: $1200\sin(8^\circ) \approx 167$ ft/s

23. (a) Since $|\cos\theta| \le 1$, we have $|\mathbf{u}\cdot\mathbf{v}| = |\mathbf{u}|\,|\mathbf{v}|\,|\cos\theta| \le |\mathbf{u}|\,|\mathbf{v}|\,(1) = |\mathbf{u}|\,|\mathbf{v}|$.

(b) We have equality precisely when $|\cos\theta| = 1$ or when one or both of $\mathbf{u}$ and $\mathbf{v}$ is $\mathbf{0}$. In the case of nonzero vectors, we have equality when $\theta = 0$ or π, i.e., when the vectors are parallel.

25. $P(x_1, y_1) = P\left(x_1, \frac{c}{b} - \frac{a}{b}x_1\right)$ and $Q(x_2, y_2) = Q\left(x_2, \frac{c}{b} - \frac{a}{b}x_2\right)$ are any two points P and Q on the line with $b \ne 0$ $\Rightarrow \overrightarrow{PQ} = (x_2 - x_1)\mathbf{i} + \frac{a}{b}(x_1 - x_2)\mathbf{j} \Rightarrow \overrightarrow{PQ}\cdot\mathbf{v} = \left[(x_2 - x_1)\mathbf{i} + \frac{a}{b}(x_1 - x_2)\mathbf{j}\right]\cdot(a\mathbf{i} + b\mathbf{j}) = a(x_2 - x_1) + b\left(\frac{a}{b}\right)(x_1 - x_2)$ $= 0 \Rightarrow \mathbf{v}$ is perpendicular to $\overrightarrow{PQ}$ for $b \ne 0$. If $b = 0$, then $\mathbf{v} = a\mathbf{i}$ is perpendicular to the vertical line $ax = c$. Alternatively, the slope of $\mathbf{v}$ is $\frac{b}{a}$ and the slope of the line $ax + by = c$ is $-\frac{a}{b}$, so the slopes are negative reciprocals $\Rightarrow$ the vector $\mathbf{v}$ and the line are perpendicular.

27. $\mathbf{v} = \mathbf{i} + 2\mathbf{j}$ is perpendicular to the line $x + 2y = c$;
$P(2, 1)$ on the line $\Rightarrow 2 + 2 = c \Rightarrow x + 2y = 4$

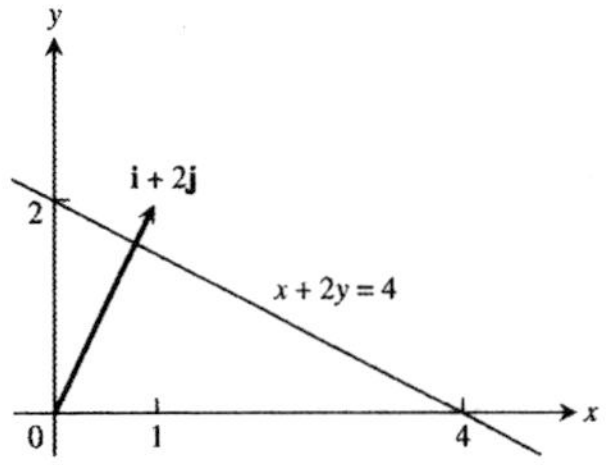

29. $\mathbf{v} = -2\mathbf{i} + \mathbf{j}$ is perpendicular to the line $-2x + y = c$;
$P(-2, -7)$ on the line $\Rightarrow (-2)(-2) - 7 = c$
$\Rightarrow -2x + y = -3$

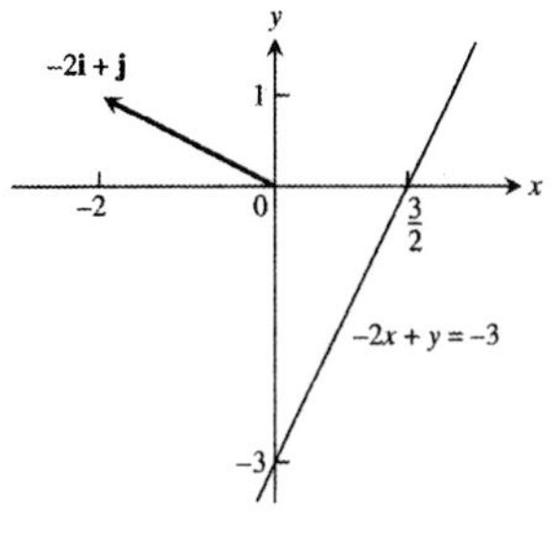

31. $\mathbf{v} = \mathbf{i} - \mathbf{j}$ is parallel to the line $-x - y = c$;
$P(-2, 1)$ on the line $\Rightarrow -(-2) - 1 = c \Rightarrow -x - y = 1$
or $x + y = -1$.

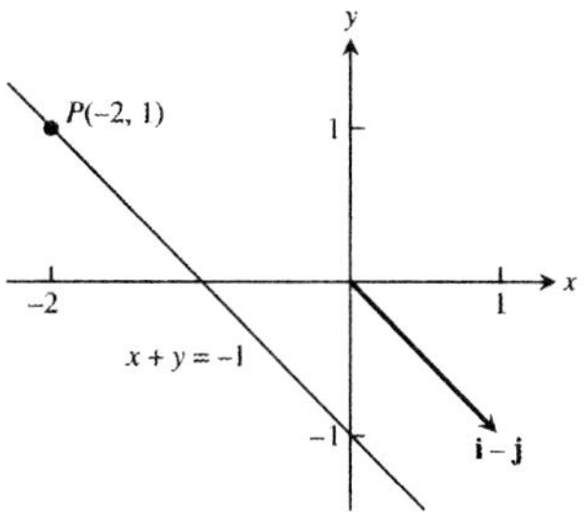

33. $\mathbf{v} = -\mathbf{i} - 2\mathbf{j}$ is parallel to the line $-2x + y = c$;
$P(1, 2)$ on the line $\Rightarrow -2(1) + 2 = c \Rightarrow -2x - y = 0$
or $2x - y = 0$.

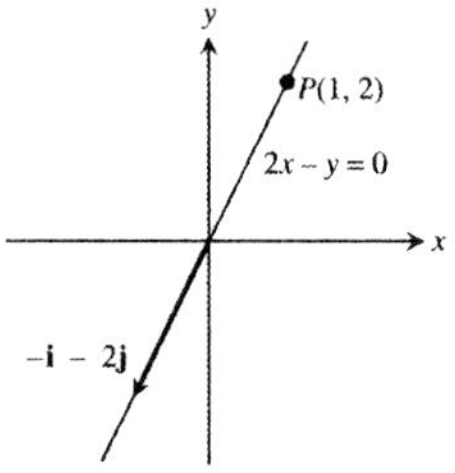

35. $P(0, 0)$, $Q(1, 1)$ and $\mathbf{F} = 5\mathbf{j} \Rightarrow \overrightarrow{PQ} = \mathbf{i} + \mathbf{j}$ and $\mathbf{W} = \mathbf{F} \cdot \overrightarrow{PQ} = (5\mathbf{j}) \cdot (\mathbf{i} + \mathbf{j}) = 5\ \mathrm{N \cdot m} = 5\ \mathrm{J}$

In Exercises 37-40 we use the fact that $\mathbf{n} = a\mathbf{i} + b\mathbf{j}$ is normal to the line $ax + by = c$.

37. $\mathbf{n}_1 = 3\mathbf{i} + \mathbf{j}$ and $\mathbf{n}_2 = 2\mathbf{i} - \mathbf{j} \Rightarrow \theta = \cos^{-1}\left(\frac{\mathbf{n}_1 \cdot \mathbf{n}_2}{|\mathbf{n}_1|\,|\mathbf{n}_2|}\right) = \cos^{-1}\left(\frac{6-1}{\sqrt{10}\,\sqrt{5}}\right) = \cos^{-1}\left(\frac{1}{\sqrt{2}}\right) = \frac{\pi}{4}$

39. $\mathbf{n}_1 = \sqrt{3}\mathbf{i} - \mathbf{j}$ and $\mathbf{n}_2 = \mathbf{i} - \sqrt{3}\mathbf{j} \Rightarrow \theta = \cos^{-1}\left(\frac{\mathbf{n}_1 \cdot \mathbf{n}_2}{|\mathbf{n}_1|\,|\mathbf{n}_2|}\right) = \cos^{-1}\left(\frac{\sqrt{3}+\sqrt{3}}{\sqrt{4}\,\sqrt{4}}\right) = \cos^{-1}\left(\frac{\sqrt{3}}{2}\right) = \frac{\pi}{6}$

9.4 THE CROSS PRODUCT

1. $\mathbf{u} \times \mathbf{v} = \begin{vmatrix} \mathbf{i} & \mathbf{j} & \mathbf{k} \\ 2 & -2 & -1 \\ 1 & 0 & -1 \end{vmatrix} = 3\left(\frac{2}{3}\mathbf{i} + \frac{1}{3}\mathbf{j} + \frac{2}{3}\mathbf{k}\right) \Rightarrow$ length $= 3$ and the direction is $\frac{2}{3}\mathbf{i} + \frac{1}{3}\mathbf{j} + \frac{2}{3}\mathbf{k}$;
$\mathbf{v} \times \mathbf{u} = -(\mathbf{u} \times \mathbf{v}) = -3\left(\frac{2}{3}\mathbf{i} + \frac{1}{3}\mathbf{j} + \frac{2}{3}\mathbf{k}\right) \Rightarrow$ length $= 3$ and the direction is $-\frac{2}{3}\mathbf{i} - \frac{1}{3}\mathbf{j} - \frac{2}{3}\mathbf{k}$

3. $\mathbf{u} \times \mathbf{v} = \begin{vmatrix} \mathbf{i} & \mathbf{j} & \mathbf{k} \\ 2 & -2 & 4 \\ -1 & 1 & -2 \end{vmatrix} = \mathbf{0} \Rightarrow$ length $= 0$ and has no direction
$\mathbf{v} \times \mathbf{u} = -(\mathbf{u} \times \mathbf{v}) = \mathbf{0} \Rightarrow$ length $= 0$ and has no direction

5. $\mathbf{u} \times \mathbf{v} = \begin{vmatrix} \mathbf{i} & \mathbf{j} & \mathbf{k} \\ 2 & 0 & 0 \\ 0 & -3 & 0 \end{vmatrix} = -6(\mathbf{k}) \Rightarrow$ length $= 6$ and the direction is $-\mathbf{k}$
$\mathbf{v} \times \mathbf{u} = -(\mathbf{u} \times \mathbf{v}) = 6(\mathbf{k}) \Rightarrow$ length $= 6$ and the direction is $\mathbf{k}$

7. $\mathbf{u} \times \mathbf{v} = \begin{vmatrix} \mathbf{i} & \mathbf{j} & \mathbf{k} \\ -8 & -2 & -4 \\ 2 & 2 & 1 \end{vmatrix} = 6\mathbf{i} - 12\mathbf{k} \Rightarrow$ length $= 6\sqrt{5}$ and the direction is $\frac{1}{\sqrt{5}}\mathbf{i} - \frac{2}{\sqrt{5}}\mathbf{k}$
$\mathbf{v} \times \mathbf{u} = -(\mathbf{u} \times \mathbf{v}) = -(6\mathbf{i} - 12\mathbf{k}) \Rightarrow$ length $= 6\sqrt{5}$ and the direction is $-\frac{1}{\sqrt{5}}\mathbf{i} + \frac{2}{\sqrt{5}}\mathbf{k}$

9. $\mathbf{u}\times\mathbf{v}=\begin{vmatrix}\mathbf{i}&\mathbf{j}&\mathbf{k}\\1&0&0\\0&1&0\end{vmatrix}=\mathbf{k}$

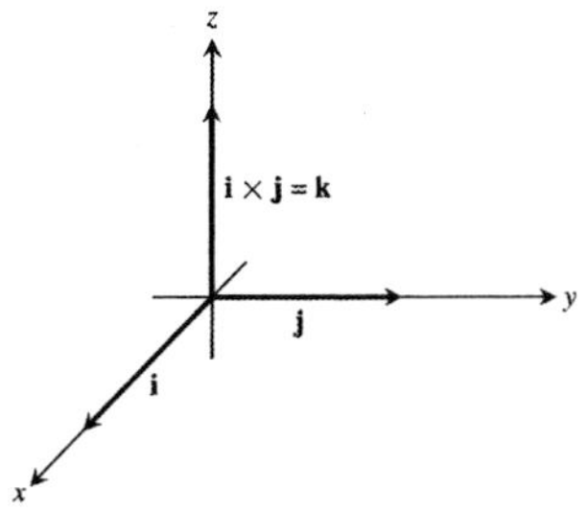

11. $\mathbf{u}\times\mathbf{v}=\begin{vmatrix}\mathbf{i}&\mathbf{j}&\mathbf{k}\\1&0&-1\\0&1&1\end{vmatrix}=\mathbf{i}-\mathbf{j}+\mathbf{k}$

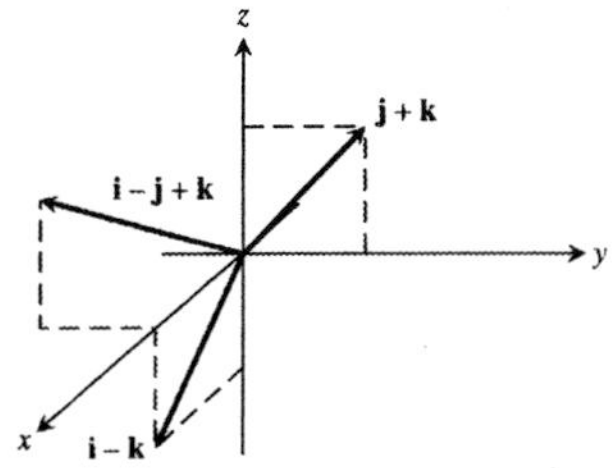

13. $\mathbf{u}\times\mathbf{v}=\begin{vmatrix}\mathbf{i}&\mathbf{j}&\mathbf{k}\\1&1&0\\1&-1&0\end{vmatrix}=-2\mathbf{k}$

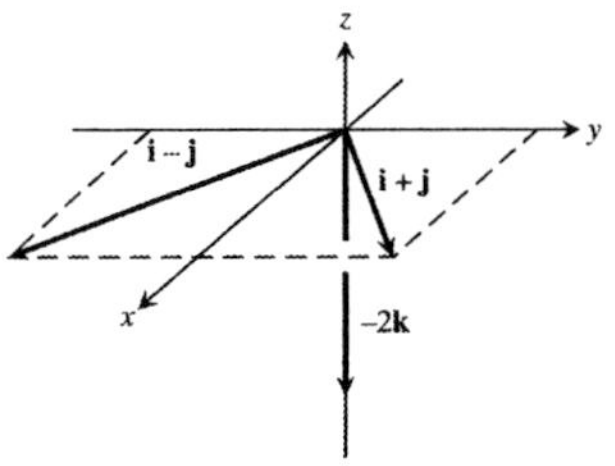

15. (a) $\overrightarrow{PQ}\times\overrightarrow{PR}=\begin{vmatrix}\mathbf{i}&\mathbf{j}&\mathbf{k}\\1&1&-3\\-1&3&-1\end{vmatrix}=8\mathbf{i}+4\mathbf{j}+4\mathbf{k}\Rightarrow \text{Area}=\frac{1}{2}\left|\overrightarrow{PQ}\times\overrightarrow{PR}\right|=\frac{1}{2}\sqrt{64+16+16}=2\sqrt{6}$

(b) $\mathbf{u}=\pm\frac{\overrightarrow{PQ}\times\overrightarrow{PR}}{\left|\overrightarrow{PQ}\times\overrightarrow{PR}\right|}=\pm\frac{1}{\sqrt{6}}(2\mathbf{i}+\mathbf{j}+\mathbf{k})$

17. (a) $\overrightarrow{PQ}\times\overrightarrow{PR}=\begin{vmatrix}\mathbf{i}&\mathbf{j}&\mathbf{k}\\1&1&1\\1&1&0\end{vmatrix}=-\mathbf{i}+\mathbf{j}\Rightarrow \text{Area}=\frac{1}{2}\left|\overrightarrow{PQ}\times\overrightarrow{PR}\right|=\frac{1}{2}\sqrt{1+1}=\frac{\sqrt{2}}{2}$

(b) $\mathbf{u}=\pm\frac{\overrightarrow{PQ}\times\overrightarrow{PR}}{\left|\overrightarrow{PQ}\times\overrightarrow{PR}\right|}=\pm\frac{1}{\sqrt{2}}(-\mathbf{i}+\mathbf{j})=\pm\frac{1}{\sqrt{2}}(\mathbf{i}-\mathbf{j})$

19. If $\mathbf{u}=a_1\mathbf{i}+a_2\mathbf{j}+a_3\mathbf{k}$, $\mathbf{v}=b_1\mathbf{i}+b_2\mathbf{j}+b_3\mathbf{k}$, and $\mathbf{w}=c_1\mathbf{i}+c_2\mathbf{j}+c_3\mathbf{k}$, then $\mathbf{u}\cdot(\mathbf{v}\times\mathbf{w})=\begin{vmatrix}a_1&a_2&a_3\\b_1&b_2&b_3\\c_1&c_2&c_3\end{vmatrix}$,

$\mathbf{v}\cdot(\mathbf{w}\times\mathbf{u})=\begin{vmatrix}b_1&b_2&b_3\\c_1&c_2&c_3\\a_1&a_2&a_3\end{vmatrix}$ and $\mathbf{w}\cdot(\mathbf{u}\times\mathbf{v})=\begin{vmatrix}c_1&c_2&c_3\\a_1&a_2&a_3\\b_1&b_2&b_3\end{vmatrix}$ which all have the same value, since the interchanging of two pair of rows in a determinant does not change its value $\Rightarrow$ the volume is

$|(\mathbf{u}\times\mathbf{v})\cdot\mathbf{w}|=\text{abs}\begin{vmatrix}2&0&0\\0&2&0\\0&0&2\end{vmatrix}=8$

21. $|(\mathbf{u}\times\mathbf{v})\cdot\mathbf{w}|=\text{abs}\begin{vmatrix}2&1&0\\2&-1&1\\1&0&2\end{vmatrix}=|-7|=7$ (for details about verification, see Exercise 19)

23. (a) $\mathbf{u}\cdot\mathbf{v} = -6$, $\mathbf{u}\cdot\mathbf{w} = -81$, $\mathbf{v}\cdot\mathbf{w} = 18 \Rightarrow$ none

(b) $\mathbf{u}\times\mathbf{v} = \begin{vmatrix} \mathbf{i} & \mathbf{j} & \mathbf{k} \\ 5 & -1 & 1 \\ 0 & 1 & -5 \end{vmatrix} \neq \mathbf{0}$, $\mathbf{u}\times\mathbf{w} = \begin{vmatrix} \mathbf{i} & \mathbf{j} & \mathbf{k} \\ 5 & -1 & 1 \\ -15 & 3 & -3 \end{vmatrix} = \mathbf{0}$, $\mathbf{v}\times\mathbf{w} = \begin{vmatrix} \mathbf{i} & \mathbf{j} & \mathbf{k} \\ 0 & 1 & -5 \\ -15 & 3 & -3 \end{vmatrix} \neq \mathbf{0}$

$\Rightarrow$ **u** and **w** are parallel

25. $\left|\overrightarrow{PQ}\times\mathbf{F}\right| = \left|\overrightarrow{PQ}\right| |\mathbf{F}| \sin(60^\circ) = \frac{2}{3}\cdot 30\cdot\frac{\sqrt{3}}{2}$ ft · lb $= 10\sqrt{3}$ ft · lb

27. (a) true, $|\mathbf{u}| = \sqrt{a_1^2 + a_2^2 + a_3^2} = \sqrt{\mathbf{u}\cdot\mathbf{u}}$

(b) not always true, $\mathbf{u}\cdot\mathbf{u} = |\mathbf{u}|^2$

(c) true, $\mathbf{u}\times\mathbf{0} = \begin{vmatrix} \mathbf{i} & \mathbf{j} & \mathbf{k} \\ a_1 & a_2 & a_3 \\ 0 & 0 & 0 \end{vmatrix} = 0\mathbf{i} + 0\mathbf{j} + 0\mathbf{k} = \mathbf{0}$ and $\mathbf{0}\times\mathbf{u} = \begin{vmatrix} \mathbf{i} & \mathbf{j} & \mathbf{k} \\ 0 & 0 & 0 \\ a_1 & a_2 & a_3 \end{vmatrix} = 0\mathbf{i} + 0\mathbf{j} + 0\mathbf{k} = \mathbf{0}$

(d) true, $\mathbf{u}\times(-\mathbf{u}) = \begin{vmatrix} \mathbf{i} & \mathbf{j} & \mathbf{k} \\ a_1 & a_2 & a_3 \\ -a_1 & -a_2 & -a_3 \end{vmatrix} = (-a_2a_3 + a_2a_3)\mathbf{i} - (-a_1a_3 + a_1a_3)\mathbf{j} + (-a_1a_2 + a_1a_2)\mathbf{k} = \mathbf{0}$

(e) not always true, $\mathbf{i}\times\mathbf{j} = \mathbf{k} \neq -\mathbf{k} = \mathbf{j}\times\mathbf{i}$ for example

(f) true, distributive property of the cross product

(g) true, $(\mathbf{u}\times\mathbf{v})\cdot\mathbf{v} = \mathbf{u}\cdot(\mathbf{v}\times\mathbf{v}) = \mathbf{u}\cdot\mathbf{0} = 0$

(h) true, the volume of a parallelpiped with **u**, **v**, and **w** along the three edges is $(\mathbf{u}\times\mathbf{v})\cdot\mathbf{w} = (\mathbf{v}\times\mathbf{w})\cdot\mathbf{u} = \mathbf{u}\cdot(\mathbf{v}\times\mathbf{w})$, since the dot product is commutative.

29. (a) $\text{proj}_\mathbf{v}\,\mathbf{u} = \left(\frac{\mathbf{u}\cdot\mathbf{v}}{|\mathbf{v}||\mathbf{v}|}\right)\mathbf{v}$ (b) $\pm(\mathbf{u}\times\mathbf{v})$ (c) $\pm((\mathbf{u}\times\mathbf{v})\times\mathbf{w})$ (d) $|(\mathbf{u}\times\mathbf{v})\cdot\mathbf{w}|$

(e) $(\mathbf{u}\times\mathbf{v})\times(\mathbf{u}\times\mathbf{w})$ (f) $|\mathbf{u}|\,\frac{\mathbf{v}}{|\mathbf{v}|}$

31. (a) yes, $\mathbf{u}\times\mathbf{v}$ and **w** are both vectors (b) no, **u** is a vector but $\mathbf{v}\cdot\mathbf{w}$ is a scalar

(c) yes, **u** and $\mathbf{u}\times\mathbf{w}$ are both vectors (d) no, **u** is a vector but $\mathbf{v}\cdot\mathbf{w}$ is a scalar

33. No, **v** need not equal **w**. For example, $\mathbf{i}+\mathbf{j} \neq -\mathbf{i}+\mathbf{j}$, but $\mathbf{i}\times(\mathbf{i}+\mathbf{j}) = \mathbf{i}\times\mathbf{i} + \mathbf{i}\times\mathbf{j} = \mathbf{0} + \mathbf{k} = \mathbf{k}$ and $\mathbf{i}\times(-\mathbf{i}+\mathbf{j}) = -\mathbf{i}\times\mathbf{i} + \mathbf{i}\times\mathbf{j} = \mathbf{0} + \mathbf{k} = \mathbf{k}$.

35. $\overrightarrow{AB} = -\mathbf{i}+\mathbf{j}$ and $\overrightarrow{AD} = -\mathbf{i}-\mathbf{j} \Rightarrow \overrightarrow{AB}\times\overrightarrow{AD} = \begin{vmatrix} \mathbf{i} & \mathbf{j} & \mathbf{k} \\ -1 & 1 & 0 \\ -1 & -1 & 0 \end{vmatrix} = 2\mathbf{k} \Rightarrow$ area $= \left|\overrightarrow{AB}\times\overrightarrow{AD}\right| = 2$

37. $\overrightarrow{AB} = 3\mathbf{i}-2\mathbf{j}$ and $\overrightarrow{AD} = 5\mathbf{i}+\mathbf{j} \Rightarrow \overrightarrow{AB}\times\overrightarrow{AD} = \begin{vmatrix} \mathbf{i} & \mathbf{j} & \mathbf{k} \\ 3 & -2 & 0 \\ 5 & 1 & 0 \end{vmatrix} = 13\mathbf{k} \Rightarrow$ area $= \left|\overrightarrow{AB}\times\overrightarrow{AD}\right| = 13$

39. $\overrightarrow{AB} = -2\mathbf{i}+3\mathbf{j}$ and $\overrightarrow{AC} = 3\mathbf{i}+\mathbf{j} \Rightarrow \overrightarrow{AB}\times\overrightarrow{AC} = \begin{vmatrix} \mathbf{i} & \mathbf{j} & \mathbf{k} \\ -2 & 3 & 0 \\ 3 & 1 & 0 \end{vmatrix} = -11\mathbf{k} \Rightarrow$ area $= \frac{1}{2}\left|\overrightarrow{AB}\times\overrightarrow{AC}\right| = \frac{11}{2}$

41. $\overrightarrow{AB} = 6\mathbf{i}-5\mathbf{j}$ and $\overrightarrow{AC} = 11\mathbf{i}-5\mathbf{j} \Rightarrow \overrightarrow{AB}\times\overrightarrow{AC} = \begin{vmatrix} \mathbf{i} & \mathbf{j} & \mathbf{k} \\ 6 & -5 & 0 \\ 11 & -5 & 0 \end{vmatrix} = 25\mathbf{k} \Rightarrow$ area $= \frac{1}{2}\left|\overrightarrow{AB}\times\overrightarrow{AC}\right| = \frac{25}{2}$

9.5 LINES AND PLANES IN SPACE

1. The direction $\mathbf{i} + \mathbf{j} + \mathbf{k}$ and $P(3, -4, -1) \Rightarrow x = 3 + t, y = -4 + t, z = -1 + t$

3. The direction $\overrightarrow{PQ} = 5\mathbf{i} + 5\mathbf{j} - 5\mathbf{k}$ and $P(-2, 0, 3) \Rightarrow x = -2 + 5t, y = 5t, z = 3 - 5t$

5. The direction $2\mathbf{j} + \mathbf{k}$ and $P(0, 0, 0) \Rightarrow x = 0, y = 2t, z = t$

7. The direction $\mathbf{k}$ and $P(1, 1, 1) \Rightarrow x = 1, y = 1, z = 1 + t$

9. The direction $\mathbf{i} + 2\mathbf{j} + 2\mathbf{k}$ and $P(0, -7, 0) \Rightarrow x = t, y = -7 + 2t, z = 2t$

11. The direction $\mathbf{i}$ and $P(0, 0, 0) \Rightarrow x = t, y = 0, z = 0$

13. The direction $\overrightarrow{PQ} = \mathbf{i} + \mathbf{j} + \frac{3}{2}\mathbf{k}$ and $P(0, 0, 0) \Rightarrow x = t$, $y = t, z = \frac{3}{2}t$, where $0 \le t \le 1$

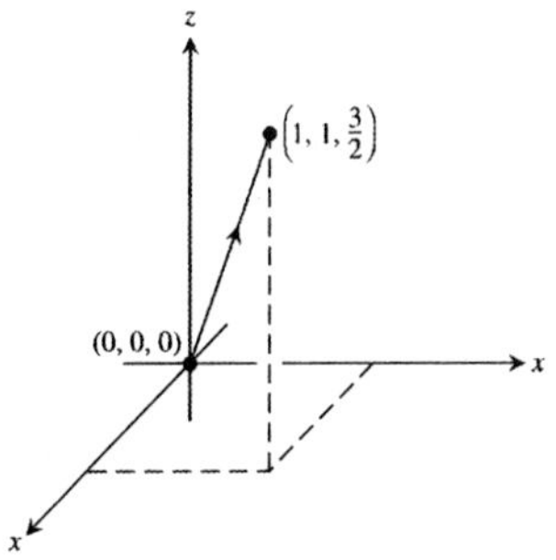

15. The direction $\overrightarrow{PQ} = \mathbf{j}$ and $P(1, 1, 0) \Rightarrow x = 1, y = 1 + t$, $z = 0$, where $-1 \le t \le 0$

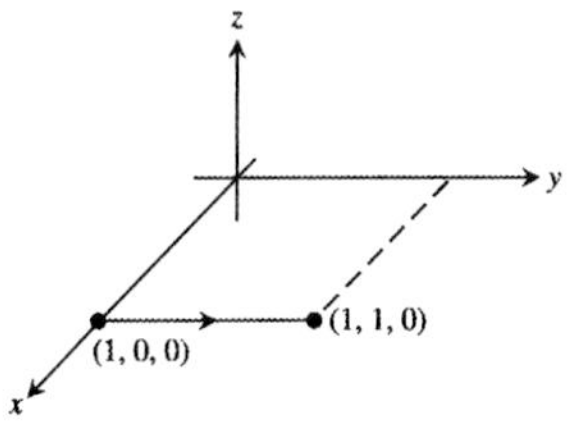

17. The direction $\overrightarrow{PQ} = -2\mathbf{j}$ and $P(0, 1, 1) \Rightarrow x = 0$, $y = 1 - 2t, z = 1$, where $0 \le t \le 1$

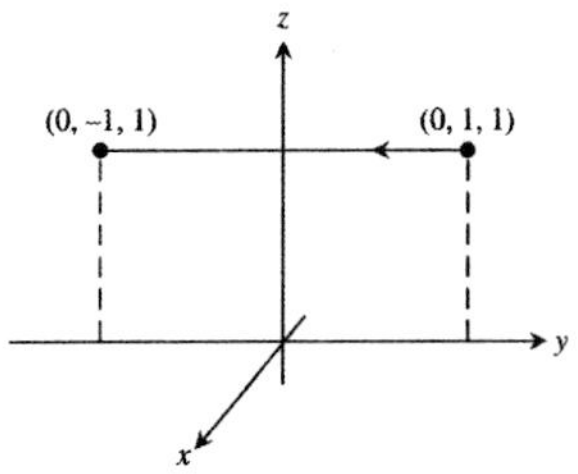

19. The direction $\overrightarrow{PQ} = -2\mathbf{i} + 2\mathbf{j} - 2\mathbf{k}$ and $P(2, 0, 2)$ $\Rightarrow x = 2 - 2t, y = 2t, z = 2 - 2t$, where $0 \le t \le 1$

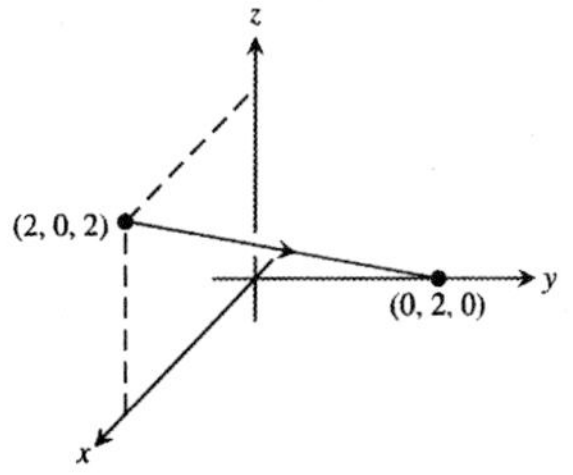

21. $3(x - 0) + (-2)(y - 2) + (-1)(z + 1) = 0 \Rightarrow 3x - 2y - z = -3$

23. $\overrightarrow{PQ} = \mathbf{i} - \mathbf{j} + 3\mathbf{k}$, $\overrightarrow{PS} = -\mathbf{i} - 3\mathbf{j} + 2\mathbf{k} \Rightarrow \overrightarrow{PQ} \times \overrightarrow{PS} = \begin{vmatrix} \mathbf{i} & \mathbf{j} & \mathbf{k} \\ 1 & -1 & 3 \\ -1 & -3 & 2 \end{vmatrix} = 7\mathbf{i} - 5\mathbf{j} - 4\mathbf{k}$ is normal to the plane

$\Rightarrow 7(x-2) + (-5)(y-0) + (-4)(z-2) = 0 \Rightarrow 7x - 5y - 4z = 6$

25. $\mathbf{n} = \mathbf{i} + 3\mathbf{j} + 4\mathbf{k}$, $P(2,4,5) = (1)(x-2) + (3)(y-4) + (4)(z-5) = 0 \Rightarrow x + 3y + 4z = 34$

27. $\begin{cases} x = 2t + 1 = s + 2 \\ y = 3t + 2 = 2s + 4 \end{cases} \Rightarrow \begin{cases} 2t - s = 1 \\ 3t - 2s = 2 \end{cases} \Rightarrow \begin{cases} 4t - 2s = 2 \\ 3t - 2s = 2 \end{cases} \Rightarrow t = 0$ and $s = -1$; then $z = 4t + 3 = -4s - 1$

$\Rightarrow 4(0) + 3 = (-4)(-1) - 1$ is satisfied $\Rightarrow$ the lines do intersect when $t = 0$ and $s = -1 \Rightarrow$ the point of intersection is $x = 1$, $y = 2$, and $z = 3$ or $P(1,2,3)$. A vector normal to the plane determined by these lines is

$\mathbf{n}_1 \times \mathbf{n}_2 = \begin{vmatrix} \mathbf{i} & \mathbf{j} & \mathbf{k} \\ 2 & 3 & 4 \\ 1 & 2 & -4 \end{vmatrix} = -20\mathbf{i} + 12\mathbf{j} + \mathbf{k}$, where $\mathbf{n}_1$ and $\mathbf{n}_2$ are directions of the lines $\Rightarrow$ the plane containing the lines is represented by$(-20)(x-1) + (12)(y-2) + (1)(z-3) = 0 \Rightarrow -20x + 12y + z = 7$.

29. The cross product of $\mathbf{i} + \mathbf{j} - \mathbf{k}$ and $-4\mathbf{i} + 2\mathbf{j} - 2\mathbf{k}$ has the same direction as the normal to the plane

$\Rightarrow \mathbf{n} = \begin{vmatrix} \mathbf{i} & \mathbf{j} & \mathbf{k} \\ 1 & 1 & -1 \\ -4 & 2 & -2 \end{vmatrix} = 6\mathbf{j} + 6\mathbf{k}$. Select a point on either line, such as $P(-1,2,1)$. Since the lines are given to intersect, the desired plane is $0(x+1) + 6(y-2) + 6(z-1) = 0 \Rightarrow 6y + 6z = 18 \Rightarrow y + z = 3$.

31. $\mathbf{n}_1 \times \mathbf{n}_2 = \begin{vmatrix} \mathbf{i} & \mathbf{j} & \mathbf{k} \\ 2 & 1 & -1 \\ 1 & 2 & 1 \end{vmatrix} = 3\mathbf{i} - 3\mathbf{j} + 3\mathbf{k}$ is a vector in the direction of the line of intersection of the planes

$\Rightarrow 3(x-2) + (-3)(y-1) + 3(z+1) = 0 \Rightarrow 3x - 3y + 3z = 0 \Rightarrow x - y + z = 0$ is the desired plane containing $P_0(2,1,-1)$

33. $S(0,0,12)$, $P(0,0,0)$ and $\mathbf{v} = 4\mathbf{i} - 2\mathbf{j} + 2\mathbf{k} \Rightarrow \overrightarrow{PS} \times \mathbf{v} = \begin{vmatrix} \mathbf{i} & \mathbf{j} & \mathbf{k} \\ 0 & 0 & 12 \\ 4 & -2 & 2 \end{vmatrix} = 24\mathbf{i} + 48\mathbf{j} = 24(\mathbf{i} + 2\mathbf{j})$

$\Rightarrow d = \frac{|\overrightarrow{PS} \times \mathbf{v}|}{|\mathbf{v}|} = \frac{24\sqrt{1+4}}{\sqrt{16+4+4}} = \frac{24\sqrt{5}}{\sqrt{24}} = \sqrt{5 \cdot 24} = 2\sqrt{30}$ is the distance from S to the line

35. $S(2,1,3)$, $P(2,1,3)$ and $\mathbf{v} = 2\mathbf{i} + 6\mathbf{j} \Rightarrow \overrightarrow{PS} \times \mathbf{v} = \mathbf{0} \Rightarrow d = \frac{|\overrightarrow{PS} \times \mathbf{v}|}{|\mathbf{v}|} = \frac{0}{\sqrt{40}} = 0$ is the distance from S to the line (i.e., the point S lies on the line)

37. $S(3,-1,4)$, $P(4,3,-5)$ and $\mathbf{v} = -\mathbf{i} + 2\mathbf{j} + 3\mathbf{k} \Rightarrow \overrightarrow{PS} \times \mathbf{v} = \begin{vmatrix} \mathbf{i} & \mathbf{j} & \mathbf{k} \\ -1 & -4 & 9 \\ -1 & 2 & 3 \end{vmatrix} = -30\mathbf{i} - 6\mathbf{j} - 6\mathbf{k}$

$\Rightarrow d = \frac{|\overrightarrow{PS} \times \mathbf{v}|}{|\mathbf{v}|} = \frac{\sqrt{900+36+36}}{\sqrt{1+4+9}} = \frac{\sqrt{972}}{\sqrt{14}} = \frac{\sqrt{486}}{\sqrt{7}} = \frac{\sqrt{81 \cdot 6}}{\sqrt{7}} = \frac{9\sqrt{42}}{7}$ is the distance from S to the line

39. $S(2,-3,4)$, $x + 2y + 2z = 13$ and $P(13,0,0)$ is on the plane $\Rightarrow \overrightarrow{PS} = -11\mathbf{i} - 3\mathbf{j} + 4\mathbf{k}$ and $\mathbf{n} = \mathbf{i} + 2\mathbf{j} + 2\mathbf{k}$

$\Rightarrow d = \left|\overrightarrow{PS} \cdot \frac{\mathbf{n}}{|\mathbf{n}|}\right| = \left|\frac{-11-6+8}{\sqrt{1+4+4}}\right| = \left|\frac{-9}{\sqrt{9}}\right| = 3$

41. $S(0,1,1)$, $4y + 3z = -12$ and $P(0,-3,0)$ is on the plane $\Rightarrow \overrightarrow{PS} = 4\mathbf{j} + \mathbf{k}$ and $\mathbf{n} = 4\mathbf{j} + 3\mathbf{k}$

$\Rightarrow d = \left|\overrightarrow{PS} \cdot \frac{\mathbf{n}}{|\mathbf{n}|}\right| = \left|\frac{16+3}{\sqrt{16+9}}\right| = \frac{19}{5}$

43. S(0, −1, 0), 2x + y + 2z = 4 and P(2, 0, 0) is on the plane $\Rightarrow \overrightarrow{PS} = -2\mathbf{i} - \mathbf{j}$ and $\mathbf{n} = 2\mathbf{i} + \mathbf{j} + 2\mathbf{k}$
$\Rightarrow d = \left| \overrightarrow{PS} \cdot \frac{\mathbf{n}}{|\mathbf{n}|} \right| = \left| \frac{-4-1+0}{\sqrt{4+1+4}} \right| = \frac{5}{3}$

45. The point P(1, 0, 0) is on the first plane and S(10, 0, 0) is a point on the second plane $\Rightarrow \overrightarrow{PS} = 9\mathbf{i}$, and $\mathbf{n} = \mathbf{i} + 2\mathbf{j} + 6\mathbf{k}$ is normal to the first plane $\Rightarrow$ the distance from S to the first plane is $d = \left| \overrightarrow{PS} \cdot \frac{\mathbf{n}}{|\mathbf{n}|} \right|$
$= \left| \frac{9}{\sqrt{1+4+36}} \right| = \frac{9}{\sqrt{41}}$, which is also the distance between the planes.

47. $\mathbf{n}_1 = \mathbf{i} + \mathbf{j}$ and $\mathbf{n}_2 = 2\mathbf{i} + \mathbf{j} - 2\mathbf{k} \Rightarrow \theta = \cos^{-1}\left(\frac{\mathbf{n}_1 \cdot \mathbf{n}_2}{|\mathbf{n}_1|\,|\mathbf{n}_2|}\right) = \cos^{-1}\left(\frac{2+1}{\sqrt{2}\,\sqrt{9}}\right) = \cos^{-1}\left(\frac{1}{\sqrt{2}}\right) = \frac{\pi}{4}$

49. $\mathbf{n}_1 = 2\mathbf{i} + 2\mathbf{j} + 2\mathbf{k}$ and $\mathbf{n}_2 = 2\mathbf{i} - 2\mathbf{j} - \mathbf{k} \Rightarrow \theta = \cos^{-1}\left(\frac{\mathbf{n}_1 \cdot \mathbf{n}_2}{|\mathbf{n}_1|\,|\mathbf{n}_2|}\right) = \cos^{-1}\left(\frac{4-4-2}{\sqrt{12}\,\sqrt{9}}\right) = \cos^{-1}\left(\frac{-1}{3\sqrt{3}}\right) \approx 1.76$ rad

51. $\mathbf{n}_1 = 2\mathbf{i} + 2\mathbf{j} - \mathbf{k}$ and $\mathbf{n}_2 = \mathbf{i} + 2\mathbf{j} + \mathbf{k} \Rightarrow \theta = \cos^{-1}\left(\frac{\mathbf{n}_1 \cdot \mathbf{n}_2}{|\mathbf{n}_1|\,|\mathbf{n}_2|}\right) = \cos^{-1}\left(\frac{2+4-1}{\sqrt{9}\,\sqrt{6}}\right) = \cos^{-1}\left(\frac{5}{3\sqrt{6}}\right) \approx 0.82$ rad

53. $2x - y + 3z = 6 \Rightarrow 2(1-t) - (3t) + 3(1+t) = 6 \Rightarrow -2t + 5 = 6 \Rightarrow t = -\frac{1}{2} \Rightarrow x = \frac{3}{2}, y = -\frac{3}{2}$ and $z = \frac{1}{2}$
$\Rightarrow \left(\frac{3}{2}, -\frac{3}{2}, \frac{1}{2}\right)$ is the point

55. $x + y + z = 2 \Rightarrow (1 + 2t) + (1 + 5t) + (3t) = 2 \Rightarrow 10t + 2 = 2 \Rightarrow t = 0 \Rightarrow x = 1, y = 1$ and $z = 0 \Rightarrow (1, 1, 0)$ is the point

57. $\mathbf{n}_1 = \mathbf{i} + \mathbf{j} + \mathbf{k}$ and $\mathbf{n}_2 = \mathbf{i} + \mathbf{j} \Rightarrow \mathbf{n}_1 \times \mathbf{n}_2 = \begin{vmatrix} \mathbf{i} & \mathbf{j} & \mathbf{k} \\ 1 & 1 & 1 \\ 1 & 1 & 0 \end{vmatrix} = -\mathbf{i} + \mathbf{j}$, the direction of the desired line; (1, 1, −1) is on both planes $\Rightarrow$ the desired line is x = 1 − t, y = 1 + t, z = −1

59. $\mathbf{n}_1 = \mathbf{i} - 2\mathbf{j} + 4\mathbf{k}$ and $\mathbf{n}_2 = \mathbf{i} + \mathbf{j} - 2\mathbf{k} \Rightarrow \mathbf{n}_1 \times \mathbf{n}_2 = \begin{vmatrix} \mathbf{i} & \mathbf{j} & \mathbf{k} \\ 1 & -2 & 4 \\ 1 & 1 & -2 \end{vmatrix} = 6\mathbf{j} + 3\mathbf{k}$, the direction of the desired line; (4, 3, 1) is on both planes $\Rightarrow$ the desired line is x = 4, y = 3 + 6t, z = 1 + 3t

61. <u>L1 & L2</u>: $x = 3 + 2t = 1 + 4s$ and $y = -1 + 4t = 1 + 2s \Rightarrow \begin{cases} 2t - 4s = -2 \\ 4t - 2s = 2 \end{cases} \Rightarrow \begin{cases} 2t - 4s = -2 \\ 2t - s = 1 \end{cases}$
$\Rightarrow -3s = -3 \Rightarrow s = 1$ and $t = 1 \Rightarrow$ on L1, z = 1 and on L2, z = 1 $\Rightarrow$ L1 and L2 intersect at (5, 3, 1).
<u>L2 & L3</u>: The direction of L2 is $\frac{1}{6}(4\mathbf{i} + 2\mathbf{j} + 4\mathbf{k}) = \frac{1}{3}(2\mathbf{i} + \mathbf{j} + 2\mathbf{k})$ which is the same as the direction $\frac{1}{3}(2\mathbf{i} + \mathbf{j} + 2\mathbf{k})$ of L3; hence L2 and L3 are parallel.
<u>L1 & L3</u>: $x = 3 + 2t = 3 + 2r$ and $y = -1 + 4t = 2 + r \Rightarrow \begin{cases} 2t - 2r = 0 \\ 4t - r = 3 \end{cases} \Rightarrow \begin{cases} t - r = 0 \\ 4t - r = 3 \end{cases} \Rightarrow 3t = 3$
$\Rightarrow t = 1$ and $r = 1 \Rightarrow$ on L1, z = 2 while on L3, z = 0 $\Rightarrow$ L1 and L2 do not intersect. The direction of L1 is $\frac{1}{\sqrt{21}}(2\mathbf{i} + 4\mathbf{j} - \mathbf{k})$ while the direction of L3 is $\frac{1}{3}(2\mathbf{i} + \mathbf{j} + 2\mathbf{k})$ and neither is a multiple of the other; hence L1 and L3 are skew.

63. $x = 2 + 2t, y = -4 - t, z = 7 + 3t; x = -2 - t, y = -2 + \frac{1}{2}t, z = 1 - \frac{3}{2}t$

65. $x = 0 \Rightarrow t = -\frac{1}{2}, y = -\frac{1}{2}, z = -\frac{3}{2} \Rightarrow \left(0, -\frac{1}{2}, -\frac{3}{2}\right); y = 0 \Rightarrow t = -1, x = -1, z = -3 \Rightarrow (-1, 0, -3); z = 0$
$\Rightarrow t = 0, x = 1, y = -1 \Rightarrow (1, -1, 0)$

67. With substitution of the line into the plane we have $2(1-2t)+(2+5t)-(-3t)=8 \Rightarrow 2-4t+2+5t+3t=8$ $\Rightarrow 4t+4=8 \Rightarrow t=1 \Rightarrow$ the point $(-1,7,-3)$ is contained in both the line and plane, so they are not parallel.

69. There are many possible answers. One is found as follows: eliminate t to get $t=x-1=2-y=\frac{z-3}{2}$ $\Rightarrow x-1=2-y$ and $2-y=\frac{z-3}{2} \Rightarrow x+y=3$ and $2y+z=7$ are two such planes.

71. The points $(a,0,0)$, $(0,b,0)$ and $(0,0,c)$ are the x, y, and z intercepts of the plane. Since a, b, and c are all nonzero, the plane must intersect all three coordinate axes and cannot pass through the origin. Thus, $\frac{x}{a}+\frac{y}{b}+\frac{z}{c}=1$ describes all planes <u>except</u> those through the origin or parallel to a coordinate axis.

73. (a) $\overrightarrow{EP}=c\overrightarrow{EP_1} \Rightarrow -x_0\mathbf{i}+y\mathbf{j}+z\mathbf{k}=c\left[(x_1-x_0)\mathbf{i}+y_1\mathbf{j}+z_1\mathbf{k}\right] \Rightarrow -x_0=c(x_1-x_0)$, $y=cy_1$ and $z=cz_1$, where c is a positive real number

(b) At $x_1=0 \Rightarrow c=1 \Rightarrow y=y_1$ and $z=z_1$; at $x_1=x_0 \Rightarrow x_0=0, y=0, z=0$; $\lim\limits_{x_0\to\infty} c=\lim\limits_{x_0\to\infty}\frac{-x_0}{x_1-x_0}$ $=\lim\limits_{x_0\to\infty}\frac{-1}{-1}=1 \Rightarrow c\to 1$ so that $y\to y_1$ and $z\to z_1$

9.6 CYLINDERS AND QUADRIC SURFACES

1. d, ellipsoid

3. a, cylinder

5. l, hyperbolic paraboloid

7. b, cylinder

9. k, hyperbolic paraboloid

11. h, cone

13. $x^2+y^2=4$

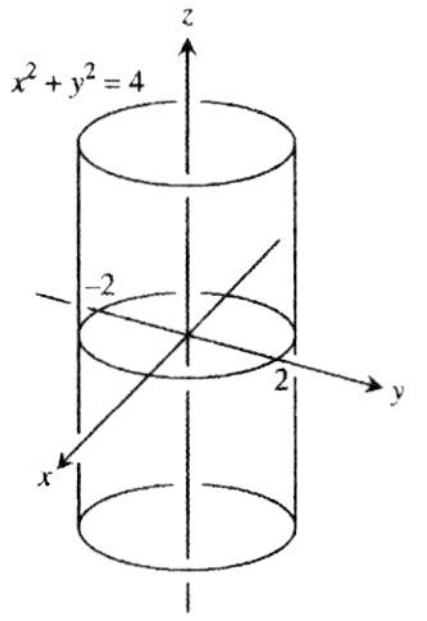

15. $x^2+4z^2=16$

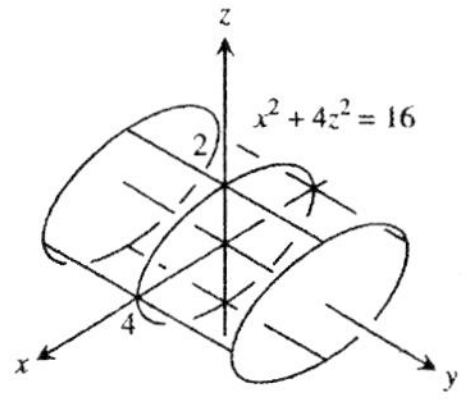

17. $9x^2+y^2+z^2=9$

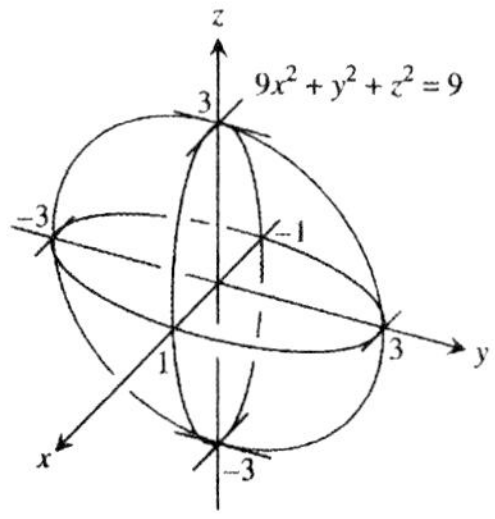

19. $4x^2+9y^2+4z^2=36$

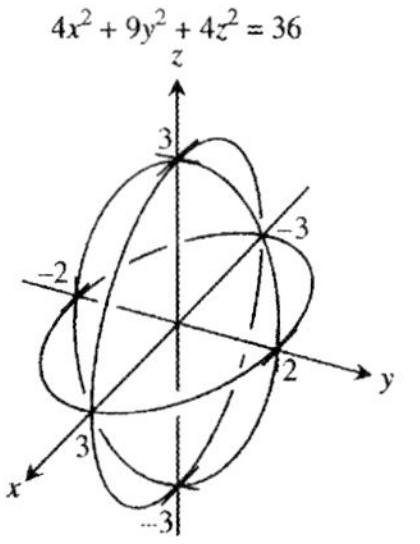

21. $x^2+4y^2=z$

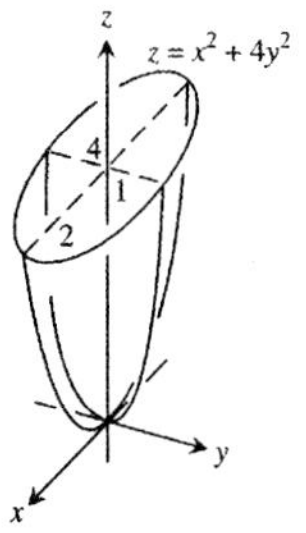

23. $x=4-4y^2-z^2$

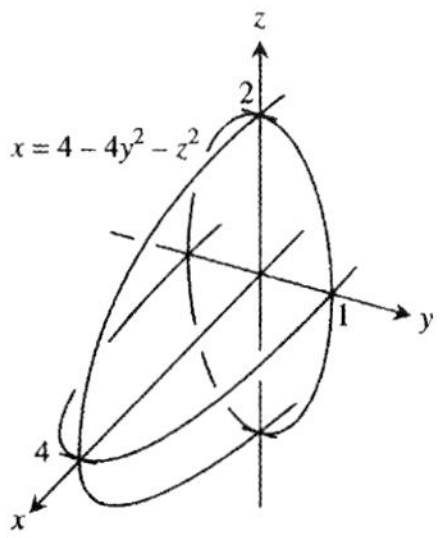

25. $x^2 + y^2 = z^2$

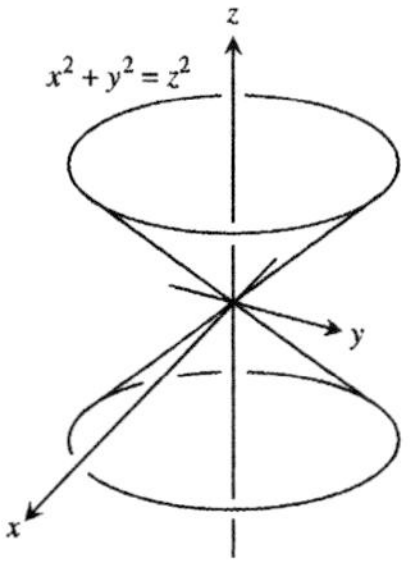

27. $x^2 + y^2 - z^2 = 1$

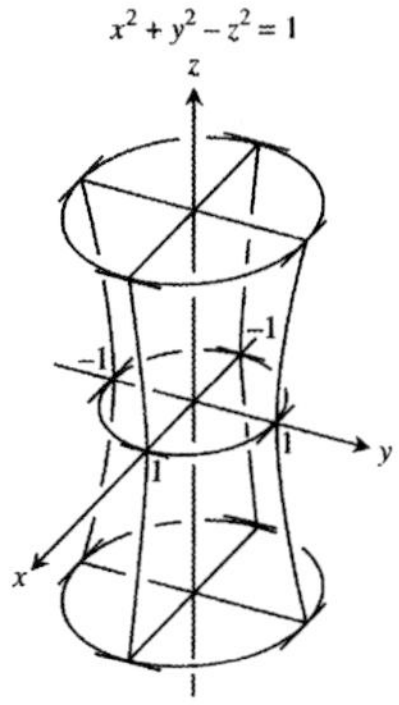

29. $z^2 - x^2 - y^2 = 1$

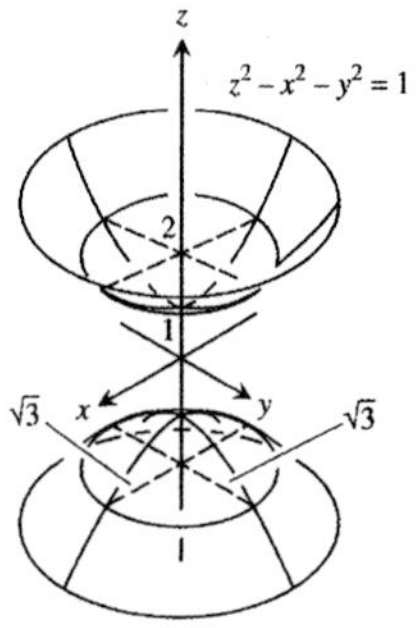

31. $y^2 - x^2 = z$

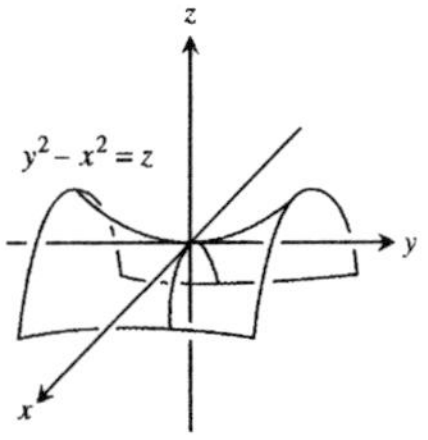

33. $z = 1 + y^2 - x^2$

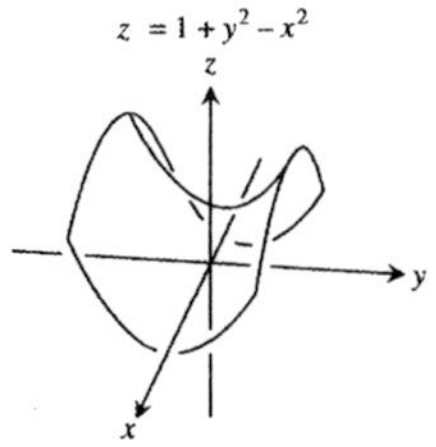

35. $y = -(x^2 + z^2)$

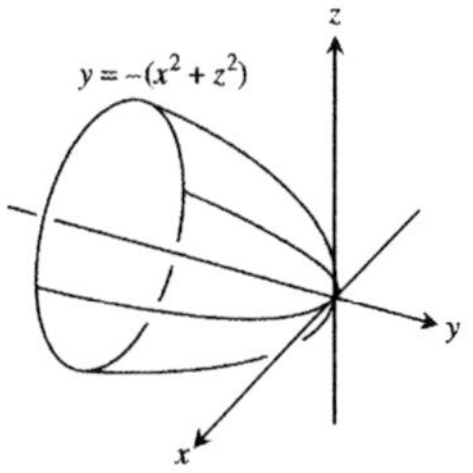

37. $x^2 + y^2 - z^2 = 4$

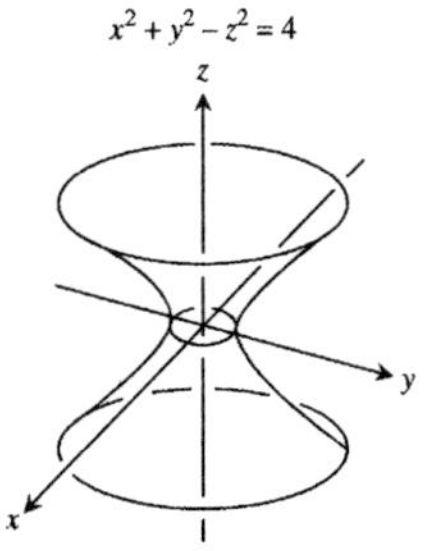

39. $x^2 + z^2 = 1$

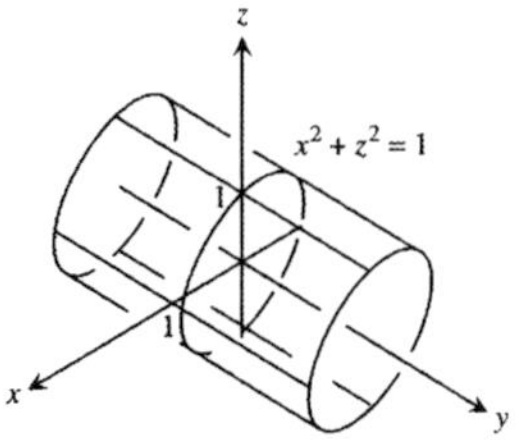

41. $z = -(x^2 + y^2)$

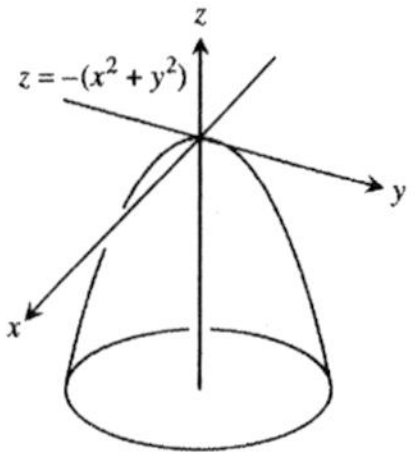

43. $4y^2 + z^2 - 4x^2 = 4$

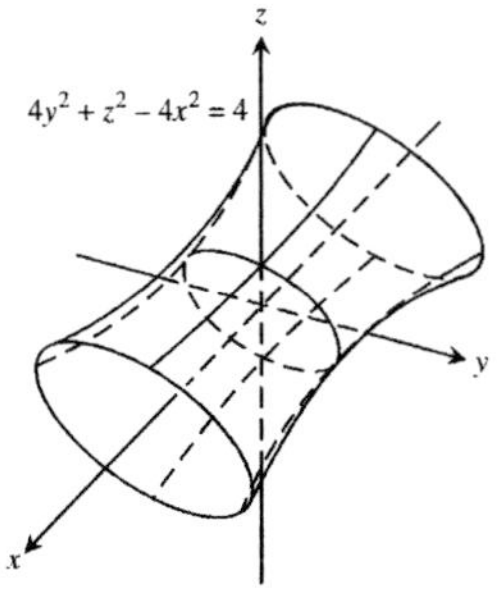

45. (a) If $x^2 + \frac{y^2}{4} + \frac{z^2}{9} = 1$ and $z = c$, then $x^2 + \frac{y^2}{4} = \frac{9-c^2}{9} \Rightarrow \frac{x^2}{\left(\frac{9-c^2}{9}\right)} + \frac{y^2}{\left[\frac{4(9-c^2)}{9}\right]} = 1 \Rightarrow A = ab\pi$

$= \pi\left(\frac{\sqrt{9-c^2}}{3}\right)\left(\frac{2\sqrt{9-c^2}}{3}\right) = \frac{2\pi(9-c^2)}{9}$

(b) From part (a), each slice has the area $\frac{2\pi(9-z^2)}{9}$, where $-3 \le z \le 3$. Thus $V = 2\int_0^3 \frac{2\pi}{9}(9-z^2)\,dz$

$= \frac{4\pi}{9}\int_0^3 (9-z^2)\,dz = \frac{4\pi}{9}\left[9z - \frac{z^3}{3}\right]_0^3 = \frac{4\pi}{9}(27-9) = 8\pi$

(c) $\frac{x^2}{a^2} + \frac{y^2}{b^2} + \frac{z^2}{c^2} = 1 \Rightarrow \frac{x^2}{\left[\frac{a^2(c^2-z^2)}{c^2}\right]} + \frac{y^2}{\left[\frac{b^2(c^2-z^2)}{c^2}\right]} = 1 \Rightarrow A = \pi\left(\frac{a\sqrt{c^2-z^2}}{c}\right)\left(\frac{b\sqrt{c^2-z^2}}{c}\right)$

$\Rightarrow V = 2\int_0^c \frac{\pi ab}{c^2}(c^2 - z^2)\,dz = \frac{2\pi ab}{c^2}\left[c^2 z - \frac{z^3}{3}\right]_0^c = \frac{2\pi ab}{c^2}\left(\frac{2}{3}c^3\right) = \frac{4\pi abc}{3}$. Note that if $r = a = b = c$, then $V = \frac{4\pi r^3}{3}$, which is the volume of a sphere.

CHAPTER 9 PRACTICE AND ADDITIONAL EXERCISES

1. (a) $3\langle -3, 4\rangle - 4\langle 2, -5\rangle = \langle -9-8, 12+20\rangle = \langle -17, 32\rangle$
   (b) $\sqrt{17^2 + 32^2} = \sqrt{1313}$

3. (a) $\langle -2(-3), -2(4)\rangle = \langle 6, -8\rangle$
   (b) $\sqrt{6^2 + (-8)^2} = 10$

5. $\frac{\pi}{6}$ radians below the negative x-axis: $\left\langle -\frac{\sqrt{3}}{2}, -\frac{1}{2}\right\rangle$ [assuming counterclockwise].

7. $2\left(\frac{1}{\sqrt{4^2+1^2}}\right)(4\mathbf{i} - \mathbf{j}) = \left(\frac{8}{\sqrt{17}}\mathbf{i} - \frac{2}{\sqrt{17}}\mathbf{j}\right)$

9. length $= \left|\sqrt{2}\mathbf{i} + \sqrt{2}\mathbf{j}\right| = \sqrt{2+2} = 2$, $\sqrt{2}\mathbf{i} + \sqrt{2}\mathbf{j} = 2\left(\frac{1}{\sqrt{2}}\mathbf{i} + \frac{1}{\sqrt{2}}\mathbf{j}\right) \Rightarrow$ the direction is $\frac{1}{\sqrt{2}}\mathbf{i} + \frac{1}{\sqrt{2}}\mathbf{j}$

11. $t = \frac{\pi}{2} \Rightarrow \mathbf{v} = (-2\sin\frac{\pi}{2})\mathbf{i} + \left(2\cos\frac{\pi}{2}\right)\mathbf{j} = -2\mathbf{i}$; length $= |-2\mathbf{i}| = \sqrt{4+0} = 2$; $-2\mathbf{i} = 2(-\mathbf{i}) \Rightarrow$ the direction is $-\mathbf{i}$

13. length $= |2\mathbf{i} - 3\mathbf{j} + 6\mathbf{k}| = \sqrt{4+9+36} = 7$, $2\mathbf{i} - 3\mathbf{j} + 6\mathbf{k} = 7\left(\frac{2}{7}\mathbf{i} - \frac{3}{7}\mathbf{j} + \frac{6}{7}\mathbf{k}\right) \Rightarrow$ the direction is $\frac{2}{7}\mathbf{i} - \frac{3}{7}\mathbf{j} + \frac{6}{7}\mathbf{k}$

15. $2\frac{\mathbf{v}}{|\mathbf{v}|} = 2 \cdot \frac{4\mathbf{i} - \mathbf{j} + 4\mathbf{k}}{\sqrt{4^2 + (-1)^2 + 4^2}} = 2 \cdot \frac{4\mathbf{i} - \mathbf{j} + 4\mathbf{k}}{\sqrt{33}} = \frac{8}{\sqrt{33}}\mathbf{i} - \frac{2}{\sqrt{33}}\mathbf{j} + \frac{8}{\sqrt{33}}\mathbf{k}$

17. $|\mathbf{v}| = \sqrt{1+1} = \sqrt{2}$, $|\mathbf{u}| = \sqrt{4+1+4} = 3$, $\mathbf{v}\cdot\mathbf{u} = 3$, $\mathbf{u}\cdot\mathbf{v} = 3$, $\mathbf{v}\times\mathbf{u} = \begin{vmatrix} \mathbf{i} & \mathbf{j} & \mathbf{k} \\ 1 & 1 & 0 \\ 2 & 1 & -2 \end{vmatrix} = -2\mathbf{i} + 2\mathbf{j} - \mathbf{k}$,

    $\mathbf{u}\times\mathbf{v} = -(\mathbf{v}\times\mathbf{u}) = 2\mathbf{i} - 2\mathbf{j} + \mathbf{k}$, $|\mathbf{v}\times\mathbf{u}| = \sqrt{4+4+1} = 3$, $\theta = \cos^{-1}\left(\frac{\mathbf{v}\cdot\mathbf{u}}{|\mathbf{v}|\,|\mathbf{u}|}\right) = \cos^{-1}\left(\frac{1}{\sqrt{2}}\right) = \frac{\pi}{4}$,

    $|\mathbf{u}|\cos\theta = \frac{3}{\sqrt{2}}$, $\text{proj}_{\mathbf{v}}\,\mathbf{u} = \left(\frac{\mathbf{v}\cdot\mathbf{u}}{|\mathbf{v}||\mathbf{v}|}\right)\mathbf{v} = \frac{3}{2}(\mathbf{i} + \mathbf{j})$

19. $\mathbf{u} = \left(\frac{\mathbf{v}\cdot\mathbf{u}}{|\mathbf{v}||\mathbf{v}|}\right)\mathbf{v} + \left[\mathbf{u} - \left(\frac{\mathbf{v}\cdot\mathbf{u}}{|\mathbf{v}||\mathbf{v}|}\right)\mathbf{v}\right] = \frac{4}{3}(2\mathbf{i} + \mathbf{j} - \mathbf{k}) + \left[(\mathbf{i} + \mathbf{j} - 5\mathbf{k}) - \frac{4}{3}(2\mathbf{i} + \mathbf{j} - \mathbf{k})\right] = \frac{4}{3}(2\mathbf{i} + \mathbf{j} - \mathbf{k}) - \frac{1}{3}(5\mathbf{i} + \mathbf{j} + 11\mathbf{k})$,
    where $\mathbf{v}\cdot\mathbf{u} = 8$ and $\mathbf{v}\cdot\mathbf{v} = 6$

21. $\mathbf{u}\times\mathbf{v} = \begin{vmatrix} \mathbf{i} & \mathbf{j} & \mathbf{k} \\ 1 & 0 & 0 \\ 1 & 1 & 0 \end{vmatrix} = \mathbf{k}$

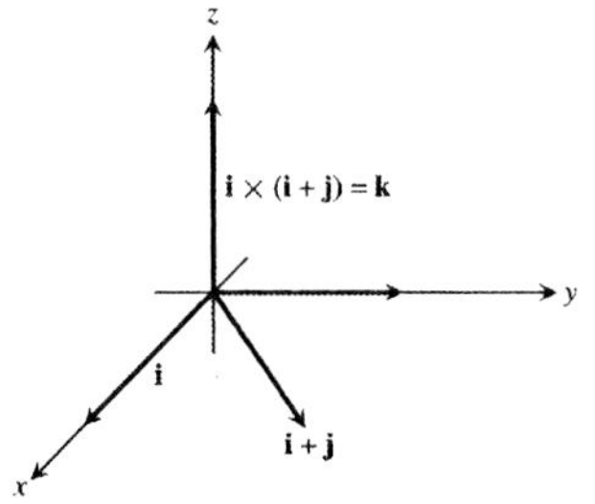

23. Let $\mathbf{v} = v_1\mathbf{i} + v_2\mathbf{j} + v_3\mathbf{k}$ and $\mathbf{w} = w_1\mathbf{i} + w_2\mathbf{j} + w_3\mathbf{k}$. Then $|\mathbf{v} - 2\mathbf{w}|^2 = |(v_1\mathbf{i} + v_2\mathbf{j} + v_3\mathbf{k}) - 2(w_1\mathbf{i} + w_2\mathbf{j} + w_3\mathbf{k})|^2$
$= |(v_1 - 2w_1)\mathbf{i} + (v_2 - 2w_2)\mathbf{j} + (v_3 - 2w_3)\mathbf{k}|^2 = \left(\sqrt{(v_1 - 2w_1)^2 + (v_2 - 2w_2)^2 + (v_3 - 2w_3)^2}\right)^2$
$= (v_1^2 + v_2^2 + v_3^2) - 4(v_1w_1 + v_2w_2 + v_3w_3) + 4\left(w_1^2 + w_2^2 + w_3^2\right) = |\mathbf{v}|^2 - 4\mathbf{v}\cdot\mathbf{w} + 4|\mathbf{w}|^2$
$= |\mathbf{v}|^2 - 4|\mathbf{v}||\mathbf{w}|\cos\theta + 4|\mathbf{w}|^2 = 4 - 4(2)(3)\left(\cos\frac{\pi}{3}\right) + 36 = 40 - 24\left(\frac{1}{2}\right) = 40 - 12 = 28 \Rightarrow |\mathbf{v} - 2\mathbf{w}| = \sqrt{28}$
$= 2\sqrt{7}$

25. (a) area $= |\mathbf{u} \times \mathbf{v}| = \text{abs}\begin{vmatrix} \mathbf{i} & \mathbf{j} & \mathbf{k} \\ 1 & 1 & -1 \\ 2 & 1 & 1 \end{vmatrix} = |2\mathbf{i} - 3\mathbf{j} - \mathbf{k}| = \sqrt{4 + 9 + 1} = \sqrt{14}$

(b) volume $= \mathbf{u}\cdot(\mathbf{v} \times \mathbf{w}) = \begin{vmatrix} 1 & 1 & -1 \\ 2 & 1 & 1 \\ -1 & -2 & 3 \end{vmatrix} = 1(3 + 2) + 1(-1 - 6) - 1(-4 + 1) = 1$

27. The desired vector is $\mathbf{n} \times \mathbf{v}$ or $\mathbf{v} \times \mathbf{n}$ since $\mathbf{n} \times \mathbf{v}$ is perpendicular to both $\mathbf{n}$ and $\mathbf{v}$ and, therefore, also parallel to the plane.

29. The line L passes through the point $P(0, 0, -1)$ parallel to $\mathbf{v} = -\mathbf{i} + \mathbf{j} + \mathbf{k}$. With $\overrightarrow{PS} = 2\mathbf{i} + 2\mathbf{j} + \mathbf{k}$ and
$\overrightarrow{PS} \times \mathbf{v} = \begin{vmatrix} \mathbf{i} & \mathbf{j} & \mathbf{k} \\ 2 & 2 & 1 \\ -1 & 1 & 1 \end{vmatrix} = (2 - 1)\mathbf{i} + (-1 - 2)\mathbf{j} + (2 + 2)\mathbf{k} = \mathbf{i} - 3\mathbf{j} + 4\mathbf{k}$, we find the distance
$d = \frac{|\overrightarrow{PS} \times \mathbf{v}|}{|\mathbf{v}|} = \frac{\sqrt{1 + 9 + 16}}{\sqrt{1 + 1 + 1}} = \frac{\sqrt{26}}{\sqrt{3}} = \frac{\sqrt{78}}{3}$.

31. Parametric equations for the line are $x = 1 - 3t$, $y = 2$, $z = 3 + 7t$.

33. The point $P(4, 0, 0)$ lies on the plane $x - y = 4$, and $\overrightarrow{PS} = (6 - 4)\mathbf{i} + 0\mathbf{j} + (-6 + 0)\mathbf{k} = 2\mathbf{i} - 6\mathbf{k}$ with $\mathbf{n} = \mathbf{i} - \mathbf{j}$
$\Rightarrow d = \frac{|\mathbf{n}\cdot\overrightarrow{PS}|}{|\mathbf{n}|} = \left|\frac{2 + 0 + 0}{\sqrt{1 + 1 + 0}}\right| = \frac{2}{\sqrt{2}} = \sqrt{2}$.

35. $P(3, -2, 1)$ and $\mathbf{n} = 2\mathbf{i} + \mathbf{j} + \mathbf{k} \Rightarrow (2)(x - 3) + (1)(y - (-2)) + (1)(z - 1) = 0 \Rightarrow 2x + y + z = 5$

37. $P(1, -1, 2)$, $Q(2, 1, 3)$ and $R(-1, 2, -1) \Rightarrow \overrightarrow{PQ} = \mathbf{i} + 2\mathbf{j} + \mathbf{k}$, $\overrightarrow{PR} = -2\mathbf{i} + 3\mathbf{j} - 3\mathbf{k}$ and $\overrightarrow{PQ} \times \overrightarrow{PR}$
$= \begin{vmatrix} \mathbf{i} & \mathbf{j} & \mathbf{k} \\ 1 & 2 & 1 \\ -2 & 3 & -3 \end{vmatrix} = -9\mathbf{i} + \mathbf{j} + 7\mathbf{k}$ is normal to the plane $\Rightarrow (-9)(x - 1) + (1)(y + 1) + (7)(z - 2) = 0$
$\Rightarrow -9x + y + 7z = 4$

39. $\left(0, -\frac{1}{2}, -\frac{3}{2}\right)$, since $t = -\frac{1}{2}$, $y = -\frac{1}{2}$ and $z = -\frac{3}{2}$ when $x = 0$; $(-1, 0, -3)$, since $t = -1$, $x = -1$ and $z = -3$ when $y = 0$; $(1, -1, 0)$, since $t = 0$, $x = 1$ and $y = -1$ when $z = 0$

41. $\mathbf{n}_1 = \mathbf{i}$ and $\mathbf{n}_2 = \mathbf{i} + \mathbf{j} + \sqrt{2}\mathbf{k} \Rightarrow$ the desired angle is $\cos^{-1}\left(\frac{\mathbf{n}_1\cdot\mathbf{n}_2}{|\mathbf{n}_1||\mathbf{n}_2|}\right) = \cos^{-1}\left(\frac{1}{2}\right) = \frac{\pi}{3}$

43. The direction of the line is $\mathbf{n}_1 \times \mathbf{n}_2 = \begin{vmatrix} \mathbf{i} & \mathbf{j} & \mathbf{k} \\ 1 & 2 & 1 \\ 1 & -1 & 2 \end{vmatrix} = 5\mathbf{i} - \mathbf{j} - 3\mathbf{k}$. Since the point $(-5, 3, 0)$ is on both planes, the desired line is $x = -5 + 5t$, $y = 3 - t$, $z = -3t$.

45. (a) The corresponding normals are $\mathbf{n}_1 = 3\mathbf{i} + 6\mathbf{k}$ and $\mathbf{n}_2 = 2\mathbf{i} + 2\mathbf{j} - \mathbf{k}$ and since $\mathbf{n}_1 \cdot \mathbf{n}_2$
$= (3)(2) + (0)(2) + (6)(-1) = 6 + 0 - 6 = 0$, we have that the planes are orthogonal

(b) The line of intersection is parallel to $\mathbf{n}_1 \times \mathbf{n}_2 = \begin{vmatrix} \mathbf{i} & \mathbf{j} & \mathbf{k} \\ 3 & 0 & 6 \\ 2 & 2 & -1 \end{vmatrix} = -12\mathbf{i} + 15\mathbf{j} + 6\mathbf{k}$. Now to find a point in the intersection, solve $\begin{cases} 3x + 6z = 1 \\ 2x + 2y - z = 3 \end{cases} \Rightarrow \begin{cases} 3x + 6z = 1 \\ 12x + 12y - 6z = 18 \end{cases} \Rightarrow 15x + 12y = 19 \Rightarrow x = 0$ and $y = \frac{19}{12}$ $\Rightarrow \left(0, \frac{19}{12}, \frac{1}{6}\right)$ is a point on the line we seek. Therefore, the line is $x = -12t$, $y = \frac{19}{12} + 15t$ and $z = \frac{1}{6} + 6t$.

47. Yes; $\mathbf{v} \cdot \mathbf{n} = (2\mathbf{i} - 4\mathbf{j} + \mathbf{k}) \cdot (2\mathbf{i} + \mathbf{j} + 0\mathbf{k}) = 2 \cdot 2 - 4 \cdot 1 + 1 \cdot 0 = 0 \Rightarrow$ the vector is orthogonal to the plane's normal $\Rightarrow$ $\mathbf{v}$ is parallel to the plane

49. A normal to the plane is $\mathbf{n} = \overrightarrow{AB} \times \overrightarrow{AC} = \begin{vmatrix} \mathbf{i} & \mathbf{j} & \mathbf{k} \\ 2 & 0 & -1 \\ 2 & -1 & 0 \end{vmatrix} = -\mathbf{i} - 2\mathbf{j} - 2\mathbf{k} \Rightarrow$ the distance is $d = \left|\frac{\overrightarrow{AP} \cdot \mathbf{n}}{\mathbf{n}}\right|$ $= \left|\frac{(\mathbf{i} + 4\mathbf{j}) \cdot (-\mathbf{i} - 2\mathbf{j} - 2\mathbf{k})}{\sqrt{1+4+4}}\right| = \left|\frac{-1-8+0}{3}\right| = 3$

51. $\mathbf{n} = 2\mathbf{i} - \mathbf{j} - \mathbf{k}$ is normal to the plane $\Rightarrow \mathbf{n} \times \mathbf{v} = \begin{vmatrix} \mathbf{i} & \mathbf{j} & \mathbf{k} \\ 2 & -1 & -1 \\ 1 & 1 & 1 \end{vmatrix} = 0\mathbf{i} - 3\mathbf{j} + 3\mathbf{k} = -3\mathbf{j} + 3\mathbf{k}$ is orthogonal to $\mathbf{v}$ and parallel to the plane

53. A vector parallel to the line of intersection is $\mathbf{v} = \mathbf{n}_1 \times \mathbf{n}_2 = \begin{vmatrix} \mathbf{i} & \mathbf{j} & \mathbf{k} \\ 1 & 2 & 1 \\ 1 & -1 & 2 \end{vmatrix} = 5\mathbf{i} - \mathbf{j} - 3\mathbf{k}$ $\Rightarrow |\mathbf{v}| = \sqrt{25 + 1 + 9} = \sqrt{35} \Rightarrow 2\left(\frac{\mathbf{v}}{|\mathbf{v}|}\right) = \frac{2}{\sqrt{35}}(5\mathbf{i} - \mathbf{j} - 3\mathbf{k})$ is the desired vector.

55. The line is represented by $x = 3 + 2t$, $y = 2 - t$, and $z = 1 + 2t$. It meets the plane $2x - y + 2z = -2$ when $2(3 + 2t) - (2 - t) + 2(1 + 2t) = -2 \Rightarrow t = -\frac{8}{9} \Rightarrow$ the point is $\left(\frac{11}{9}, \frac{26}{9}, -\frac{7}{9}\right)$.

57. The intersection occurs when $(3 + 2t) + 3(2t) - t = -4 \Rightarrow t = -1 \Rightarrow$ the point is $(1, -2, -1)$. The required line must be perpendicular to both the given line and to the normal, and hence is parallel to $\begin{vmatrix} \mathbf{i} & \mathbf{j} & \mathbf{k} \\ 2 & 2 & 1 \\ 1 & 3 & -1 \end{vmatrix}$ $= -5\mathbf{i} + 3\mathbf{j} + 4\mathbf{k} \Rightarrow$ the line is represented by $x = 1 - 5t$, $y = -2 + 3t$, and $z = -1 + 4t$.

59. The vector $\overrightarrow{AB} \times \overrightarrow{CD} = \begin{vmatrix} \mathbf{i} & \mathbf{j} & \mathbf{k} \\ 3 & -2 & 4 \\ \frac{26}{5} & 0 & -\frac{26}{5} \end{vmatrix} = \frac{26}{5}(2\mathbf{i} + 7\mathbf{j} + 2\mathbf{k})$ is normal to the plane and $A(-2, 0, -3)$ lies on the plane $\Rightarrow 2(x + 2) + 7(y - 0) + 2(z - (-3)) = 0 \Rightarrow 2x + 7y + 2z + 10 = 0$ is an equation of the plane.

61. $\overrightarrow{AB} = -2\mathbf{i} + \mathbf{j} + \mathbf{k}$, $\overrightarrow{CD} = \mathbf{i} + 4\mathbf{j} - \mathbf{k}$, and $\overrightarrow{AC} = 2\mathbf{i} + \mathbf{j} \Rightarrow \mathbf{n} = \begin{vmatrix} \mathbf{i} & \mathbf{j} & \mathbf{k} \\ -2 & 1 & 1 \\ 1 & 4 & -1 \end{vmatrix} = -5\mathbf{i} - \mathbf{j} - 9\mathbf{k} \Rightarrow$ the distance is $d = \left|\frac{(2\mathbf{i} + \mathbf{j}) \cdot (-5\mathbf{i} - \mathbf{j} - 9\mathbf{k})}{\sqrt{25 + 1 + 81}}\right| = \frac{11}{\sqrt{107}}$

63. $x^2 + y^2 + z^2 = 4$
sphere

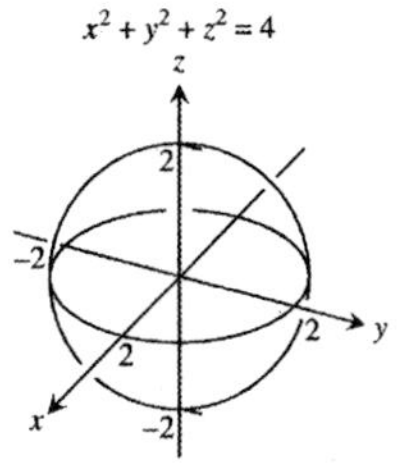

65. $4x^2 + 4y^2 + z^2 = 4$
ellipsoid

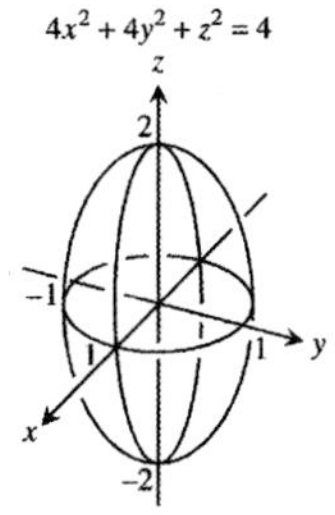

67. $z = -(x^2 + y^2)$
elliptical paraboloid

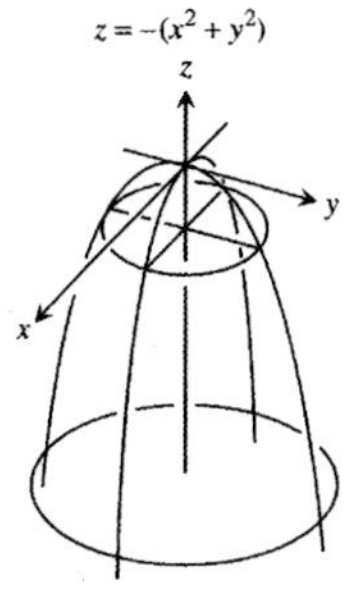

69. $x^2 + y^2 = z^2$
elliptical cone

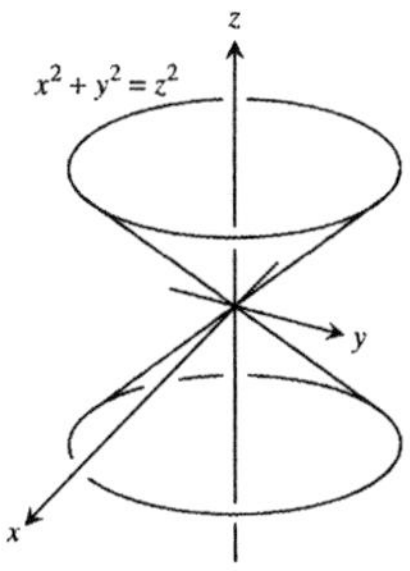

71. $x^2 + y^2 - z^2 = 4$
hyperboloid of one sheet

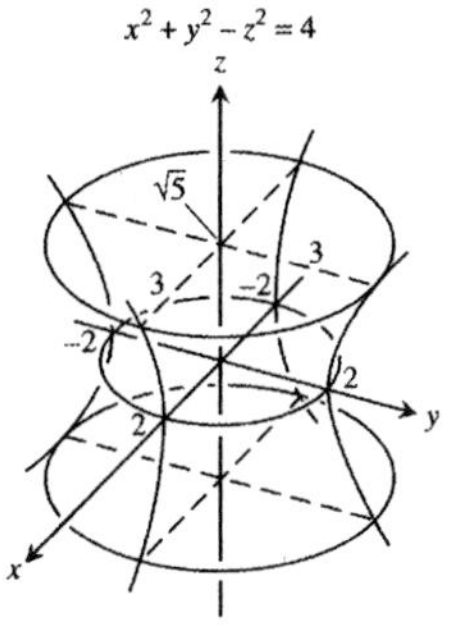

73. $y^2 - x^2 - z^2 = 1$
hyperboloid of two sheets

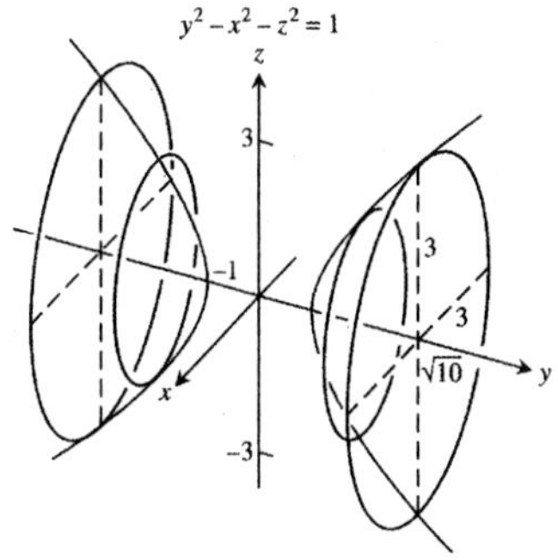

NOTES:

CHAPTER 10 VECTOR-VALUED FUNCTIONS AND MOTION IN SPACE

10.1 VECTOR FUNCTIONS AND THEIR DERIVATIVES

1. $x = t + 1$ and $y = t^2 - 1 \Rightarrow y = (x-1)^2 - 1 = x^2 - 2x$; $\mathbf{v} = \frac{d\mathbf{r}}{dt} = \mathbf{i} + 2t\mathbf{j} \Rightarrow \mathbf{a} = \frac{d\mathbf{v}}{dt} = 2\mathbf{j} \Rightarrow \mathbf{v} = \mathbf{i} + 2\mathbf{j}$ and $\mathbf{a} = 2\mathbf{j}$ at $t = 1$

3. $x = e^t$ and $y = \frac{2}{9}e^{2t} \Rightarrow y = \frac{2}{9}x^2$; $\mathbf{v} = \frac{d\mathbf{r}}{dt} = e^t\mathbf{i} + \frac{4}{9}e^{2t}\mathbf{j} \Rightarrow \mathbf{a} = e^t\mathbf{i} + \frac{8}{9}e^{2t}\mathbf{j} \Rightarrow \mathbf{v} = 3\mathbf{i} + 4\mathbf{j}$ and $\mathbf{a} = 3\mathbf{i} + 8\mathbf{j}$ at $t = \ln 3$

5. $\mathbf{v} = \frac{d\mathbf{r}}{dt} = (\cos t)\mathbf{i} - (\sin t)\mathbf{j}$ and $\mathbf{a} = \frac{d\mathbf{v}}{dt} = -(\sin t)\mathbf{i} - (\cos t)\mathbf{j}$
$\Rightarrow$ for $t = \frac{\pi}{4}$, $\mathbf{v}\left(\frac{\pi}{4}\right) = \frac{\sqrt{2}}{2}\mathbf{i} - \frac{\sqrt{2}}{2}\mathbf{j}$ and
$\mathbf{a}\left(\frac{\pi}{4}\right) = -\frac{\sqrt{2}}{2}\mathbf{i} - \frac{\sqrt{2}}{2}\mathbf{j}$; for $t = \frac{\pi}{2}$, $\mathbf{v}\left(\frac{\pi}{2}\right) = -\mathbf{j}$ and
$\mathbf{a}\left(\frac{\pi}{2}\right) = -\mathbf{i}$

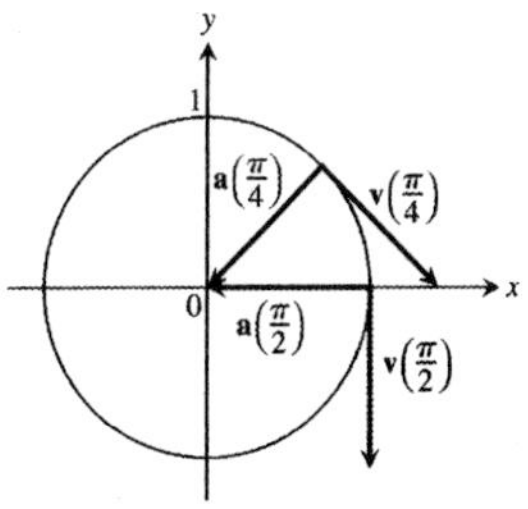

7. $\mathbf{v} = \frac{d\mathbf{r}}{dt} = (1 - \cos t)\mathbf{i} + (\sin t)\mathbf{j}$ and $\mathbf{a} = \frac{d\mathbf{v}}{dt}$
$= (\sin t)\mathbf{i} + (\cos t)\mathbf{j} \Rightarrow$ for $t = \pi$, $\mathbf{v}(\pi) = 2\mathbf{i}$ and $\mathbf{a}(\pi) = -\mathbf{j}$;
for $t = \frac{3\pi}{2}$, $\mathbf{v}\left(\frac{3\pi}{2}\right) = \mathbf{i} - \mathbf{j}$ and $\mathbf{a}\left(\frac{3\pi}{2}\right) = -\mathbf{i}$

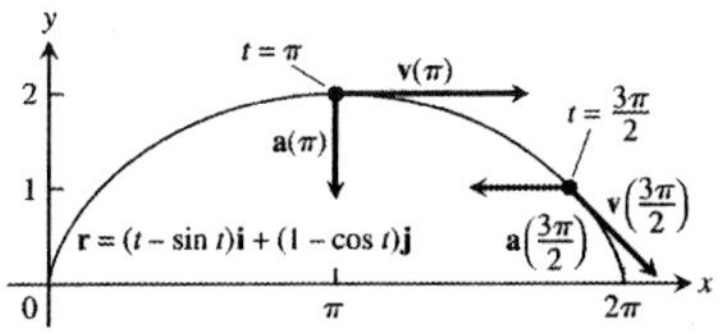

9. $\mathbf{r} = (t+1)\mathbf{i} + (t^2 - 1)\mathbf{j} + 2t\mathbf{k} \Rightarrow \mathbf{v} = \frac{d\mathbf{r}}{dt} = \mathbf{i} + 2t\mathbf{j} + 2\mathbf{k} \Rightarrow \mathbf{a} = \frac{d^2\mathbf{r}}{dt^2} = 2\mathbf{j}$; Speed: $|\mathbf{v}(1)| = \sqrt{1^2 + (2(1))^2 + 2^2} = 3$;
Direction: $\frac{\mathbf{v}(1)}{|\mathbf{v}(1)|} = \frac{\mathbf{i} + 2(1)\mathbf{j} + 2\mathbf{k}}{3} = \frac{1}{3}\mathbf{i} + \frac{2}{3}\mathbf{j} + \frac{2}{3}\mathbf{k} \Rightarrow \mathbf{v}(1) = 3\left(\frac{1}{3}\mathbf{i} + \frac{2}{3}\mathbf{j} + \frac{2}{3}\mathbf{k}\right)$

11. $\mathbf{r} = (2\cos t)\mathbf{i} + (3\sin t)\mathbf{j} + 4t\mathbf{k} \Rightarrow \mathbf{v} = \frac{d\mathbf{r}}{dt} = (-2\sin t)\mathbf{i} + (3\cos t)\mathbf{j} + 4\mathbf{k} \Rightarrow \mathbf{a} = \frac{d^2\mathbf{r}}{dt^2} = (-2\cos t)\mathbf{i} - (3\sin t)\mathbf{j}$;
Speed: $\left|\mathbf{v}\left(\frac{\pi}{2}\right)\right| = \sqrt{\left(-2\sin\frac{\pi}{2}\right)^2 + \left(3\cos\frac{\pi}{2}\right)^2 + 4^2} = 2\sqrt{5}$; Direction: $\frac{\mathbf{v}\left(\frac{\pi}{2}\right)}{\left|\mathbf{v}\left(\frac{\pi}{2}\right)\right|}$
$= \left(-\frac{2}{2\sqrt{5}}\sin\frac{\pi}{2}\right)\mathbf{i} + \left(\frac{3}{2\sqrt{5}}\cos\frac{\pi}{2}\right)\mathbf{j} + \frac{4}{2\sqrt{5}}\mathbf{k} = -\frac{1}{\sqrt{5}}\mathbf{i} + \frac{2}{\sqrt{5}}\mathbf{k} \Rightarrow \mathbf{v}\left(\frac{\pi}{2}\right) = 2\sqrt{5}\left(-\frac{1}{\sqrt{5}}\mathbf{i} + \frac{2}{\sqrt{5}}\mathbf{k}\right)$

13. $\mathbf{r} = (2\ln(t+1))\mathbf{i} + t^2\mathbf{j} + \frac{t^2}{2}\mathbf{k} \Rightarrow \mathbf{v} = \frac{d\mathbf{r}}{dt} = \left(\frac{2}{t+1}\right)\mathbf{i} + 2t\mathbf{j} + t\mathbf{k} \Rightarrow \mathbf{a} = \frac{d^2\mathbf{r}}{dt^2} = \left[\frac{-2}{(t+1)^2}\right]\mathbf{i} + 2\mathbf{j} + \mathbf{k}$;
Speed: $|\mathbf{v}(1)| = \sqrt{\left(\frac{2}{1+1}\right)^2 + (2(1))^2 + 1^2} = \sqrt{6}$; Direction: $\frac{\mathbf{v}(1)}{|\mathbf{v}(1)|} = \frac{\left(\frac{2}{1+1}\right)\mathbf{i} + 2(1)\mathbf{j} + (1)\mathbf{k}}{\sqrt{6}}$
$= \frac{1}{\sqrt{6}}\mathbf{i} + \frac{2}{\sqrt{6}}\mathbf{j} + \frac{1}{\sqrt{6}}\mathbf{k} \Rightarrow \mathbf{v}(1) = \sqrt{6}\left(\frac{1}{\sqrt{6}}\mathbf{i} + \frac{2}{\sqrt{6}}\mathbf{j} + \frac{1}{\sqrt{6}}\mathbf{k}\right)$

15. $\mathbf{v} = 3\mathbf{i} + \sqrt{3}\mathbf{j} + 2t\mathbf{k}$ and $\mathbf{a} = 2\mathbf{k} \Rightarrow \mathbf{v}(0) = 3\mathbf{i} + \sqrt{3}\mathbf{j}$ and $\mathbf{a}(0) = 2\mathbf{k} \Rightarrow |\mathbf{v}(0)| = \sqrt{3^2 + \left(\sqrt{3}\right)^2 + 0^2} = \sqrt{12}$ and $|\mathbf{a}(0)| = \sqrt{2^2} = 2$; $\mathbf{v}(0) \cdot \mathbf{a}(0) = 0 \Rightarrow \cos\theta = 0 \Rightarrow \theta = \frac{\pi}{2}$

17. $\mathbf{v} = \left(\frac{2t}{t^2+1}\right)\mathbf{i} + \left(\frac{1}{t^2+1}\right)\mathbf{j} + t(t^2+1)^{-1/2}\mathbf{k}$ and $\mathbf{a} = \left[\frac{-2t^2+2}{(t^2+1)^2}\right]\mathbf{i} - \left[\frac{2t}{(t^2+1)^2}\right]\mathbf{j} + \left[\frac{1}{(t^2+1)^{3/2}}\right]\mathbf{k} \Rightarrow \mathbf{v}(0) = \mathbf{j}$ and $\mathbf{a}(0) = 2\mathbf{i} + \mathbf{k} \Rightarrow |\mathbf{v}(0)| = 1$ and $|\mathbf{a}(0)| = \sqrt{2^2 + 1^2} = \sqrt{5}$; $\mathbf{v}(0) \cdot \mathbf{a}(0) = 0 \Rightarrow \cos\theta = 0 \Rightarrow \theta = \frac{\pi}{2}$

19. $\mathbf{r}(t) = (\sin t)\mathbf{i} + (t^2 - \cos t)\,\mathbf{j} + e^t\mathbf{k} \Rightarrow \mathbf{v}(t) = (\cos t)\mathbf{i} + (2t + \sin t)\mathbf{j} + e^t\mathbf{k}\,;\ t_0 = 0 \Rightarrow \mathbf{v}(t_0) = \mathbf{i} + \mathbf{k}$ and $\mathbf{r}(t_0) = P_0 = (0, -1, 1) \Rightarrow x = 0 + t = t,\ y = -1$, and $z = 1 + t$ are parametric equations of the tangent line

21. $\mathbf{r}(t) = (a \sin t)\mathbf{i} + (a \cos t)\,\mathbf{j} + bt\mathbf{k} \Rightarrow \mathbf{v}(t) = (a \cos t)\mathbf{i} - (a \sin t)\mathbf{j} + b\mathbf{k}\,;\ t_0 = 2\pi \Rightarrow \mathbf{v}(t_0) = a\mathbf{i} + b\mathbf{k}$ and $\mathbf{r}(t_0) = P_0 = (0, a, 2b\pi) \Rightarrow x = 0 + at = at,\ y = a$, and $z = 2\pi b + bt$ are parametric equations of the tangent line

23. (a) $\mathbf{v}(t) = -(\sin t)\mathbf{i} + (\cos t)\mathbf{j} \Rightarrow \mathbf{a}(t) = -(\cos t)\mathbf{i} - (\sin t)\mathbf{j}\,;$
 (i) $|\mathbf{v}(t)| = \sqrt{(-\sin t)^2 + (\cos t)^2} = 1 \Rightarrow$ constant speed;
 (ii) $\mathbf{v} \cdot \mathbf{a} = (\sin t)(\cos t) - (\cos t)(\sin t) = 0 \Rightarrow$ yes, orthogonal;
 (iii) counterclockwise movement;
 (iv) yes, $\mathbf{r}(0) = \mathbf{i} + 0\mathbf{j}$
 (b) $\mathbf{v}(t) = -(2 \sin 2t)\mathbf{i} + (2 \cos 2t)\mathbf{j} \Rightarrow \mathbf{a}(t) = -(4 \cos 2t)\mathbf{i} - (4 \sin 2t)\mathbf{j};$
 (i) $|\mathbf{v}(t)| = \sqrt{4 \sin^2 2t + 4 \cos^2 2t} = 2 \Rightarrow$ constant speed;
 (ii) $\mathbf{v} \cdot \mathbf{a} = 8 \sin 2t \cos 2t - 8 \cos 2t \sin 2t = 0 \Rightarrow$ yes, orthogonal;
 (iii) counterclockwise movement;
 (iv) yes, $\mathbf{r}(0) = \mathbf{i} + 0\mathbf{j}$
 (c) $\mathbf{v}(t) = -\sin\left(t - \frac{\pi}{2}\right)\mathbf{i} + \cos\left(t - \frac{\pi}{2}\right)\mathbf{j} \Rightarrow \mathbf{a}(t) = -\cos\left(t - \frac{\pi}{2}\right)\mathbf{i} - \sin\left(t - \frac{\pi}{2}\right)\mathbf{j}\,;$
 (i) $|\mathbf{v}(t)| = \sqrt{\sin^2\left(t - \frac{\pi}{2}\right) + \cos^2\left(t - \frac{\pi}{2}\right)} = 1 \Rightarrow$ constant speed;
 (ii) $\mathbf{v} \cdot \mathbf{a} = \sin\left(t - \frac{\pi}{2}\right)\cos\left(t - \frac{\pi}{2}\right) - \cos\left(t - \frac{\pi}{2}\right)\sin\left(t - \frac{\pi}{2}\right) = 0 \Rightarrow$ yes, orthogonal;
 (iii) counterclockwise movement;
 (iv) no, $\mathbf{r}(0) = 0\mathbf{i} - \mathbf{j}$ instead of $\mathbf{i} + 0\mathbf{j}$
 (d) $\mathbf{v}(t) = -(\sin t)\mathbf{i} - (\cos t)\mathbf{j} \Rightarrow \mathbf{a}(t) = -(\cos t)\mathbf{i} + (\sin t)\mathbf{j}\,;$
 (i) $|\mathbf{v}(t)| = \sqrt{(-\sin t)^2 + (-\cos t)^2} = 1 \Rightarrow$ constant speed;
 (ii) $\mathbf{v} \cdot \mathbf{a} = (\sin t)(\cos t) - (\cos t)(\sin t) = 0 \Rightarrow$ yes, orthogonal;
 (iii) clockwise movement;
 (iv) yes, $\mathbf{r}(0) = \mathbf{i} - 0\mathbf{j}$
 (e) $\mathbf{v}(t) = -(2t \sin t)\mathbf{i} + (2t \cos t)\mathbf{j} \Rightarrow \mathbf{a}(t) = -(2 \sin t + 2t \cos t)\mathbf{i} + (2 \cos t - 2t \sin t)\mathbf{j}\,;$
 (i) $|\mathbf{v}(t)| = \sqrt{[-(2t \sin t)]^2 + (2t \cos t)^2} = \sqrt{4t^2(\sin^2 t + \cos^2 t)} = 2|t| = 2t,\ t \ge 0$
 $\Rightarrow$ variable speed;
 (ii) $\mathbf{v} \cdot \mathbf{a} = 4\left(t \sin^2 t + t^2 \sin t \cos t\right) + 4\left(t \cos^2 t - t^2 \cos t \sin t\right) = 4t \neq 0$ in general
 $\Rightarrow$ not orthogonal in general;
 (iii) counterclockwise movement;
 (iv) yes, $\mathbf{r}(0) = \mathbf{i} + 0\mathbf{j}$

25. The velocity vector is tangent to the graph of $y^2 = 2x$ at the point $(2, 2)$, has length 5, and a positive $\mathbf{i}$ component. Now, $y^2 = 2x \Rightarrow 2y\,\frac{dy}{dx} = 2 \Rightarrow \left.\frac{dy}{dx}\right|_{(2,2)} = \frac{2}{2\cdot 2} = \frac{1}{2} \Rightarrow$ the tangent vector lies in the direction of the vector $\mathbf{i} + \frac{1}{2}\mathbf{j} \Rightarrow$ the velocity vector is $\mathbf{v} = \frac{5}{\sqrt{1+\frac{1}{4}}}\left(\mathbf{i} + \frac{1}{2}\mathbf{j}\right) = \frac{5}{\left(\frac{\sqrt{5}}{2}\right)}\left(\mathbf{i} + \frac{1}{2}\mathbf{j}\right) = 2\sqrt{5}\mathbf{i} + \sqrt{5}\mathbf{j}$

27. $\frac{d}{dt}(\mathbf{v} \cdot \mathbf{v}) = \mathbf{v} \cdot \frac{d\mathbf{v}}{dt} + \frac{d\mathbf{v}}{dt} \cdot \mathbf{v} = 2\mathbf{v} \cdot \frac{d\mathbf{v}}{dt} = 2 \cdot 0 = 0 \Rightarrow \mathbf{v} \cdot \mathbf{v}$ is a constant $\Rightarrow |\mathbf{v}| = \sqrt{\mathbf{v} \cdot \mathbf{v}}$ is constant

29. (a) $\mathbf{u} = f(t)\mathbf{i} + g(t)\mathbf{j} + h(t)\mathbf{k} \Rightarrow c\mathbf{u} = cf(t)\mathbf{i} + cg(t)\mathbf{j} + ch(t)\mathbf{k} \Rightarrow \frac{d}{dt}(c\mathbf{u}) = c\,\frac{df}{dt}\,\mathbf{i} + c\,\frac{dg}{dt}\,\mathbf{j} + c\,\frac{dh}{dt}\,\mathbf{k}$
 $= c\left(\frac{df}{dt}\,\mathbf{i} + \frac{dg}{dt}\,\mathbf{j} + \frac{dh}{dt}\,\mathbf{k}\right) = c\,\frac{d\mathbf{u}}{dt}$
 (b) $f\mathbf{u} = ff(t)\mathbf{i} + fg(t)\mathbf{j} + fh(t)\mathbf{k} \Rightarrow \frac{d}{dt}(f\mathbf{u}) = \left[\frac{df}{dt}\,f(t) + f\,\frac{df}{dt}\right]\mathbf{i} + \left[\frac{df}{dt}\,g(t) + f\,\frac{dg}{dt}\right]\mathbf{j} + \left[\frac{df}{dt}\,h(t) + f\,\frac{dh}{dt}\right]\mathbf{k}$
 $= \frac{df}{dt}\,[f(t)\mathbf{i} + g(t)\mathbf{j} + h(t)\mathbf{k}] + f\left[\frac{df}{dt}\,\mathbf{i} + \frac{dg}{dt}\,\mathbf{j} + \frac{dh}{dt}\,\mathbf{k}\right] = \frac{df}{dt}\,\mathbf{u} + f\,\frac{d\mathbf{u}}{dt}$

31. Suppose $\mathbf{r}$ is continuous at $t = t_0$. Then $\lim_{t \to t_0} \mathbf{r}(t) = \mathbf{r}(t_0) \Leftrightarrow \lim_{t \to t_0} [f(t)\mathbf{i} + g(t)\mathbf{j} + h(t)\mathbf{k}]$

$= f(t_0)\mathbf{i} + g(t_0)\mathbf{j} + h(t_0)\mathbf{k} \Leftrightarrow \lim_{t \to t_0} f(t) = f(t_0),\ \lim_{t \to t_0} g(t) = g(t_0)$, and $\lim_{t \to t_0} h(t) = h(t_0) \Leftrightarrow$ f, g, and h are continuous at $t = t_0$.

33. $\mathbf{r}'(t_0)$ exists $\Rightarrow f'(t_0)\mathbf{i} + g'(t_0)\mathbf{j} + h'(t_0)\mathbf{k}$ exists $\Rightarrow f'(t_0), g'(t_0), h'(t_0)$ all exist $\Rightarrow$ f, g, and h are continuous at $t = t_0 \Rightarrow \mathbf{r}(t)$ is continuous at $t = t_0$

10.2 INTEGRALS OF VECTOR FUNCTIONS

1. $\int_0^1 [t^3\mathbf{i} + 7\mathbf{j} + (t+1)\mathbf{k}]\,dt = \left[\frac{t^4}{4}\right]_0^1 \mathbf{i} + [7t]_0^1 \mathbf{j} + \left[\frac{t^2}{2} + t\right]_0^1 \mathbf{k} = \frac{1}{4}\mathbf{i} + 7\mathbf{j} + \frac{3}{2}\mathbf{k}$

3. $\int_{-\pi/4}^{\pi/4} [(\sin t)\mathbf{i} + (1 + \cos t)\mathbf{j} + (\sec^2 t)\,\mathbf{k}]\,dt = [-\cos t]_{-\pi/4}^{\pi/4}\mathbf{i} + [t + \sin t]_{-\pi/4}^{\pi/4}\mathbf{j} + [\tan t]_{-\pi/4}^{\pi/4}\mathbf{k} = \left(\frac{\pi + 2\sqrt{2}}{2}\right)\mathbf{j} + 2\mathbf{k}$

5. $\int_1^4 \left(\frac{1}{t}\mathbf{i} + \frac{1}{5-t}\mathbf{j} + \frac{1}{2t}\mathbf{k}\right)dt = = [\ln t]_1^4 \mathbf{i} + [-\ln(5-t)]_1^4 \mathbf{j} + \left[\frac{1}{2}\ln t\right]_1^4 \mathbf{k} = (\ln 4)\mathbf{i} + (\ln 4)\mathbf{j} + (\ln 2)\mathbf{k}$

7. $\mathbf{r} = \int(-t\mathbf{i} - t\mathbf{j} - t\mathbf{k})\,dt = -\frac{t^2}{2}\mathbf{i} - \frac{t^2}{2}\mathbf{j} - \frac{t^2}{2}\mathbf{k} + \mathbf{C}$; $\mathbf{r}(0) = 0\mathbf{i} - 0\mathbf{j} - 0\mathbf{k} + \mathbf{C} = \mathbf{i} + 2\mathbf{j} + 3\mathbf{k} \Rightarrow \mathbf{C} = \mathbf{i} + 2\mathbf{j} + 3\mathbf{k}$
$\Rightarrow \mathbf{r} = \left(-\frac{t^2}{2} + 1\right)\mathbf{i} + \left(-\frac{t^2}{2} + 2\right)\mathbf{j} + \left(-\frac{t^2}{2} + 3\right)\mathbf{k}$

9. $\mathbf{r} = \int \left[\left(\frac{3}{2}(t+1)^{1/2}\right)\mathbf{i} + e^{-t}\mathbf{j} + \left(\frac{1}{t+1}\right)\mathbf{k}\right]dt = (t+1)^{3/2}\,\mathbf{i} - e^{-t}\,\mathbf{j} + \ln(t+1)\mathbf{k} + \mathbf{C}$;
$\mathbf{r}(0) = (0+1)^{3/2}\,\mathbf{i} - e^{-0}\,\mathbf{j} + \ln(0+1)\mathbf{k} + \mathbf{C} = \mathbf{k} \Rightarrow \mathbf{C} = -\mathbf{i} + \mathbf{j} + \mathbf{k}$
$\Rightarrow \mathbf{r} = \left[(t+1)^{3/2} - 1\right]\mathbf{i} + (1 - e^{-t})\,\mathbf{j} + [1 + \ln(t+1)]\mathbf{k}$

11. $\frac{d\mathbf{r}}{dt} = \int(-32\mathbf{k})\,dt = -32t\mathbf{k} + \mathbf{C}_1$; $\frac{d\mathbf{r}}{dt}(0) = 8\mathbf{i} + 8\mathbf{j} \Rightarrow -32(0)\mathbf{k} + \mathbf{C}_1 = 8\mathbf{i} + 8\mathbf{j} \Rightarrow \mathbf{C}_1 = 8\mathbf{i} + 8\mathbf{j}$
$\Rightarrow \frac{d\mathbf{r}}{dt} = 8\mathbf{i} + 8\mathbf{j} - 32t\mathbf{k}$; $\mathbf{r} = \int(8\mathbf{i} + 8\mathbf{j} - 32t\mathbf{k})\,dt = 8t\mathbf{i} + 8t\mathbf{j} - 16t^2\mathbf{k} + \mathbf{C}_2$; $\mathbf{r}(0) = 100\mathbf{k}$
$\Rightarrow 8(0)\mathbf{i} + 8(0)\mathbf{j} - 16(0)^2\mathbf{k} + \mathbf{C}_2 = 100\mathbf{k} \Rightarrow \mathbf{C}_2 = 100\mathbf{k} \Rightarrow \mathbf{r} = 8t\mathbf{i} + 8t\mathbf{j} + (100 - 16t^2)\,\mathbf{k}$

13. $\frac{d\mathbf{v}}{dt} = \mathbf{a} = 3\mathbf{i} - \mathbf{j} + \mathbf{k} \Rightarrow \mathbf{v}(t) = 3t\mathbf{i} - t\mathbf{j} + t\mathbf{k} + \mathbf{C}_1$; the particle travels in the direction of the vector $(4-1)\mathbf{i} + (1-2)\mathbf{j} + (4-3)\mathbf{k} = 3\mathbf{i} - \mathbf{j} + \mathbf{k}$ (since it travels in a straight line), and at time $t = 0$ it has speed
$2 \Rightarrow \mathbf{v}(0) = \frac{2}{\sqrt{9+1+1}}(3\mathbf{i} - \mathbf{j} + \mathbf{k}) = \mathbf{C}_1 \Rightarrow \frac{d\mathbf{r}}{dt} = \mathbf{v}(t) = \left(3t + \frac{6}{\sqrt{11}}\right)\mathbf{i} - \left(t + \frac{2}{\sqrt{11}}\right)\mathbf{j} + \left(t + \frac{2}{\sqrt{11}}\right)\mathbf{k}$
$\Rightarrow \mathbf{r}(t) = \left(\frac{3}{2}t^2 + \frac{6}{\sqrt{11}}t\right)\mathbf{i} - \left(\frac{1}{2}t^2 + \frac{2}{\sqrt{11}}t\right)\mathbf{j} + \left(\frac{1}{2}t^2 + \frac{2}{\sqrt{11}}t\right)\mathbf{k} + \mathbf{C}_2$; $\mathbf{r}(0) = \mathbf{i} + 2\mathbf{j} + 3\mathbf{k} = \mathbf{C}_2$
$\Rightarrow \mathbf{r}(t) = \left(\frac{3}{2}t^2 + \frac{6}{\sqrt{11}}t + 1\right)\mathbf{i} - \left(\frac{1}{2}t^2 + \frac{2}{\sqrt{11}}t - 2\right)\mathbf{j} + \left(\frac{1}{2}t^2 + \frac{2}{\sqrt{11}}t + 3\right)\mathbf{k}$
$= \left(\frac{1}{2}t^2 + \frac{2}{\sqrt{11}}t\right)(3\mathbf{i} - \mathbf{j} + \mathbf{k}) + (\mathbf{i} + 2\mathbf{j} + 3\mathbf{k})$

15. $x = (v_0 \cos\alpha)t \Rightarrow (21\text{ km})\left(\frac{1000\text{ m}}{1\text{ km}}\right) = (840\text{ m/s})(\cos 60°)t \Rightarrow t = \frac{21{,}000\text{ m}}{(840\text{ m/s})(\cos 60°)} = 50$ seconds

17. (a) $t = \frac{2v_0 \sin\alpha}{g} = \frac{2(500\text{ m/s})(\sin 45°)}{9.8\text{ m/s}^2} \approx 72.2$ seconds; $R = \frac{v_0^2}{g}\sin 2\alpha = \frac{(500\text{ m/s})^2}{9.8\text{ m/s}^2}(\sin 90°) \approx 25{,}510.2$ m
(b) $x = (v_0 \cos\alpha)t \Rightarrow 5000\text{ m} = (500\text{ m/s})(\cos 45°)t \Rightarrow t = \frac{5000\text{ m}}{(500\text{ m/s})(\cos 45°)} \approx 14.14$ s; thus,
$y = (v_0 \sin\alpha)t - \frac{1}{2}gt^2 \Rightarrow y \approx (500\text{ m/s})(\sin 45°)(14.14\text{ s}) - \frac{1}{2}(9.8\text{ m/s}^2)(14.14\text{ s})^2 \approx 4020$ m
(c) $y_{max} = \frac{(v_0 \sin\alpha)^2}{2g} = \frac{((500\text{ m/s})(\sin 45°))^2}{2(9.8\text{ m/s}^2)} \approx 6378$ m

19. (a) $R = \frac{v_0^2}{g}\sin 2\alpha \Rightarrow 10\text{ m} = \left(\frac{v_0^2}{9.8\text{ m/s}^2}\right)(\sin 90^\circ) \Rightarrow v_0^2 = 98\text{ m}^2\text{s}^2 \Rightarrow v_0 \approx 9.9\text{ m/s};$

(b) $6\text{m} \approx \frac{(9.9\text{ m/s})^2}{9.8\text{ m/s}^2}(\sin 2\alpha) \Rightarrow \sin 2\alpha \approx 0.59999 \Rightarrow 2\alpha \approx 36.87^\circ \text{ or } 143.12^\circ \Rightarrow \alpha \approx 18.4^\circ \text{ or } 71.6^\circ$

21. $R = \frac{v_0^2}{g}\sin 2\alpha \Rightarrow 16{,}000\text{ m} = \frac{(400\text{ m/s})^2}{9.8\text{ m/s}^2}\sin 2\alpha \Rightarrow \sin 2\alpha = 0.98 \Rightarrow 2\alpha \approx 78.5^\circ \text{ or } 2\alpha \approx 101.5^\circ \Rightarrow \alpha \approx 39.3^\circ \text{ or } 50.7^\circ$

23. $\frac{d\mathbf{r}}{dt} = \int(-g\mathbf{j})\,dt = -gt\mathbf{j} + \mathbf{C}_1$ and $\frac{d\mathbf{r}}{dt}(0) = (v_0\cos\alpha)\mathbf{i} + (v_0\sin\alpha)\mathbf{j} \Rightarrow -g(0)\mathbf{j} + \mathbf{C}_1 = (v_0\cos\alpha)\mathbf{i} + (v_0\sin\alpha)\mathbf{j}$
$\Rightarrow \mathbf{C}_1 = (v_0\cos\alpha)\mathbf{i} + (v_0\sin\alpha)\mathbf{j} \Rightarrow \frac{d\mathbf{r}}{dt} = (v_0\cos\alpha)\mathbf{i} + (v_0\sin\alpha - gt)\mathbf{j}$; $\mathbf{r} = \int[(v_0\cos\alpha)\mathbf{i} + (v_0\sin\alpha - gt)\mathbf{j}]\,dt$
$= (v_0t\cos\alpha)\mathbf{i} + \left(v_0t\sin\alpha - \frac{1}{2}gt^2\right)\mathbf{j} + \mathbf{C}_2$ and $\mathbf{r}(0) = x_0\mathbf{i} + y_0\mathbf{j} \Rightarrow [v_0(0)\cos\alpha]\mathbf{i} + \left[v_0(0)\sin\alpha - \frac{1}{2}g(0)^2\right]\mathbf{j} + \mathbf{C}_2$
$= x_0\mathbf{i} + y_0\mathbf{j} \Rightarrow \mathbf{C}_2 = x_0\mathbf{i} + y_0\mathbf{j} \Rightarrow \mathbf{r} = (x_0 + v_0t\cos\alpha)\mathbf{i} + \left(y_0 + v_0t\sin\alpha - \frac{1}{2}gt^2\right)\mathbf{j} \Rightarrow x = x_0 + v_0t\cos\alpha$ and
$y = y_0 + v_0t\sin\alpha - \frac{1}{2}gt^2$

25. (a) $\int_a^b k\mathbf{r}(t)\,dt = \int_a^b [kf(t)\mathbf{i} + kg(t)\mathbf{j} + kh(t)\mathbf{k}]\,dt = \int_a^b [kf(t)]\,dt\,\mathbf{i} + \int_a^b [kg(t)]\,dt\,\mathbf{j} + \int_a^b [kh(t)]\,dt\,\mathbf{k}$
$= k\left(\int_a^b f(t)\,dt\,\mathbf{i} + \int_a^b g(t)\,dt\,\mathbf{j} + \int_a^b h(t)\,dt\,\mathbf{k}\right) = k\int_a^b \mathbf{r}(t)\,dt$

(b) $\int_a^b [\mathbf{r}_1(t) \pm \mathbf{r}_2(t)]\,dt = \int_a^b ([f_1(t)\mathbf{i} + g_1(t)\mathbf{j} + h_1(t)\mathbf{k}] \pm [f_2(t)\mathbf{i} + g_2(t)\mathbf{j} + h_2(t)\mathbf{k}])\,dt$
$= \int_a^b ([f_1(t) \pm f_2(t)]\,\mathbf{i} + [g_1(t) \pm g_2(t)]\,\mathbf{j} + [h_1(t) \pm h_2(t)]\,\mathbf{k})\,dt$
$= \int_a^b [f_1(t) \pm f_2(t)]\,dt\,\mathbf{i} + \int_a^b [g_1(t) \pm g_2(t)]\,dt\,\mathbf{j} + \int_a^b [h_1(t) \pm h_2(t)]\,dt\,\mathbf{k}$
$= \left[\int_a^b f_1(t)\,dt\,\mathbf{i} \pm \int_a^b f_2(t)\,dt\,\mathbf{i}\right] + \left[\int_a^b g_1(t)\,dt\,\mathbf{j} \pm \int_a^b g_2(t)\,dt\,\mathbf{j}\right] + \left[\int_a^b h_1(t)\,dt\,\mathbf{k} \pm \int_a^b h_2(t)\,dt\,\mathbf{k}\right]$
$= \int_a^b \mathbf{r}_1(t)\,dt \pm \int_a^b \mathbf{r}_2(t)\,dt$

(c) Let $\mathbf{C} = c_1\mathbf{i} + c_2\mathbf{j} + c_3\mathbf{k}$. Then $\int_a^b \mathbf{C}\cdot\mathbf{r}(t)\,dt = \int_a^b [c_1f(t) + c_2g(t) + c_3h(t)]\,dt$
$= c_1\int_a^b f(t)\,dt + c_2\int_a^b g(t)\,dt + c_3\int_a^b h(t)\,dt = \mathbf{C}\cdot\int_a^b \mathbf{r}(t)\,dt;$
$\int_a^b \mathbf{C}\times\mathbf{r}(t)\,dt = \int_a^b [c_2h(t) - c_3g(t)]\,\mathbf{i} + [c_3f(t) - c_1h(t)]\,\mathbf{j} + [c_1g(t) - c_2f(t)]\,\mathbf{k}\,dt$
$= \left[c_2\int_a^b h(t)\,dt - c_3\int_a^b g(t)\,dt\right]\mathbf{i} + \left[c_3\int_a^b f(t)\,dt - c_1\int_a^b h(t)\,dt\right]\mathbf{j} + \left[c_1\int_a^b g(t)\,dt - c_2\int_a^b f(t)\,dt\right]\mathbf{k}$
$= \mathbf{C}\times\int_a^b \mathbf{r}(t)\,dt$

27. (a) If $\mathbf{R}_1(t)$ and $\mathbf{R}_2(t)$ have identical derivatives on I, then $\frac{d\mathbf{R}_1}{dt} = \frac{df_1}{dt}\mathbf{i} + \frac{dg_1}{dt}\mathbf{j} + \frac{dh_1}{dt}\mathbf{k} = \frac{df_2}{dt}\mathbf{i} + \frac{dg_2}{dt}\mathbf{j} + \frac{dh_2}{dt}\mathbf{k}$
$= \frac{d\mathbf{R}_2}{dt} \Rightarrow \frac{df_1}{dt} = \frac{df_2}{dt}, \frac{dg_1}{dt} = \frac{dg_2}{dt}, \frac{dh_1}{dt} = \frac{dh_2}{dt} \Rightarrow f_1(t) = f_2(t) + c_1, g_1(t) = g_2(t) + c_2, h_1(t) = h_2(t) + c_3$
$\Rightarrow f_1(t)\mathbf{i} + g_1(t)\mathbf{j} + h_1(t)\mathbf{k} = [f_2(t) + c_1]\mathbf{i} + [g_2(t) + c_2]\mathbf{j} + [h_2(t) + c_3]\mathbf{k} \Rightarrow \mathbf{R}_1(t) = \mathbf{R}_2(t) + \mathbf{C}$, where
$\mathbf{C} = c_1\mathbf{i} + c_2\mathbf{j} + c_3\mathbf{k}$.

(b) Let $\mathbf{R}(t)$ be an antiderivative of $\mathbf{r}(t)$ on I. Then $\mathbf{R}'(t) = \mathbf{r}(t)$. If $\mathbf{U}(t)$ is an antiderivative of $\mathbf{r}(t)$ on I, then $\mathbf{U}'(t) = \mathbf{r}(t)$. Thus $\mathbf{U}'(t) = \mathbf{R}'(t)$ on I $\Rightarrow \mathbf{U}(t) = \mathbf{R}(t) + \mathbf{C}$.

29. (a) (Assuming that "x" is zero at the point of impact:)
$\mathbf{r}(t) = (x(t))\mathbf{i} + (y(t))\mathbf{j}$; where $x(t) = (35\cos 27^\circ)t$ and $y(t) = 4 + (35\sin 27^\circ)t - 16t^2$.

(b) $y_{max} = \frac{(v_0\sin\alpha)^2}{2g} + 4 = \frac{(35\sin 27^\circ)^2}{64} + 4 \approx 7.945$ feet, which is reached at $t = \frac{v_0\sin\alpha}{g} = \frac{35\sin 27^\circ}{32} \approx 0.497$ seconds.

(c) For the time, solve $y = 4 + (35\sin 27^\circ)t - 16t^2 = 0$ for t, using the quadratic formula
$t = \frac{35\sin 27^\circ + \sqrt{(-35\sin 27^\circ)^2 + 256}}{32} \approx 1.201$ sec. Then the range is about $x(1.201) = (35\cos 27^\circ)(1.201)$
≈ 37.453 feet.

(d) For the time, solve $y = 4 + (35 \sin 27°)t - 16t^2 = 7$ for t, using the quadratic formula

$t = \frac{35 \sin 27° + \sqrt{(-35 \sin 27°)^2 - 192}}{32} \approx 0.254$ and 0.740 seconds. At those times the ball is about $x(0.254) = (35 \cos 27°)(0.254) \approx 7.921$ feet and $x(0.740) = (35 \cos 27°)(0.740) \approx 23.077$ feet the impact point, or about $37.453 - 7.921 \approx 29.532$ feet and $37.453 - 23.077 \approx 14.376$ feet from the landing spot.

(e) Yes. It changes things because the ball won't clear the net ($y_{max} \approx 7.945$).

10.3 ARC LENGTH IN SPACE

1. $\mathbf{r} = (2 \cos t)\mathbf{i} + (2 \sin t)\mathbf{j} + \sqrt{5}t\mathbf{k} \Rightarrow \mathbf{v} = (-2 \sin t)\mathbf{i} + (2 \cos t)\mathbf{j} + \sqrt{5}\mathbf{k}$

$\Rightarrow |\mathbf{v}| = \sqrt{(-2 \sin t)^2 + (2 \cos t)^2 + \left(\sqrt{5}\right)^2} = \sqrt{4 \sin^2 t + 4 \cos^2 t + 5} = 3; \mathbf{T} = \frac{\mathbf{v}}{|\mathbf{v}|}$

$= \left(-\frac{2}{3} \sin t\right) \mathbf{i} + \left(\frac{2}{3} \cos t\right) \mathbf{j} + \frac{\sqrt{5}}{3} \mathbf{k}$ and Length $= \int_0^{\pi} |\mathbf{v}|\, dt = \int_0^{\pi} 3\, dt = [3t]_0^{\pi} = 3\pi$

3. $\mathbf{r} = t\mathbf{i} + \frac{2}{3} t^{3/2} \mathbf{k} \Rightarrow \mathbf{v} = \mathbf{i} + t^{1/2} \mathbf{k} \Rightarrow |\mathbf{v}| = \sqrt{1^2 + (t^{1/2})^2} = \sqrt{1 + t}; \mathbf{T} = \frac{\mathbf{v}}{|\mathbf{v}|} = \frac{1}{\sqrt{1+t}} \mathbf{i} + \frac{\sqrt{t}}{\sqrt{1+t}} \mathbf{k}$

and Length $= \int_0^8 \sqrt{1 + t}\, dt = \left[\frac{2}{3} (1 + t)^{3/2}\right]_0^8 = \frac{52}{3}$

5. $\mathbf{r} = (\cos^3 t)\, \mathbf{j} + (\sin^3 t)\, \mathbf{k} \Rightarrow \mathbf{v} = (-3 \cos^2 t \sin t)\, \mathbf{j} + (3 \sin^2 t \cos t)\, \mathbf{k} \Rightarrow |\mathbf{v}|$

$= \sqrt{(-3 \cos^2 t \sin t)^2 + (3 \sin^2 t \cos t)^2} = \sqrt{(9 \cos^2 t \sin^2 t)(\cos^2 t + \sin^2 t)} = 3\, |\cos t \sin t|$;

$\mathbf{T} = \frac{\mathbf{v}}{|\mathbf{v}|} = \frac{-3 \cos^2 t \sin t}{3\, |\cos t \sin t|}\, \mathbf{j} + \frac{3 \sin^2 t \cos t}{3\, |\cos t \sin t|}\, \mathbf{k} = (-\cos t)\mathbf{j} + (\sin t)\mathbf{k}$, if $0 \le t \le \frac{\pi}{2}$, and

Length $= \int_0^{\pi/2} 3\, |\cos t \sin t|\, dt = \int_0^{\pi/2} 3 \cos t \sin t\, dt = \int_0^{\pi/2} \frac{3}{2} \sin 2t\, dt = \left[-\frac{3}{4} \cos 2t\right]_0^{\pi/2} = \frac{3}{2}$

7. $\mathbf{r} = (t \cos t)\mathbf{i} + (t \sin t)\mathbf{j} + \frac{2\sqrt{2}}{3} t^{3/2}\mathbf{k} \Rightarrow \mathbf{v} = (\cos t - t \sin t)\mathbf{i} + (\sin t + t \cos t)\mathbf{j} + \left(\sqrt{2}\, t^{1/2}\right) \mathbf{k}$

$\Rightarrow |\mathbf{v}| = \sqrt{(\cos t - t \sin t)^2 + (\sin t + t \cos t)^2 + \left(\sqrt{2t}\right)^2} = \sqrt{1 + t^2 + 2t} = \sqrt{(t + 1)^2} = |t + 1| = t + 1$, if $t \ge 0$;

$\mathbf{T} = \frac{\mathbf{v}}{|\mathbf{v}|} = \left(\frac{\cos t - t \sin t}{t+1}\right) \mathbf{i} + \left(\frac{\sin t + t \cos t}{t+1}\right) \mathbf{j} + \left(\frac{\sqrt{2}\, t^{1/2}}{t+1}\right) \mathbf{k}$ and Length $= \int_0^{\pi} (t + 1)\, dt = \left[\frac{t^2}{2} + t\right]_0^{\pi} = \frac{\pi^2}{2} + \pi$

9. Let $P(t_0)$ denote the point. Then $\mathbf{v} = (5 \cos t)\mathbf{i} - (5 \sin t)\mathbf{j} + 12\mathbf{k}$ and $26\pi = \int_0^{t_0} \sqrt{25 \cos^2 t + 25 \sin^2 t + 144}\, dt$

$= \int_0^{t_0} 13\, dt = 13t_0 \Rightarrow t_0 = 2\pi$, and the point is $P(2\pi) = (5 \sin 2\pi, 5 \cos 2\pi, 24\pi) = (0, 5, 24\pi)$

11. $\mathbf{r} = (4 \cos t)\mathbf{i} + (4 \sin t)\mathbf{j} + 3t\mathbf{k} \Rightarrow \mathbf{v} = (-4 \sin t)\mathbf{i} + (4 \cos t)\mathbf{j} + 3\mathbf{k} \Rightarrow |\mathbf{v}| = \sqrt{(-4 \sin t)^2 + (4 \cos t)^2 + 3^2}$

$= \sqrt{25} = 5 \Rightarrow s(t) = \int_0^t 5\, d\tau = 5t \Rightarrow$ Length $= s\left(\frac{\pi}{2}\right) = \frac{5\pi}{2}$

13. $\mathbf{r} = (e^t \cos t)\, \mathbf{i} + (e^t \sin t)\, \mathbf{j} + e^t\mathbf{k} \Rightarrow \mathbf{v} = (e^t \cos t - e^t \sin t)\, \mathbf{i} + (e^t \sin t + e^t \cos t)\, \mathbf{j} + e^t\mathbf{k}$

$\Rightarrow |\mathbf{v}| = \sqrt{(e^t \cos t - e^t \sin t)^2 + (e^t \sin t + e^t \cos t)^2 + (e^t)^2} = = \sqrt{3e^{2t}} = \sqrt{3}\, e^t \Rightarrow s(t) = \int_0^t \sqrt{3}\, e^{\tau}\, d\tau$

$= \sqrt{3}\, e^t - \sqrt{3} \Rightarrow$ Length $= s(0) - s(-\ln 4) = 0 - \left(\sqrt{3}\, e^{-\ln 4} - \sqrt{3}\right) = \frac{3\sqrt{3}}{4}$

15. $\mathbf{r} = \left(\sqrt{2}t\right) \mathbf{i} + \left(\sqrt{2}t\right) \mathbf{j} + (1 - t^2)\, \mathbf{k} \Rightarrow \mathbf{v} = \sqrt{2}\mathbf{i} + \sqrt{2}\mathbf{j} - 2t\mathbf{k} \Rightarrow |\mathbf{v}| = \sqrt{\left(\sqrt{2}\right)^2 + \left(\sqrt{2}\right)^2 + (-2t)^2} = \sqrt{4 + 4t^2}$

$= 2\sqrt{1 + t^2} \Rightarrow$ Length $= \int_0^1 2\sqrt{1 + t^2}\, dt = \left[2\left(\frac{t}{2}\sqrt{1 + t^2} + \frac{1}{2} \ln\left(t + \sqrt{1 + t^2}\right)\right)\right]_0^1 = \sqrt{2} + \ln\left(1 + \sqrt{2}\right)$

17. $\angle PQB = \angle QOB = t$ and $PQ = \text{arc}(AQ) = t$ since PQ = length of the unwound string = length of arc (AQ); thus $x = OB + BC = OB + DP = \cos t + t \sin t$, and $y = PC = QB - QD = \sin t - t\cos t$

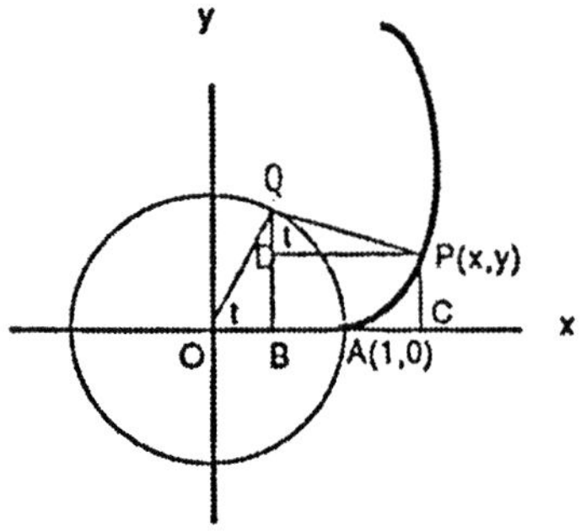

19. $\mathbf{v} = \frac{d}{dt}(x_0 + t u_1)\mathbf{i} + \frac{d}{dt}(y_0 + t u_2)\mathbf{j} + \frac{d}{dt}(z_0 + t u_3)\mathbf{k} = u_1\mathbf{i} + u_2\mathbf{j} + u_3\mathbf{k} = \mathbf{u}$, so $s(t) = \int_0^t |\mathbf{v}|dt = \int_0^t |\mathbf{u}|d\tau = \int_0^t 1\, d\tau = t$

10.4 CURVATURE OF A CURVE

1. $\mathbf{r} = t\mathbf{i} + \ln(\cos t)\mathbf{j} \Rightarrow \mathbf{v} = \mathbf{i} + \left(\frac{-\sin t}{\cos t}\right)\mathbf{j} = \mathbf{i} - (\tan t)\mathbf{j} \Rightarrow |\mathbf{v}| = \sqrt{1^2 + (-\tan t)^2} = \sqrt{\sec^2 t} = |\sec t| = \sec t$, since $-\frac{\pi}{2} < t < \frac{\pi}{2} \Rightarrow \mathbf{T} = \frac{\mathbf{v}}{|\mathbf{v}|} = \left(\frac{1}{\sec t}\right)\mathbf{i} - \left(\frac{\tan t}{\sec t}\right)\mathbf{j} = (\cos t)\mathbf{i} - (\sin t)\mathbf{j}$; $\frac{d\mathbf{T}}{dt} = (-\sin t)\mathbf{i} - (\cos t)\mathbf{j}$
$\Rightarrow \left|\frac{d\mathbf{T}}{dt}\right| = \sqrt{(-\sin t)^2 + (-\cos t)^2} = 1 \Rightarrow \mathbf{N} = \frac{\left(\frac{d\mathbf{T}}{dt}\right)}{\left|\frac{d\mathbf{T}}{dt}\right|} = (-\sin t)\mathbf{i} - (\cos t)\mathbf{j}$; $\kappa = \frac{1}{|\mathbf{v}|} \cdot \left|\frac{d\mathbf{T}}{dt}\right| = \frac{1}{\sec t} \cdot 1 = \cos t$.

3. $\mathbf{r} = (2t+3)\mathbf{i} + (5 - t^2)\mathbf{j} \Rightarrow \mathbf{v} = 2\mathbf{i} - 2t\mathbf{j} \Rightarrow |\mathbf{v}| = \sqrt{2^2 + (-2t)^2} = 2\sqrt{1+t^2} \Rightarrow \mathbf{T} = \frac{\mathbf{v}}{|\mathbf{v}|} = \frac{2}{2\sqrt{1+t^2}}\mathbf{i} + \frac{-2t}{2\sqrt{1+t^2}}\mathbf{j}$
$= \frac{1}{\sqrt{1+t^2}}\mathbf{i} - \frac{t}{\sqrt{1+t^2}}\mathbf{j}$; $\frac{d\mathbf{T}}{dt} = \frac{-t}{\left(\sqrt{1+t^2}\right)^3}\mathbf{i} - \frac{1}{\left(\sqrt{1+t^2}\right)^3}\mathbf{j} \Rightarrow \left|\frac{d\mathbf{T}}{dt}\right| = \sqrt{\left(\frac{-t}{\left(\sqrt{1+t^2}\right)^3}\right)^2 + \left(-\frac{1}{\left(\sqrt{1+t^2}\right)^3}\right)^2}$
$= \sqrt{\frac{1}{(1+t^2)^2}} = \frac{1}{1+t^2} \Rightarrow \mathbf{N} = \frac{\left(\frac{d\mathbf{T}}{dt}\right)}{\left|\frac{d\mathbf{T}}{dt}\right|} = \frac{-t}{\sqrt{1+t^2}}\mathbf{i} - \frac{1}{\sqrt{1+t^2}}\mathbf{j}$; $\kappa = \frac{1}{|\mathbf{v}|} \cdot \left|\frac{d\mathbf{T}}{dt}\right| = \frac{1}{2\sqrt{1+t^2}} \cdot \frac{1}{1+t^2} = \frac{1}{2(1+t^2)^{3/2}}$

5. (a) $\kappa(x) = \frac{1}{|\mathbf{v}(x)|} \cdot \left|\frac{d\mathbf{T}(x)}{dt}\right|$. Now, $\mathbf{v} = \mathbf{i} + f'(x)\mathbf{j} \Rightarrow |\mathbf{v}(x)| = \sqrt{1 + [f'(x)]^2} \Rightarrow \mathbf{T} = \frac{\mathbf{v}}{|\mathbf{v}|}$
$= \left(1 + [f'(x)]^2\right)^{-1/2}\mathbf{i} + f'(x)\left(1 + [f'(x)]^2\right)^{-1/2}\mathbf{j}$. Thus $\frac{d\mathbf{T}}{dt}(x) = \frac{-f'(x)f''(x)}{\left(1+[f'(x)]^2\right)^{3/2}}\mathbf{i} + \frac{f''(x)}{\left(1+[f'(x)]^2\right)^{3/2}}\mathbf{j}$
$\Rightarrow \left|\frac{d\mathbf{T}(x)}{dt}\right| = \sqrt{\left[\frac{-f'(x)f''(x)}{\left(1+[f'(x)]^2\right)^{3/2}}\right]^2 + \left(\frac{f''(x)}{\left(1+[f'(x)]^2\right)^{3/2}}\right)^2} = \sqrt{\frac{[f''(x)]^2\left(1+[f'(x)]^2\right)}{\left(1+[f'(x)]^2\right)^3}} = \frac{|f''(x)|}{\left|1+[f'(x)]^2\right|}$
Thus $\kappa(x) = \frac{1}{(1+[f'(x)]^2)^{1/2}} \cdot \frac{|f''(x)|}{|1+[f'(x)]^2|} = \frac{|f''(x)|}{\left(1+[f'(x)]^2\right)^{3/2}}$

(b) $y = \ln(\cos x) \Rightarrow \frac{dy}{dx} = \left(\frac{1}{\cos x}\right)(-\sin x) = -\tan x \Rightarrow \frac{d^2y}{dx^2} = -\sec^2 x \Rightarrow \kappa = \frac{|-\sec^2 x|}{[1+(-\tan x)^2]^{3/2}} = \frac{\sec^2 x}{|\sec^3 x|}$
$= \frac{1}{\sec x} = \cos x$, since $-\frac{\pi}{2} < x < \frac{\pi}{2}$

(c) Note that $f''(x) = 0$ at an inflection point.

7. (a) $\mathbf{r}(t) = f(t)\mathbf{i} + g(t)\mathbf{j} \Rightarrow \mathbf{v} = f'(t)\mathbf{i} + g'(t)\mathbf{j}$ is tangent to the curve at the point $(f(t), g(t))$;
$\mathbf{n} \cdot \mathbf{v} = [-g'(t)\mathbf{i} + f'(t)\mathbf{j}] \cdot [f'(t)\mathbf{i} + g'(t)\mathbf{j}] = -g'(t)f'(t) + f'(t)g'(t) = 0$; $-\mathbf{n} \cdot \mathbf{v} = -(\mathbf{n} \cdot \mathbf{v}) = 0$; thus, $\mathbf{n}$ and $-\mathbf{n}$ are both normal to the curve at the point

(b) $\mathbf{r}(t) = t\mathbf{i} + e^{2t}\mathbf{j} \Rightarrow \mathbf{v} = \mathbf{i} + 2e^{2t}\mathbf{j} \Rightarrow \mathbf{n} = -2e^{2t}\mathbf{i} + \mathbf{j}$ points toward the concave side of the curve; $\mathbf{N} = \frac{\mathbf{n}}{|\mathbf{n}|}$ and $|\mathbf{n}| = \sqrt{4e^{4t} + 1} \Rightarrow \mathbf{N} = \frac{-2e^{2t}}{\sqrt{1+4e^{4t}}}\mathbf{i} + \frac{1}{\sqrt{1+4e^{4t}}}\mathbf{j}$

(c) $\mathbf{r}(t) = \sqrt{4-t^2}\,\mathbf{i} + t\mathbf{j} \Rightarrow \mathbf{v} = \frac{-t}{\sqrt{4-t^2}}\mathbf{i} + \mathbf{j} \Rightarrow \mathbf{n} = -\mathbf{i} - \frac{t}{\sqrt{4-t^2}}\mathbf{j}$ points toward the concave side of the curve;
$\mathbf{N} = \frac{\mathbf{n}}{|\mathbf{n}|}$ and $|\mathbf{n}| = \sqrt{1 + \frac{t^2}{4-t^2}} = \frac{2}{\sqrt{4-t^2}} \Rightarrow \mathbf{N} = -\frac{1}{2}\left(\sqrt{4-t^2}\,\mathbf{i} + t\mathbf{j}\right)$

9. $\mathbf{r} = (3\sin t)\mathbf{i} + (3\cos t)\mathbf{j} + 4t\mathbf{k} \Rightarrow \mathbf{v} = (3\cos t)\mathbf{i} + (-3\sin t)\mathbf{j} + 4\mathbf{k} \Rightarrow |\mathbf{v}| = \sqrt{(3\cos t)^2 + (-3\sin t)^2 + 4^2}$
$= \sqrt{25} = 5 \Rightarrow \mathbf{T} = \frac{\mathbf{v}}{|\mathbf{v}|} = \left(\frac{3}{5}\cos t\right)\mathbf{i} - \left(\frac{3}{5}\sin t\right)\mathbf{j} + \frac{4}{5}\mathbf{k} \Rightarrow \frac{d\mathbf{T}}{dt} = \left(-\frac{3}{5}\sin t\right)\mathbf{i} - \left(\frac{3}{5}\cos t\right)\mathbf{j}$
$\Rightarrow \left|\frac{d\mathbf{T}}{dt}\right| = \sqrt{\left(-\frac{3}{5}\sin t\right)^2 + \left(-\frac{3}{5}\cos t\right)^2} = \frac{3}{5} \Rightarrow \mathbf{N} = \frac{\left(\frac{d\mathbf{T}}{dt}\right)}{\left|\frac{d\mathbf{T}}{dt}\right|} = (-\sin t)\mathbf{i} - (\cos t)\mathbf{j}\,;\ \kappa = \frac{1}{5} \cdot \frac{3}{5} = \frac{3}{25}$

11. $\mathbf{r} = (e^t\cos t)\,\mathbf{i} + (e^t\sin t)\,\mathbf{j} + 2\mathbf{k} \Rightarrow \mathbf{v} = (e^t\cos t - e^t\sin t)\,\mathbf{i} + (e^t\sin t + e^t\cos t)\,\mathbf{j} \Rightarrow$
$|\mathbf{v}| = \sqrt{(e^t\cos t - e^t\sin t)^2 + (e^t\sin t + e^t\cos t)^2} = \sqrt{2e^{2t}} = e^t\sqrt{2}\,;$
$\mathbf{T} = \frac{\mathbf{v}}{|\mathbf{v}|} = \left(\frac{\cos t - \sin t}{\sqrt{2}}\right)\mathbf{i} + \left(\frac{\sin t + \cos t}{\sqrt{2}}\right)\mathbf{j} \Rightarrow \frac{d\mathbf{T}}{dt} = \left(\frac{-\sin t - \cos t}{\sqrt{2}}\right)\mathbf{i} + \left(\frac{\cos t - \sin t}{\sqrt{2}}\right)\mathbf{j}$
$\Rightarrow \left|\frac{d\mathbf{T}}{dt}\right| = \sqrt{\left(\frac{-\sin t - \cos t}{\sqrt{2}}\right)^2 + \left(\frac{\cos t - \sin t}{\sqrt{2}}\right)^2} = 1 \Rightarrow \mathbf{N} = \frac{\left(\frac{d\mathbf{T}}{dt}\right)}{\left|\frac{d\mathbf{T}}{dt}\right|} = \left(\frac{-\cos t - \sin t}{\sqrt{2}}\right)\mathbf{i} + \left(\frac{-\sin t + \cos t}{\sqrt{2}}\right)\mathbf{j}\,;$
$\kappa = \frac{1}{|\mathbf{v}|} \cdot \left|\frac{d\mathbf{T}}{dt}\right| = \frac{1}{e^t\sqrt{2}} \cdot 1 = \frac{1}{e^t\sqrt{2}}$

13. $\mathbf{r} = \left(\frac{t^3}{3}\right)\mathbf{i} + \left(\frac{t^2}{2}\right)\mathbf{j},\ t > 0 \Rightarrow \mathbf{v} = t^2\mathbf{i} + t\mathbf{j} \Rightarrow |\mathbf{v}| = \sqrt{t^4 + t^2} = t\sqrt{t^2 + 1}$, since $t > 0 \Rightarrow \mathbf{T} = \frac{\mathbf{v}}{|\mathbf{v}|}$
$= \frac{t}{\sqrt{t^2 + t}}\mathbf{i} + \frac{1}{\sqrt{t^2 + 1}}\mathbf{j} \Rightarrow \frac{d\mathbf{T}}{dt} = \frac{1}{(t^2+1)^{3/2}}\mathbf{i} - \frac{t}{(t^2+1)^{3/2}}\mathbf{j} \Rightarrow \left|\frac{d\mathbf{T}}{dt}\right| = \sqrt{\left(\frac{1}{(t^2+1)^{3/2}}\right)^2 + \left(\frac{-t}{(t^2+1)^{3/2}}\right)^2}$
$= \sqrt{\frac{1+t^2}{(t^2+1)^3}} = \frac{1}{t^2+1} \Rightarrow \mathbf{N} = \frac{\left(\frac{d\mathbf{T}}{dt}\right)}{\left|\frac{d\mathbf{T}}{dt}\right|} = \frac{1}{\sqrt{t^2+1}}\mathbf{i} - \frac{t}{\sqrt{t^2+1}}\mathbf{j}\,;\ \kappa = \frac{1}{|\mathbf{v}|} \cdot \left|\frac{d\mathbf{T}}{dt}\right| = \frac{1}{t\sqrt{t^2+1}} \cdot \frac{1}{t^2+1} = \frac{1}{t(t^2+1)^{3/2}}.$

15. $\mathbf{r} = t\mathbf{i} + \left(a\cosh\frac{t}{a}\right)\mathbf{j},\ a > 0 \Rightarrow \mathbf{v} = \mathbf{i} + \left(\sinh\frac{t}{a}\right)\mathbf{j} \Rightarrow |\mathbf{v}| = \sqrt{1 + \sinh^2\left(\frac{t}{a}\right)} = \sqrt{\cosh^2\left(\frac{t}{a}\right)} = \cosh\frac{t}{a}$
$\Rightarrow \mathbf{T} = \frac{\mathbf{v}}{|\mathbf{v}|} = \left(\operatorname{sech}\frac{t}{a}\right)\mathbf{i} + \left(\tanh\frac{t}{a}\right)\mathbf{j} \Rightarrow \frac{d\mathbf{T}}{dt} = \left(-\frac{1}{a}\operatorname{sech}\frac{t}{a}\tanh\frac{t}{a}\right)\mathbf{i} + \left(\frac{1}{a}\operatorname{sech}^2\frac{t}{a}\right)\mathbf{j}$
$\Rightarrow \left|\frac{d\mathbf{T}}{dt}\right| = \sqrt{\frac{1}{a^2}\operatorname{sech}^2\left(\frac{t}{a}\right)\tanh^2\left(\frac{t}{a}\right) + \frac{1}{a^2}\operatorname{sech}^4\left(\frac{t}{a}\right)} = \frac{1}{a}\operatorname{sech}\left(\frac{t}{a}\right) \Rightarrow \mathbf{N} = \frac{\left(\frac{d\mathbf{T}}{dt}\right)}{\left|\frac{d\mathbf{T}}{dt}\right|} = \left(-\tanh\frac{t}{a}\right)\mathbf{i} + \left(\operatorname{sech}\frac{t}{a}\right)\mathbf{j}\,;$
$\kappa = \frac{1}{|\mathbf{v}|} \cdot \left|\frac{d\mathbf{T}}{dt}\right| = \frac{1}{\cosh\frac{t}{a}} \cdot \frac{1}{a}\operatorname{sech}\left(\frac{t}{a}\right) = \frac{1}{a}\operatorname{sech}^2\left(\frac{t}{a}\right).$

17. $y = ax^2 \Rightarrow y' = 2ax \Rightarrow y'' = 2a$; from Exercise 5(a), $\kappa(x) = \frac{|2a|}{(1+4a^2x^2)^{3/2}} = |2a|\,(1 + 4a^2x^2)^{-3/2}$
$\Rightarrow \kappa'(x) = -\frac{3}{2}\,|2a|\,(1 + 4a^2x^2)^{-5/2}\,(8a^2x)$; thus, $\kappa'(x) = 0 \Rightarrow x = 0$. Now, $\kappa'(x) > 0$ for $x < 0$ and $\kappa'(x) < 0$ for $x > 0$ so that $\kappa(x)$ has an absolute maximum at $x = 0$ which is the vertex of the parabola. Since $x = 0$ is the only critical point for $\kappa(x)$, the curvature has no minimum value.

19. $\kappa = \frac{a}{a^2 + b^2} \Rightarrow \frac{d\kappa}{da} = \frac{-a^2 + b^2}{(a^2+b^2)^2}\,;\ \frac{d\kappa}{da} = 0 \Rightarrow -a^2 + b^2 = 0 \Rightarrow a = \pm b \Rightarrow a = b$ since $a, b \ge 0$. Now, $\frac{d\kappa}{da} > 0$ if $a < b$ and $\frac{d\kappa}{da} < 0$ if $a > b \Rightarrow \kappa$ is at a maximum for $a = b$ and $\kappa(b) = \frac{b}{b^2 + b^2} = \frac{1}{2b}$ is the maximum value of κ.

21. $\mathbf{r} = t\mathbf{i} + (\sin t)\mathbf{j} \Rightarrow \mathbf{v} = \mathbf{i} + (\cos t)\mathbf{j} \Rightarrow |\mathbf{v}| = \sqrt{1^2 + (\cos t)^2} = \sqrt{1 + \cos^2 t} \Rightarrow \left|\mathbf{v}\left(\frac{\pi}{2}\right)\right| = \sqrt{1 + \cos^2\left(\frac{\pi}{2}\right)} = 1;\ \mathbf{T} = \frac{\mathbf{v}}{|\mathbf{v}|}$
$= \frac{\mathbf{i} + \cos t\,\mathbf{j}}{\sqrt{1 + \cos^2 t}} \Rightarrow \frac{d\mathbf{T}}{dt} = \frac{\sin t\cos t}{(1+\cos^2 t)^{3/2}}\mathbf{i} + \frac{-\sin t}{(1+\cos^2 t)^{3/2}}\mathbf{j} \Rightarrow \left|\frac{d\mathbf{T}}{dt}\right| = \frac{|\sin t|}{1 + \cos^2 t};\ \left|\frac{d\mathbf{T}}{dt}\right|_{t=\frac{\pi}{2}} = \frac{\left|\sin\frac{\pi}{2}\right|}{1 + \cos^2\left(\frac{\pi}{2}\right)} = \frac{1}{1} = 1$. Thus $\kappa\left(\frac{\pi}{2}\right) = \frac{1}{1} \cdot 1 = 1$
$\Rightarrow \rho = \frac{1}{1} = 1$ and the center is $\left(\frac{\pi}{2}, 0\right) \Rightarrow \left(x - \frac{\pi}{2}\right)^2 + y^2 = 1$

23. $y = x^2 \Rightarrow f'(x) = 2x$ and $f''(x) = 2$
$\Rightarrow \kappa = \frac{|2|}{(1 + (2x)^2)^{3/2}} = \frac{2}{(1+4x^2)^{3/2}}$

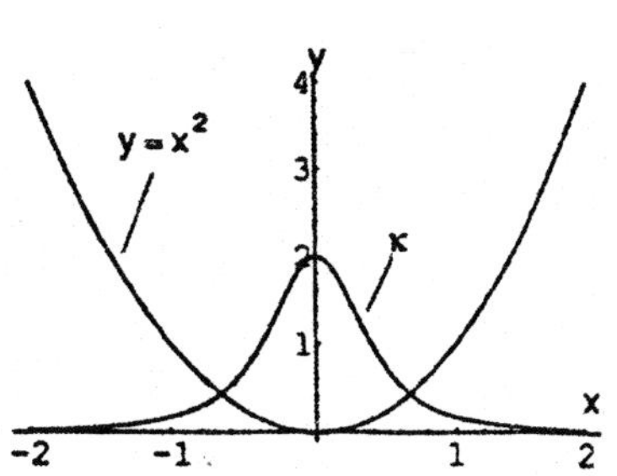

25. $y = \sin x \Rightarrow f'(x) = \cos x$ and $f''(x) = -\sin x$
$\Rightarrow \kappa = \frac{|-\sin x|}{(1+\cos^2 x)^{3/2}} = \frac{|\sin x|}{(1+\cos^2 x)^{3/2}}$

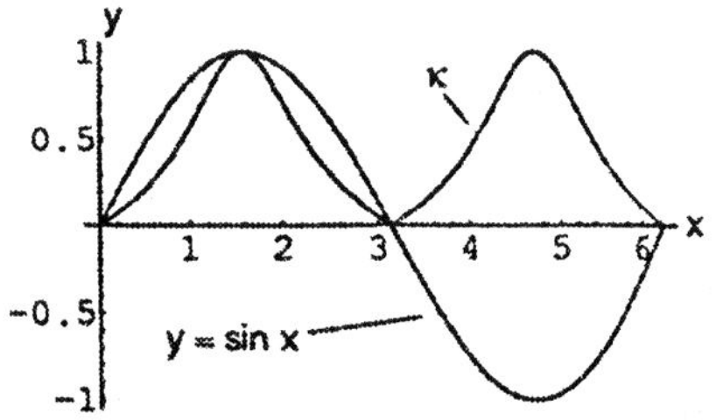

10.5 TANGENTIAL AND NORMAL COMPONENTS OF ACCELERATION

1. $\mathbf{r} = (a\cos t)\mathbf{i} + (a\sin t)\mathbf{j} + bt\mathbf{k} \Rightarrow \mathbf{v} = (-a\sin t)\mathbf{i} + (a\cos t)\mathbf{j} + b\mathbf{k} \Rightarrow |\mathbf{v}| = \sqrt{(-a\sin t)^2 + (a\cos t)^2 + b^2}$
$= \sqrt{a^2+b^2} \Rightarrow a_T = \frac{d}{dt}|\mathbf{v}| = 0;\ \mathbf{a} = (-a\cos t)\mathbf{i} + (-a\sin t)\mathbf{j} \Rightarrow |\mathbf{a}| = \sqrt{(-a\cos t)^2 + (-a\sin t)^2} = \sqrt{a^2} = |a|$
$\Rightarrow a_N = \sqrt{|\mathbf{a}|^2 - a_T^2} = \sqrt{|\mathbf{a}|^2 - 0^2} = |\mathbf{a}| = |a| \Rightarrow \mathbf{a} = (0)\mathbf{T} + |a|\,\mathbf{N} = |a|\,\mathbf{N}$

3. $\mathbf{r} = (t+1)\mathbf{i} + 2t\mathbf{j} + t^2\mathbf{k} \Rightarrow \mathbf{v} = \mathbf{i} + 2\mathbf{j} + 2t\mathbf{k} \Rightarrow |\mathbf{v}| = \sqrt{1^2+2^2+(2t)^2} = \sqrt{5+4t^2} \Rightarrow a_T = \frac{1}{2}(5+4t^2)^{-1/2}(8t)$
$= 4t(5+4t^2)^{-1/2} \Rightarrow a_T(1) = \frac{4}{\sqrt{9}} = \frac{4}{3};\ \mathbf{a} = 2\mathbf{k} \Rightarrow \mathbf{a}(1) = 2\mathbf{k} \Rightarrow |\mathbf{a}(1)| = 2 \Rightarrow a_N = \sqrt{|\mathbf{a}|^2 - a_T^2} = \sqrt{2^2 - \left(\frac{4}{3}\right)^2}$
$= \sqrt{\frac{20}{9}} = \frac{2\sqrt{5}}{3} \Rightarrow \mathbf{a}(1) = \frac{4}{3}\mathbf{T} + \frac{2\sqrt{5}}{3}\mathbf{N}$

5. $\mathbf{r} = t^2\mathbf{i} + \left(t + \frac{1}{3}t^3\right)\mathbf{j} + \left(t - \frac{1}{3}t^3\right)\mathbf{k} \Rightarrow \mathbf{v} = 2t\mathbf{i} + (1+t^2)\mathbf{j} + (1-t^2)\mathbf{k} \Rightarrow |\mathbf{v}| = \sqrt{(2t)^2 + (1+t^2)^2 + (1-t^2)^2}$
$= \sqrt{2(t^4+2t^2+1)} = \sqrt{2}(1+t^2) \Rightarrow a_T = 2t\sqrt{2} \Rightarrow a_T(0) = 0;\ \mathbf{a} = 2\mathbf{i} + 2t\mathbf{j} - 2t\mathbf{k} \Rightarrow \mathbf{a}(0) = 2\mathbf{i} \Rightarrow |\mathbf{a}(0)| = 2$
$\Rightarrow a_N = \sqrt{|\mathbf{a}|^2 - a_T^2} = \sqrt{2^2 - 0^2} = 2 \Rightarrow \mathbf{a}(0) = (0)\mathbf{T} + 2\mathbf{N} = 2\mathbf{N}$

7. $\mathbf{r} = (\cos t)\mathbf{i} + (\sin t)\mathbf{j} - \mathbf{k} \Rightarrow \mathbf{v} = (-\sin t)\mathbf{i} + (\cos t)\mathbf{j} \Rightarrow |\mathbf{v}| = \sqrt{(-\sin t)^2 + (\cos t)^2} = 1 \Rightarrow \mathbf{T} = \frac{\mathbf{v}}{|\mathbf{v}|}$
$= (-\sin t)\mathbf{i} + (\cos t)\mathbf{j} \Rightarrow \mathbf{T}\left(\frac{\pi}{4}\right) = -\frac{\sqrt{2}}{2}\mathbf{i} + \frac{\sqrt{2}}{2}\mathbf{j};\ \frac{d\mathbf{T}}{dt} = (-\cos t)\mathbf{i} - (\sin t)\mathbf{j} \Rightarrow \left|\frac{d\mathbf{T}}{dt}\right| = \sqrt{(-\cos t)^2 + (-\sin t)^2}$
$= 1 \Rightarrow \mathbf{N} = \frac{\left(\frac{d\mathbf{T}}{dt}\right)}{\left|\frac{d\mathbf{T}}{dt}\right|} = (-\cos t)\mathbf{i} - (\sin t)\mathbf{j} \Rightarrow \mathbf{N}\left(\frac{\pi}{4}\right) = -\frac{\sqrt{2}}{2}\mathbf{i} - \frac{\sqrt{2}}{2}\mathbf{j};\ \mathbf{B} = \mathbf{T} \times \mathbf{N} = \begin{vmatrix} \mathbf{i} & \mathbf{j} & \mathbf{k} \\ -\sin t & \cos t & 0 \\ -\cos t & -\sin t & 0 \end{vmatrix} = \mathbf{k}$
$\Rightarrow \mathbf{B}\left(\frac{\pi}{4}\right) = \mathbf{k}$, the normal to the osculating plane; $\mathbf{r}\left(\frac{\pi}{4}\right) = \frac{\sqrt{2}}{2}\mathbf{i} + \frac{\sqrt{2}}{2}\mathbf{j} - \mathbf{k} \Rightarrow P = \left(\frac{\sqrt{2}}{2}, \frac{\sqrt{2}}{2}, -1\right)$ lies on the osculating plane $\Rightarrow 0\left(x - \frac{\sqrt{2}}{2}\right) + 0\left(y - \frac{\sqrt{2}}{2}\right) + (z - (-1)) = 0 \Rightarrow z = -1$ is the osculating plane; $\mathbf{T}$ is normal to the normal plane $\Rightarrow \left(-\frac{\sqrt{2}}{2}\right)\left(x - \frac{\sqrt{2}}{2}\right) + \left(\frac{\sqrt{2}}{2}\right)\left(y - \frac{\sqrt{2}}{2}\right) + 0(z-(-1)) = 0 \Rightarrow -\frac{\sqrt{2}}{2}x + \frac{\sqrt{2}}{2}y = 0$
$\Rightarrow -x + y = 0$ is the normal plane; $\mathbf{N}$ is normal to the rectifying plane
$\Rightarrow \left(-\frac{\sqrt{2}}{2}\right)\left(x - \frac{\sqrt{2}}{2}\right) + \left(-\frac{\sqrt{2}}{2}\right)\left(y - \frac{\sqrt{2}}{2}\right) + 0(z-(-1)) = 0 \Rightarrow -\frac{\sqrt{2}}{2}x - \frac{\sqrt{2}}{2}y = -1 \Rightarrow x + y = \sqrt{2}$ is the rectifying plane

9. By Exercise 9 in Section 10.4, $\mathbf{T} = \left(\frac{3}{5}\cos t\right)\mathbf{i} + \left(-\frac{3}{5}\sin t\right)\mathbf{j} + \frac{4}{5}\mathbf{k}$ and $\mathbf{N} = (-\sin t)\mathbf{i} - (\cos t)\mathbf{j}$ so that $\mathbf{B} = \mathbf{T} \times \mathbf{N}$
$= \begin{vmatrix} \mathbf{i} & \mathbf{j} & \mathbf{k} \\ \frac{3}{5}\cos t & -\frac{3}{5}\sin t & \frac{4}{5} \\ -\sin t & -\cos t & 0 \end{vmatrix} = \left(\frac{4}{5}\cos t\right)\mathbf{i} - \left(\frac{4}{5}\sin t\right)\mathbf{j} - \frac{3}{5}\mathbf{k}$. Also $\mathbf{v} = (3\cos t)\mathbf{i} + (-3\sin t)\mathbf{j} + 4\mathbf{k}$
$\Rightarrow \mathbf{a} = (-3\sin t)\mathbf{i} + (-3\cos t)\mathbf{j} \Rightarrow \frac{d\mathbf{a}}{dt} = (-3\cos t)\mathbf{i} + (3\sin t)\mathbf{j}$ and $\mathbf{v} \times \mathbf{a} = \begin{vmatrix} \mathbf{i} & \mathbf{j} & \mathbf{k} \\ 3\cos t & -3\sin t & 4 \\ -3\sin t & -3\cos t & 0 \end{vmatrix}$
$= (12\cos t)\mathbf{i} - (12\sin t)\mathbf{j} - 9\mathbf{k} \Rightarrow |\mathbf{v} \times \mathbf{a}|^2 = (12\cos t)^2 + (-12\sin t)^2 + (-9)^2 = 225$. Thus

$$\tau = \frac{\begin{vmatrix} 3\cos t & -3\sin t & 4 \\ -3\sin t & -3\sin t & 0 \\ -3\cos t & 3\sin t & 0 \end{vmatrix}}{225} = \frac{4\cdot(-9\sin^2 t - 9\cos^2 t)}{225} = \frac{-36}{225} = -\frac{4}{25}$$

11. By Exercise 11 in Section 10.4, $\mathbf{T} = \left(\frac{\cos t - \sin t}{\sqrt{2}}\right)\mathbf{i} + \left(\frac{\sin t + \cos t}{\sqrt{2}}\right)\mathbf{j}$ and $\mathbf{N} = \left(\frac{-\cos t - \sin t}{\sqrt{2}}\right)\mathbf{i} + \left(\frac{-\sin t + \cos t}{\sqrt{2}}\right)\mathbf{j}$; Thus

$$\mathbf{B} = \mathbf{T} \times \mathbf{N} = \begin{vmatrix} \mathbf{i} & \mathbf{j} & \mathbf{k} \\ \frac{\cos t - \sin t}{\sqrt{2}} & \frac{\sin t + \cos t}{\sqrt{2}} & 0 \\ \frac{-\cos t - \sin t}{\sqrt{2}} & \frac{-\sin t + \cos t}{\sqrt{2}} & 0 \end{vmatrix} = \left[\left(\frac{\cos^2 t - 2\cos t \sin t + \sin^2 t}{2}\right) + \left(\frac{\sin^2 t + 2\sin t \cos t + \cos^2 t}{2}\right)\right]\mathbf{k}$$

$= \left[\left(\frac{1 - \sin(2t)}{2}\right) + \left(\frac{1 + \sin(2t)}{2}\right)\right]\mathbf{k} = \mathbf{k}$. Also, $\mathbf{v} = (e^t \cos t - e^t \sin t)\,\mathbf{i} + (e^t \sin t + e^t \cos t)\,\mathbf{j}$

$\Rightarrow \mathbf{a} = \left[e^t(-\sin t - \cos t) + e^t(\cos t - \sin t)\right]\mathbf{i} + \left[e^t(\cos t - \sin t) + e^t(\sin t + \cos t)\right]\mathbf{j} = (-2e^t \sin t)\,\mathbf{i} + (2e^t \cos t)\,\mathbf{j}$

$\Rightarrow \frac{d\mathbf{a}}{dt} = -2e^t(\cos t + \sin t)\,\mathbf{i} + 2e^t(-\sin t + \cos t)\mathbf{j}$. Thus $\mathbf{v} \times \mathbf{a} = \begin{vmatrix} \mathbf{i} & \mathbf{j} & \mathbf{k} \\ e^t(\cos t - \sin t) & e^t(\sin t + \cos t) & 0 \\ -2e^t \sin t & 2e^t \cos t & 0 \end{vmatrix} = 2e^{2t}\mathbf{k}$

$\Rightarrow |\mathbf{v} \times \mathbf{a}|^2 = (2e^{2t})^2 = 4e^{4t}$. Thus $\tau = \frac{\begin{vmatrix} e^t(\cos t - \sin t) & e^t(\sin t + \cos t) & 0 \\ -2e^t \sin t & 2e^t \cos t & 0 \\ -2e^t(\cos t + \sin t) & 2e^t(-\sin t + \cos t) & 0 \end{vmatrix}}{4e^{4t}} = 0$

13. By Exercise 13 in Section 10.4, $\mathbf{T} = \frac{t}{(t^2+1)^{1/2}}\,\mathbf{i} + \frac{1}{(t^2+1)^{1/2}}\,\mathbf{j}$ and $\mathbf{N} = \frac{1}{\sqrt{t^2+1}}\,\mathbf{i} - \frac{t}{\sqrt{t^2+1}}\,\mathbf{j}$ so that $\mathbf{B} = \mathbf{T} \times \mathbf{N}$

$= \begin{vmatrix} \mathbf{i} & \mathbf{j} & \mathbf{k} \\ \frac{t}{\sqrt{t^2+1}} & \frac{1}{\sqrt{t^2+1}} & 0 \\ \frac{1}{\sqrt{t^2+1}} & \frac{-t}{\sqrt{t^2+1}} & 0 \end{vmatrix} = -\mathbf{k}$. Also, $\mathbf{v} = t^2\mathbf{i} + t\mathbf{j} \Rightarrow \mathbf{a} = 2t\mathbf{i} + \mathbf{j} \Rightarrow \frac{d\mathbf{a}}{dt} = 2\mathbf{i}$ so that $\begin{vmatrix} t^2 & t & 0 \\ 2t & 1 & 0 \\ 2 & 0 & 0 \end{vmatrix} = 0 \Rightarrow \tau = 0$

15. By Exercise 15 in Section 10.4, $\mathbf{T} = \frac{\mathbf{v}}{|\mathbf{v}|} = \left(\operatorname{sech}\frac{t}{a}\right)\mathbf{i} + \left(\tanh\frac{t}{a}\right)\mathbf{j}$ and $\mathbf{N} = \left(-\tanh\frac{t}{a}\right)\mathbf{i} + \left(\operatorname{sech}\frac{t}{a}\right)\mathbf{j}$ so that $\mathbf{B} = \mathbf{T} \times \mathbf{N}$

$= \begin{vmatrix} \mathbf{i} & \mathbf{j} & \mathbf{k} \\ \operatorname{sech}\left(\frac{t}{a}\right) & \tanh\left(\frac{t}{a}\right) & 0 \\ -\tanh\left(\frac{t}{a}\right) & \operatorname{sech}\left(\frac{t}{a}\right) & 0 \end{vmatrix} = \mathbf{k}$. Also, $\mathbf{v} = \mathbf{i} + \left(\sinh\frac{t}{a}\right)\mathbf{j} \Rightarrow \mathbf{a} = \left(\frac{1}{a}\cosh\frac{t}{a}\right)\mathbf{j} \Rightarrow \frac{d\mathbf{a}}{dt} = \frac{1}{a^2}\sinh\left(\frac{t}{a}\right)\mathbf{j}$ so that

$\begin{vmatrix} 1 & \sinh\left(\frac{t}{a}\right) & 0 \\ 0 & \frac{1}{a}\cosh\left(\frac{t}{a}\right) & 0 \\ 0 & \frac{1}{a^2}\sinh\left(\frac{t}{a}\right) & 0 \end{vmatrix} = 0 \Rightarrow \tau = 0$

17. Yes. If the car is moving along a curved path, then $\kappa \neq 0$ and $a_N = \kappa\,|\mathbf{v}|^2 \neq 0 \Rightarrow \mathbf{a} = a_T\mathbf{T} + a_N\mathbf{N} \neq \mathbf{0}$.

19. $\mathbf{a} \perp \mathbf{v} \Rightarrow \mathbf{a} \perp \mathbf{T} \Rightarrow a_T = 0 \Rightarrow \frac{d}{dt}|\mathbf{v}| = 0 \Rightarrow |\mathbf{v}|$ is constant

21. By $\mathbf{a} = a_T\mathbf{T} + a_N\mathbf{N}$ we have $\mathbf{v} \times \mathbf{a} = \left(\frac{ds}{dt}\mathbf{T}\right) \times \left[\frac{d^2s}{dt^2}\mathbf{T} + \kappa\left(\frac{ds}{dt}\right)^2\mathbf{N}\right] = \left(\frac{ds}{dt}\frac{d^2s}{dt^2}\right)(\mathbf{T} \times \mathbf{T}) + \kappa\left(\frac{ds}{dt}\right)^3(\mathbf{T} \times \mathbf{N})$

$= \kappa\left(\frac{ds}{dt}\right)^3\mathbf{B}$. It follows that $|\mathbf{v} \times \mathbf{a}| = \kappa\left|\frac{ds}{dt}\right|^3|\mathbf{B}| = \kappa\,|\mathbf{v}|^3 \Rightarrow \kappa = \frac{|\mathbf{v} \times \mathbf{a}|}{|\mathbf{v}|^3}$

23. From Example 1, $|\mathbf{v}| = t$ and $a_N = t$ so that $a_N = \kappa\,|\mathbf{v}|^2 \Rightarrow \kappa = \frac{a_N}{|\mathbf{v}|^2} = \frac{t}{t^2} = \frac{1}{t}$, $t \neq 0 \Rightarrow \rho = \frac{1}{\kappa} = t$

25. If a plane curve is sufficiently differentiable the torsion is zero as the following argument shows:
$\mathbf{r} = f(t)\mathbf{i} + g(t)\mathbf{j} \Rightarrow \mathbf{v} = f'(t)\mathbf{i} + g'(t)\mathbf{j} \Rightarrow \mathbf{a} = f''(t)\mathbf{i} + g''(t)\mathbf{j} \Rightarrow \frac{d\mathbf{a}}{dt} = f'''(t)\mathbf{i} + g'''(t)\mathbf{j}$

$$\Rightarrow \tau = \frac{\begin{vmatrix} f'(t) & g'(t) & 0 \\ f''(t) & g''(t) & 0 \\ f'''(t) & g'''(t) & 0 \end{vmatrix}}{|\mathbf{v} \times \mathbf{a}|^2} = 0$$

27. $\mathbf{r}(t) = f(t)\mathbf{i} + g(t)\mathbf{j} + h(t)\mathbf{k} \Rightarrow \mathbf{v} = f'(t)\mathbf{i} + g'(t)\mathbf{j} + h'(t)\mathbf{k};\ \mathbf{v}\cdot\mathbf{k} = 0 \Rightarrow h'(t) = 0 \Rightarrow h(t) = C$
$\Rightarrow \mathbf{r}(t) = f(t)\mathbf{i} + g(t)\mathbf{j} + C\mathbf{k}$ and $\mathbf{r}(a) = f(a)\mathbf{i} + g(a)\mathbf{j} + C\mathbf{k} = \mathbf{0} \Rightarrow f(a) = 0,\ g(a) = 0$ and $C = 0 \Rightarrow h(t) = 0.$

10.6 VELOCITY AND ACCELERATION IN POLAR COORDINATES

1. $\frac{d\theta}{dt} = 3 = \dot{\theta} \Rightarrow \ddot{\theta} = 0,\ r = a(1 - \cos\theta) \Rightarrow \dot{r} = a\sin\theta\frac{d\theta}{dt} = 3a\sin\theta \Rightarrow \ddot{r} = 3a\cos\theta\frac{d\theta}{dt} = 9a\cos\theta$
$\mathbf{v} = (3a\sin\theta)\mathbf{u}_r + (a(1-\cos\theta))(3)\mathbf{u}_\theta = (3a\sin\theta)\mathbf{u}_r + 3a(1-\cos\theta)\mathbf{u}_\theta$
$\mathbf{a} = \left(9a\cos\theta - a(1-\cos\theta)(3)^2\right)\mathbf{u}_r + (a(1-\cos\theta)\cdot 0 + 2(3a\sin\theta)(3))\mathbf{u}_\theta$
$= (9a\cos\theta - 9a + 9a\cos\theta)\mathbf{u}_r + (18a\sin\theta)\mathbf{u}_\theta = 9a(2\cos\theta - 1)\mathbf{u}_r + (18a\sin\theta)\mathbf{u}_\theta$

3. $\frac{d\theta}{dt} = 2 = \dot{\theta} \Rightarrow \ddot{\theta} = 0,\ r = e^{a\theta} \Rightarrow \dot{r} = e^{a\theta}\cdot a\frac{d\theta}{dt} = 2a\,e^{a\theta} \Rightarrow \ddot{r} = 2a\,e^{a\theta}\cdot a\frac{d\theta}{dt} = 4a^2 e^{a\theta}$
$\mathbf{v} = \left(2a\,e^{a\theta}\right)\mathbf{u}_r + \left(e^{a\theta}\right)(2)\mathbf{u}_\theta = \left(2a\,e^{a\theta}\right)\mathbf{u}_r + \left(2e^{a\theta}\right)\mathbf{u}_\theta$
$\mathbf{a} = \left[\left(4a^2 e^{a\theta}\right) - \left(e^{a\theta}\right)(2)^2\right]\mathbf{u}_r + \left[\left(e^{a\theta}\right)(0) + 2\left(2a\,e^{a\theta}\right)(2)\right]\mathbf{u}_\theta = \left[4a^2 e^{a\theta} - 4e^{a\theta}\right]\mathbf{u}_r + \left[0 + 8a\,e^{a\theta}\right]\mathbf{u}_\theta$
$= 4e^{a\theta}(a^2 - 1)\mathbf{u}_r + \left(8a\,e^{a\theta}\right)\mathbf{u}_\theta$

5. $\theta = 2t \Rightarrow \dot{\theta} = 2 \Rightarrow \ddot{\theta} = 0,\ r = 2\cos 4t \Rightarrow \dot{r} = -8\sin 4t \Rightarrow \ddot{r} = -32\cos 4t$
$\mathbf{v} = (-8\sin 4t)\mathbf{u}_r + (2\cos 4t)(2)\mathbf{u}_\theta = -8(\sin 4t)\mathbf{u}_r + 4(\cos 4t)\mathbf{u}_\theta$
$\mathbf{a} = \left((-32\cos 4t) - (2\cos 4t)(2)^2\right)\mathbf{u}_r + ((2\cos 4t)\cdot 0 + 2(-8\sin 4t)(2))\mathbf{u}_\theta$
$= (-32\cos 4t - 8\cos 4t)\mathbf{u}_r + (0 - 32\sin 4t)\mathbf{u}_\theta = -40(\cos 4t)\mathbf{u}_r - 32(\sin 4t)\mathbf{u}_\theta$

7. $r = \frac{GM}{v^2} \Rightarrow v^2 = \frac{GM}{r} \Rightarrow v = \sqrt{\frac{GM}{r}}$ which is constant since G, M, and r (the radius of orbit) are constant

9. $T = \left(\frac{2\pi a^2}{r_0 v_0}\right)\sqrt{1 - e^2} \Rightarrow T^2 = \left(\frac{4\pi^2 a^4}{r_0^2 v_0^2}\right)(1 - e^2) = \left(\frac{4\pi^2 a^4}{r_0^2 v_0^2}\right)\left[1 - \left(\frac{r_0 v_0^2}{GM} - 1\right)^2\right]$ (from Equation 5)
$= \left(\frac{4\pi^2 a^4}{r_0^2 v_0^2}\right)\left[-\frac{r_0^2 v_0^4}{G^2M^2} + 2\left(\frac{r_0 v_0^2}{GM}\right)\right] = \left(\frac{4\pi^2 a^4}{r_0^2 v_0^2}\right)\left[\frac{2GMr_0 v_0^2 - r_0^2 v_0^4}{G^2M^2}\right] = \frac{(4\pi^2 a^4)(2GM - r_0 v_0^2)}{r_0 G^2 M^2}$
$= (4\pi^2 a^4)\left(\frac{2GM - r_0 v_0^2}{2r_0 GM}\right)\left(\frac{2}{GM}\right) = (4\pi^2 a^4)\left(\frac{1}{2a}\right)\left(\frac{2}{GM}\right)$ (from Equation 10) $\Rightarrow T^2 = \frac{4\pi^2 a^3}{GM} \Rightarrow \frac{T^2}{a^3} = \frac{4\pi^2}{GM}$

CHAPTER 10 PRACTICE AND ADDITIONAL EXERCISES

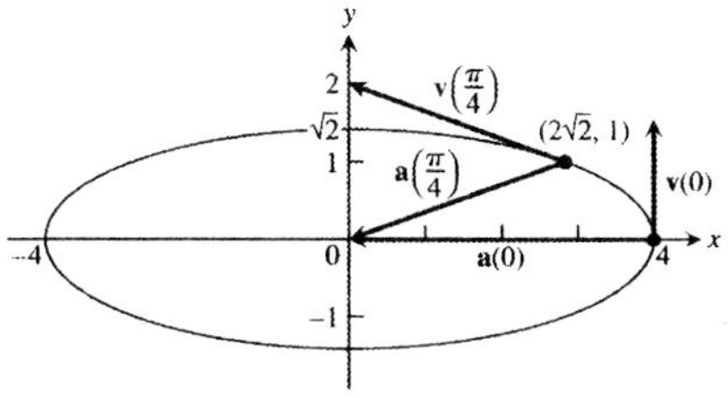

1. $\mathbf{r}(t) = (4\cos t)\mathbf{i} + \left(\sqrt{2}\sin t\right)\mathbf{j} \Rightarrow x = 4\cos t$
and $y = \sqrt{2}\sin t \Rightarrow \frac{x^2}{16} + \frac{y^2}{2} = 1;$
$\mathbf{v} = (-4\sin t)\mathbf{i} + \left(\sqrt{2}\cos t\right)\mathbf{j}$ and
$\mathbf{a} = (-4\cos t)\mathbf{i} - \left(\sqrt{2}\sin t\right)\mathbf{j};\ \mathbf{r}(0) = 4\,\mathbf{i},\ \mathbf{v}(0) = \sqrt{2}\mathbf{j},$
$\mathbf{a}(0) = -4\mathbf{i};\ \mathbf{r}\left(\frac{\pi}{4}\right) = 2\sqrt{2}\mathbf{i} + \mathbf{j},\ \mathbf{v}\left(\frac{\pi}{4}\right) = -2\sqrt{2}\mathbf{i} + \mathbf{j},$
$\mathbf{a}\left(\frac{\pi}{4}\right) = -2\sqrt{2}\mathbf{i} - \mathbf{j};\ |\mathbf{v}| = \sqrt{16\sin^2 t + 2\cos^2 t}$
$\Rightarrow a_T = \frac{d}{dt}|\mathbf{v}| = \frac{14\sin t\cos t}{\sqrt{16\sin^2 t + 2\cos^2 t}}$; at $t = 0$: $a_T = 0,\ a_N = \sqrt{|\mathbf{a}|^2 - 0} = 4,\ \mathbf{a} = 0\mathbf{T} + 4\mathbf{N} = 4\mathbf{N},\ \kappa = \frac{a_N}{|\mathbf{v}|^2} = \frac{4}{2} = 2;$
at $t = \frac{\pi}{4}$: $a_T = \frac{7}{\sqrt{8+1}} = \frac{7}{3},\ a_N = \sqrt{9 - \frac{49}{9}} = \frac{4\sqrt{2}}{3},\ \mathbf{a} = \frac{7}{3}\mathbf{T} + \frac{4\sqrt{2}}{3}\mathbf{N},\ \kappa = \frac{a_N}{|\mathbf{v}|^2} = \frac{4\sqrt{2}}{27}$

3. $\mathbf{r} = \frac{1}{\sqrt{1+t^2}}\mathbf{i} + \frac{t}{\sqrt{1+t^2}}\mathbf{j} \Rightarrow \mathbf{v} = -t(1+t^2)^{-3/2}\mathbf{i} + (1+t^2)^{-3/2}\mathbf{j} \Rightarrow |\mathbf{v}| = \sqrt{\left[-t(1+t^2)^{-3/2}\right]^2 + \left[(1+t^2)^{-3/2}\right]^2}$ $= \frac{1}{1+t^2}$. We want to maximize $|\mathbf{v}|$: $\frac{d|\mathbf{v}|}{dt} = \frac{-2t}{(1+t^2)^2}$ and $\frac{d|\mathbf{v}|}{dt} = 0 \Rightarrow \frac{-2t}{(1+t^2)^2} = 0 \Rightarrow t = 0$. For $t < 0$, $\frac{-2t}{(1+t^2)^2} > 0$; for $t > 0$, $\frac{-2t}{(1+t^2)^2} < 0 \Rightarrow |\mathbf{v}|_{max}$ occurs when $t = 0 \Rightarrow |\mathbf{v}|_{max} = 1$

5. $\mathbf{v} = 3\mathbf{i} + 4\mathbf{j}$ and $\mathbf{a} = 5\mathbf{i} + 15\mathbf{j} \Rightarrow \mathbf{v} \times \mathbf{a} = \begin{vmatrix} \mathbf{i} & \mathbf{j} & \mathbf{k} \\ 3 & 4 & 0 \\ 5 & 15 & 0 \end{vmatrix} = 25\mathbf{k} \Rightarrow |\mathbf{v} \times \mathbf{a}| = 25$; $|\mathbf{v}| = \sqrt{3^2 + 4^2} = 5$ $\Rightarrow \kappa = \frac{|\mathbf{v}\times\mathbf{a}|}{|\mathbf{v}|^3} = \frac{25}{5^3} = \frac{1}{5}$

7. $\mathbf{r} = x\mathbf{i} + y\mathbf{j} \Rightarrow \mathbf{v} = \frac{dx}{dt}\mathbf{i} + \frac{dy}{dt}\mathbf{j}$ and $\mathbf{v} \cdot \mathbf{i} = y \Rightarrow \frac{dx}{dt} = y$. Since the particle moves around the unit circle $x^2 + y^2 = 1$, $2x\frac{dx}{dt} + 2y\frac{dy}{dt} = 0 \Rightarrow \frac{dy}{dt} = -\frac{x}{y}\frac{dx}{dt} \Rightarrow \frac{dy}{dt} = -\frac{x}{y}(y) = -x$. Since $\frac{dx}{dt} = y$ and $\frac{dy}{dt} = -x$, we have $\mathbf{v} = y\mathbf{i} - x\mathbf{j} \Rightarrow$ at $(1, 0)$, $\mathbf{v} = -\mathbf{j}$ and the motion is clockwise.

9. $\frac{d\mathbf{r}}{dt}$ orthogonal to $\mathbf{r} \Rightarrow 0 = \frac{d\mathbf{r}}{dt} \cdot \mathbf{r} = \frac{1}{2}\frac{d\mathbf{r}}{dt} \cdot \mathbf{r} + \frac{1}{2}\mathbf{r} \cdot \frac{d\mathbf{r}}{dt} = \frac{1}{2}\frac{d}{dt}(\mathbf{r} \cdot \mathbf{r}) \Rightarrow \mathbf{r} \cdot \mathbf{r} = K$, a constant. If $\mathbf{r} = x\mathbf{i} + y\mathbf{j}$, where x and y are differentiable functions of t, then $\mathbf{r} \cdot \mathbf{r} = x^2 + y^2 \Rightarrow x^2 + y^2 = K$, which is the equation of a circle centered at the origin.

11. $y = y_0 + (v_0 \sin\alpha)t - \frac{1}{2}gt^2 \Rightarrow y = 6.5 + (44\text{ ft/sec})(\sin 45°)(3\text{ sec}) - \frac{1}{2}(32\text{ ft/sec}^2)(3\text{ sec})^2 = 6.5 + 66\sqrt{2} - 144$ ≈ -44.16 ft $\Rightarrow$ the shot put is on the ground. Now, $y = 0 \Rightarrow 6.5 + 22\sqrt{2}t - 16t^2 = 0 \Rightarrow t \approx 2.13$ sec (the positive root) $\Rightarrow x \approx (44\text{ ft/sec})(\cos 45°)(2.13\text{ sec}) \approx 66.27$ ft or about 66 ft, 3 in. from the stopboard

13. $\mathbf{r} = (2\cos t)\mathbf{i} + (2\sin t)\mathbf{j} + t^2\mathbf{k} \Rightarrow \mathbf{v} = (-2\sin t)\mathbf{i} + (2\cos t)\mathbf{j} + 2t\mathbf{k} \Rightarrow |\mathbf{v}| = \sqrt{(-2\sin t)^2 + (2\cos t)^2 + (2t)^2}$ $= 2\sqrt{1+t^2} \Rightarrow \text{Length} = \int_0^{\pi/4} 2\sqrt{1+t^2}\,dt = \left[t\sqrt{1+t^2} + \ln\left|t + \sqrt{1+t^2}\right|\right]_0^{\pi/4} = \frac{\pi}{4}\sqrt{1+\frac{\pi^2}{16}} + \ln\left(\frac{\pi}{4} + \sqrt{1+\frac{\pi^2}{16}}\right)$

15. $\mathbf{r} = \frac{4}{9}(1+t)^{3/2}\mathbf{i} + \frac{4}{9}(1-t)^{3/2}\mathbf{j} + \frac{1}{3}t\mathbf{k} \Rightarrow \mathbf{v} = \frac{2}{3}(1+t)^{1/2}\mathbf{i} - \frac{2}{3}(1-t)^{1/2}\mathbf{j} + \frac{1}{3}\mathbf{k}$ $\Rightarrow |\mathbf{v}| = \sqrt{\left[\frac{2}{3}(1+t)^{1/2}\right]^2 + \left[-\frac{2}{3}(1-t)^{1/2}\right]^2 + \left(\frac{1}{3}\right)^2} = 1 \Rightarrow \mathbf{T} = \frac{2}{3}(1+t)^{1/2}\mathbf{i} - \frac{2}{3}(1-t)^{1/2}\mathbf{j} + \frac{1}{3}\mathbf{k}$ $\Rightarrow \mathbf{T}(0) = \frac{2}{3}\mathbf{i} - \frac{2}{3}\mathbf{j} + \frac{1}{3}\mathbf{k}$; $\frac{d\mathbf{T}}{dt} = \frac{1}{3}(1+t)^{-1/2}\mathbf{i} + \frac{1}{3}(1-t)^{-1/2}\mathbf{j} \Rightarrow \frac{d\mathbf{T}}{dt}(0) = \frac{1}{3}\mathbf{i} + \frac{1}{3}\mathbf{j} \Rightarrow \left|\frac{d\mathbf{T}}{dt}(0)\right| = \frac{\sqrt{2}}{3}$ $\Rightarrow \mathbf{N}(0) = \frac{1}{\sqrt{2}}\mathbf{i} + \frac{1}{\sqrt{2}}\mathbf{j}$; $\mathbf{B}(0) = \mathbf{T}(0) \times \mathbf{N}(0) = \begin{vmatrix} \mathbf{i} & \mathbf{j} & \mathbf{k} \\ \frac{2}{3} & -\frac{2}{3} & \frac{1}{3} \\ \frac{1}{\sqrt{2}} & \frac{1}{\sqrt{2}} & 0 \end{vmatrix} = -\frac{1}{3\sqrt{2}}\mathbf{i} + \frac{1}{3\sqrt{2}}\mathbf{j} + \frac{4}{3\sqrt{2}}\mathbf{k}$;
$\mathbf{a} = \frac{1}{3}(1+t)^{-1/2}\mathbf{i} + \frac{1}{3}(1-t)^{-1/2}\mathbf{j} \Rightarrow \mathbf{a}(0) = \frac{1}{3}\mathbf{i} + \frac{1}{3}\mathbf{j}$ and $\mathbf{v}(0) = \frac{2}{3}\mathbf{i} - \frac{2}{3}\mathbf{j} + \frac{1}{3}\mathbf{k} \Rightarrow \mathbf{v}(0) \times \mathbf{a}(0)$ $= \begin{vmatrix} \mathbf{i} & \mathbf{j} & \mathbf{k} \\ \frac{2}{3} & -\frac{2}{3} & \frac{1}{3} \\ \frac{1}{3} & \frac{1}{3} & 0 \end{vmatrix} = -\frac{1}{9}\mathbf{i} + \frac{1}{9}\mathbf{j} + \frac{4}{9}\mathbf{k} \Rightarrow |\mathbf{v} \times \mathbf{a}| = \frac{\sqrt{2}}{3} \Rightarrow \kappa(0) = \frac{|\mathbf{v}\times\mathbf{a}|}{|\mathbf{v}|^3} = \frac{\left(\frac{\sqrt{2}}{3}\right)}{1^3} = \frac{\sqrt{2}}{3}$;
$\dot{\mathbf{a}} = -\frac{1}{6}(1+t)^{-3/2}\mathbf{i} + \frac{1}{6}(1-t)^{-3/2}\mathbf{j} \Rightarrow \dot{\mathbf{a}}(0) = -\frac{1}{6}\mathbf{i} + \frac{1}{6}\mathbf{j} \Rightarrow \tau(0) = \frac{\begin{vmatrix} \frac{2}{3} & -\frac{2}{3} & \frac{1}{3} \\ \frac{1}{3} & \frac{1}{3} & 0 \\ -\frac{1}{6} & \frac{1}{6} & 0 \end{vmatrix}}{|\mathbf{v}\times\mathbf{a}|^2} = \frac{\left(\frac{1}{3}\right)\left(\frac{2}{18}\right)}{\left(\frac{\sqrt{2}}{3}\right)^2} = \frac{1}{6}$

17. $\mathbf{r} = t\mathbf{i} + \frac{1}{2}e^{2t}\mathbf{j} \Rightarrow \mathbf{v} = \mathbf{i} + e^{2t}\mathbf{j} \Rightarrow |\mathbf{v}| = \sqrt{1+e^{4t}} \Rightarrow \mathbf{T} = \frac{1}{\sqrt{1+e^{4t}}}\mathbf{i} + \frac{e^{2t}}{\sqrt{1+e^{4t}}}\mathbf{j} \Rightarrow \mathbf{T}(\ln 2) = \frac{1}{\sqrt{17}}\mathbf{i} + \frac{4}{\sqrt{17}}\mathbf{j}$;
$\frac{d\mathbf{T}}{dt} = \frac{-2e^{4t}}{(1+e^{4t})^{3/2}}\mathbf{i} + \frac{2e^{2t}}{(1+e^{4t})^{3/2}}\mathbf{j} \Rightarrow \frac{d\mathbf{T}}{dt}(\ln 2) = \frac{-32}{17\sqrt{17}}\mathbf{i} + \frac{8}{17\sqrt{17}}\mathbf{j} \Rightarrow \mathbf{N}(\ln 2) = -\frac{4}{\sqrt{17}}\mathbf{i} + \frac{1}{\sqrt{17}}\mathbf{j}$;

$$\mathbf{B}(\ln 2) = \mathbf{T}(\ln 2) \times \mathbf{N}(\ln 2) = \begin{vmatrix} \mathbf{i} & \mathbf{j} & \mathbf{k} \\ \frac{1}{\sqrt{17}} & \frac{4}{\sqrt{17}} & 0 \\ -\frac{4}{\sqrt{17}} & \frac{1}{\sqrt{17}} & 0 \end{vmatrix} = \mathbf{k}\,;\ \mathbf{a} = 2e^{2t}\,\mathbf{j} \Rightarrow \mathbf{a}(\ln 2) = 8\mathbf{j} \text{ and } \mathbf{v}(\ln 2) = \mathbf{i} + 4\mathbf{j}$$

$$\Rightarrow \mathbf{v}(\ln 2) \times \mathbf{a}(\ln 2) = \begin{vmatrix} \mathbf{i} & \mathbf{j} & \mathbf{k} \\ 1 & 4 & 0 \\ 0 & 8 & 0 \end{vmatrix} = 8\mathbf{k} \Rightarrow |\mathbf{v} \times \mathbf{a}| = 8 \text{ and } |\mathbf{v}(\ln 2)| = \sqrt{17} \Rightarrow \kappa(\ln 2) = \frac{8}{17\sqrt{17}}\,;\ \dot{\mathbf{a}} = 4e^{2t}\,\mathbf{j}$$

$$\Rightarrow \dot{\mathbf{a}}(\ln 2) = 16\mathbf{j} \Rightarrow \tau(\ln 2) = \frac{\begin{vmatrix} 1 & 4 & 0 \\ 0 & 8 & 0 \\ 0 & 16 & 0 \end{vmatrix}}{|\mathbf{v}\times\mathbf{a}|^2} = 0$$

19. $\mathbf{r} = (2 + 3t + 3t^2)\,\mathbf{i} + (4t + 4t^2)\,\mathbf{j} - (6\cos t)\mathbf{k} \Rightarrow \mathbf{v} = (3 + 6t)\mathbf{i} + (4 + 8t)\mathbf{j} + (6\sin t)\mathbf{k}$

$\Rightarrow |\mathbf{v}| = \sqrt{(3 + 6t)^2 + (4 + 8t)^2 + (6\sin t)^2} = \sqrt{25 + 100t + 100t^2 + 36\sin^2 t}$

$\Rightarrow \frac{d|\mathbf{v}|}{dt} = \frac{1}{2}\left(25 + 100t + 100t^2 + 36\sin^2 t\right)^{-1/2}(100 + 200t + 72\sin t\cos t) \Rightarrow a_T(0) = \frac{d|\mathbf{v}|}{dt}(0) = 10;$

$\mathbf{a} = 6\mathbf{i} + 8\mathbf{j} + (6\cos t)\mathbf{k} \Rightarrow |\mathbf{a}| = \sqrt{6^2 + 8^2 + (6\cos t)^2} = \sqrt{100 + 36\cos^2 t} \Rightarrow |\mathbf{a}(0)| = \sqrt{136}$

$\Rightarrow a_N = \sqrt{|\mathbf{a}|^2 - a_T^2} = \sqrt{136 - 10^2} = \sqrt{36} = 6 \Rightarrow \mathbf{a}(0) = 10\mathbf{T} + 6\mathbf{N}$

21. $\mathbf{r} = (\sin t)\mathbf{i} + \left(\sqrt{2}\cos t\right)\mathbf{j} + (\sin t)\mathbf{k} \Rightarrow \mathbf{v} = (\cos t)\mathbf{i} - \left(\sqrt{2}\sin t\right)\mathbf{j} + (\cos t)\mathbf{k}$

$\Rightarrow |\mathbf{v}| = \sqrt{(\cos t)^2 + \left(-\sqrt{2}\sin t\right)^2 + (\cos t)^2} = \sqrt{2} \Rightarrow \mathbf{T} = \frac{\mathbf{v}}{|\mathbf{v}|} = \left(\frac{1}{\sqrt{2}}\cos t\right)\mathbf{i} - (\sin t)\mathbf{j} + \left(\frac{1}{\sqrt{2}}\cos t\right)\mathbf{k}\,;$

$\frac{d\mathbf{T}}{dt} = \left(-\frac{1}{\sqrt{2}}\sin t\right)\mathbf{i} - (\cos t)\mathbf{j} - \left(\frac{1}{\sqrt{2}}\sin t\right)\mathbf{k} \Rightarrow \left|\frac{d\mathbf{T}}{dt}\right| = \sqrt{\left(-\frac{1}{\sqrt{2}}\sin t\right)^2 + (-\cos t)^2 + \left(-\frac{1}{\sqrt{2}}\sin t\right)^2} = 1$

$$\Rightarrow \mathbf{N} = \frac{\left(\frac{d\mathbf{T}}{dt}\right)}{\left|\frac{d\mathbf{T}}{dt}\right|} = \left(-\frac{1}{\sqrt{2}}\sin t\right)\mathbf{i} - (\cos t)\mathbf{j} - \left(\frac{1}{\sqrt{2}}\sin t\right)\mathbf{k}\,;\ \mathbf{B} = \mathbf{T}\times\mathbf{N} = \begin{vmatrix} \mathbf{i} & \mathbf{j} & \mathbf{k} \\ \frac{1}{\sqrt{2}}\cos t & -\sin t & \frac{1}{\sqrt{2}}\cos t \\ -\frac{1}{\sqrt{2}}\sin t & -\cos t & -\frac{1}{\sqrt{2}}\sin t \end{vmatrix}$$

$$= \frac{1}{\sqrt{2}}\mathbf{i} - \frac{1}{\sqrt{2}}\mathbf{k}\,;\ \mathbf{a} = (-\sin t)\mathbf{i} - \left(\sqrt{2}\cos t\right)\mathbf{j} - (\sin t)\mathbf{k} \Rightarrow \mathbf{v}\times\mathbf{a} = \begin{vmatrix} \mathbf{i} & \mathbf{j} & \mathbf{k} \\ \cos t & -\sqrt{2}\sin t & \cos t \\ -\sin t & -\sqrt{2}\cos t & -\sin t \end{vmatrix}$$

$= \sqrt{2}\mathbf{i} - \sqrt{2}\mathbf{k} \Rightarrow |\mathbf{v}\times\mathbf{a}| = \sqrt{4} = 2 \Rightarrow \kappa = \frac{|\mathbf{v}\times\mathbf{a}|}{|\mathbf{v}|^3} = \frac{2}{\left(\sqrt{2}\right)^3} = \frac{1}{\sqrt{2}}\,;\ \dot{\mathbf{a}} = (-\cos t)\mathbf{i} + \left(\sqrt{2}\sin t\right)\mathbf{j} - (\cos t)\mathbf{k}$

$$\Rightarrow \tau = \frac{\begin{vmatrix} \cos t & -\sqrt{2}\sin t & \cos t \\ -\sin t & -\sqrt{2}\cos t & -\sin t \\ -\cos t & \sqrt{2}\sin t & -\cos t \end{vmatrix}}{|\mathbf{v}\times\mathbf{a}|^2} = \frac{(\cos t)\left(\sqrt{2}\right) - \left(\sqrt{2}\sin t\right)(0) + (\cos t)\left(-\sqrt{2}\right)}{4} = 0$$

23. $\mathbf{r} = e^t\mathbf{i} + (\sin t)\mathbf{j} + \ln(1 - t)\mathbf{k} \Rightarrow \mathbf{v} = e^t\mathbf{i} + (\cos t)\mathbf{j} - \left(\frac{1}{1-t}\right)\mathbf{k} \Rightarrow \mathbf{v}(0) = \mathbf{i} + \mathbf{j} - \mathbf{k}\,;\ \mathbf{r}(0) = \mathbf{i} \Rightarrow (1, 0, 0)$ is on the line

$\Rightarrow x = 1 + t$, $y = t$, and $z = -t$ are parametric equations of the line

25. $s = a\theta \Rightarrow \theta = \frac{s}{a} \Rightarrow \phi = \frac{s}{a} + \frac{\pi}{2} \Rightarrow \frac{d\phi}{ds} = \frac{1}{a} \Rightarrow \kappa = \left|\frac{1}{a}\right| = \frac{1}{a}$ since $a > 0$

CHAPTER 11 PARTIAL DERIVATIVES

11.1 FUNCTIONS OF SEVERAL VARIABLES

1. (a) Domain: all points in the xy-plane
   (b) Range: all real numbers
   (c) level curves are straight lines $y - x = c$ parallel to the line $y = x$
   (d) no boundary points
   (e) both open and closed
   (f) unbounded

3. (a) Domain: all points in the xy-plane
   (b) Range: $z \geq 0$
   (c) level curves: for $f(x, y) = 0$, the origin; for $f(x, y) = c > 0$, ellipses with center $(0, 0)$ and major and minor axes along the x- and y-axes, respectively
   (d) no boundary points
   (e) both open and closed
   (f) unbounded

5. (a) Domain: all points in the xy-plane
   (b) Range: all real numbers
   (c) level curves are hyperbolas with the x- and y-axes as asymptotes when $f(x, y) \neq 0$, and the x- and y-axes when $f(x, y) = 0$
   (d) no boundary points
   (e) both open and closed
   (f) unbounded

7. (a) Domain: all (x, y) satisfying $x^2 + y^2 < 16$
   (b) Range: $z \geq \frac{1}{4}$
   (c) level curves are circles centered at the origin with radii $r < 4$
   (d) boundary is the circle $x^2 + y^2 = 16$
   (e) open
   (f) bounded

9. (a) Domain: $(x, y) \neq (0, 0)$
   (b) Range: all real numbers
   (c) level curves are circles with center $(0, 0)$ and radii $r > 0$
   (d) boundary is the single point $(0, 0)$
   (e) open
   (f) unbounded

11. (a) Domain: all (x, y) satisfying $-1 \leq y - x \leq 1$
    (b) Range: $-\frac{\pi}{2} \leq z \leq \frac{\pi}{2}$
    (c) level curves are straight lines of the form $y - x = c$ where $-1 \leq c \leq 1$
    (d) boundary is the two straight lines $y = 1 + x$ and $y = -1 + x$
    (e) closed
    (f) unbounded

13. f

15. a

17. d

19. (a)

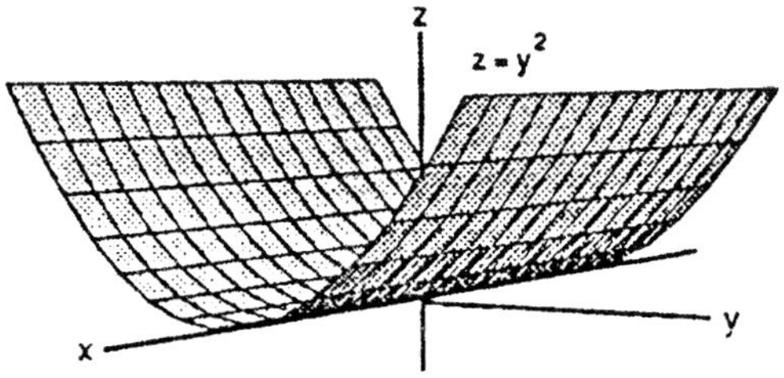

(b)

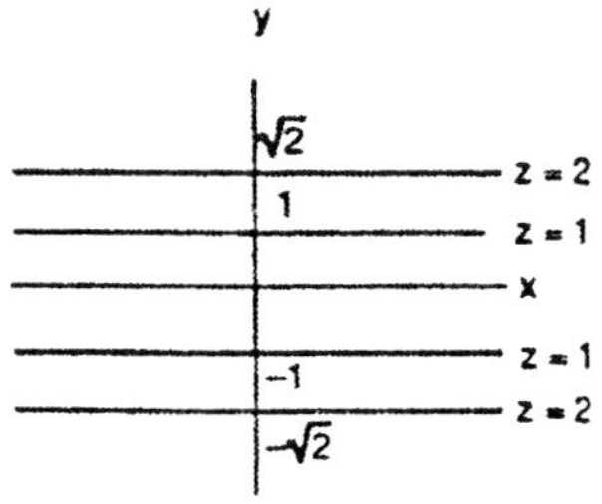

21. (a)

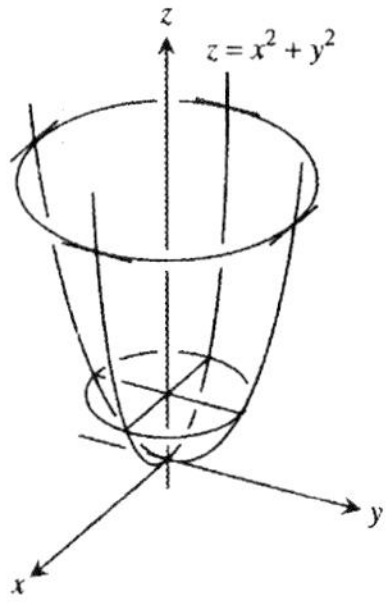

(b)

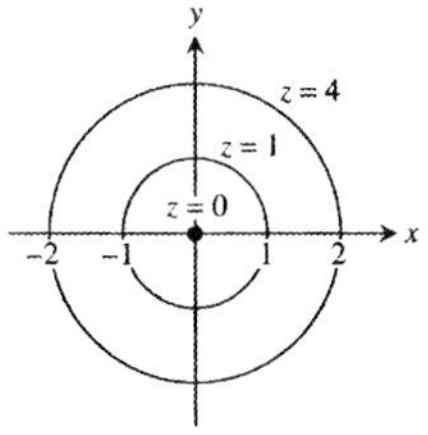

23. (a)

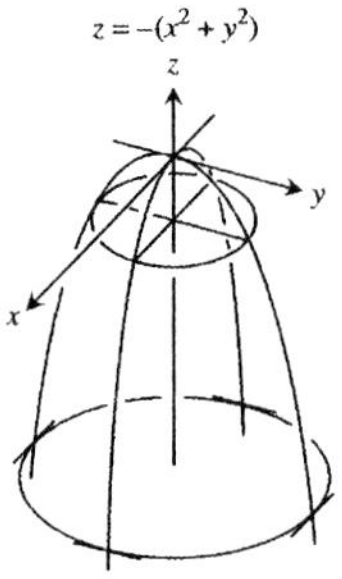

(b)

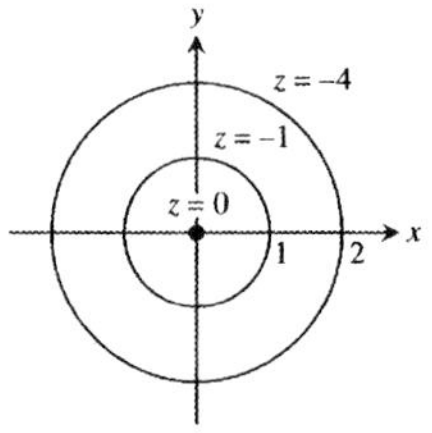

25. (a)

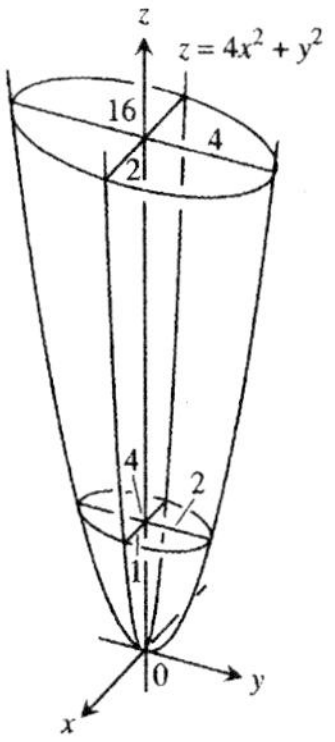

(b)

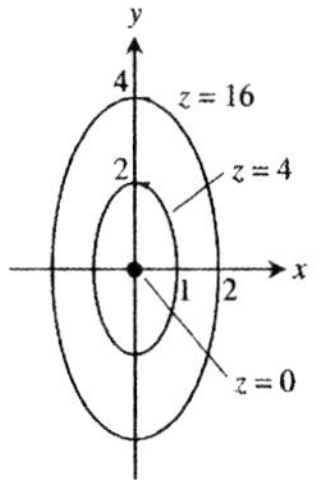

27. (a)

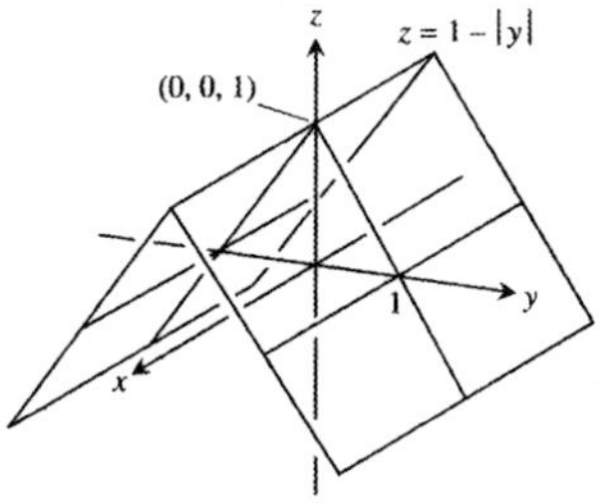

(b)

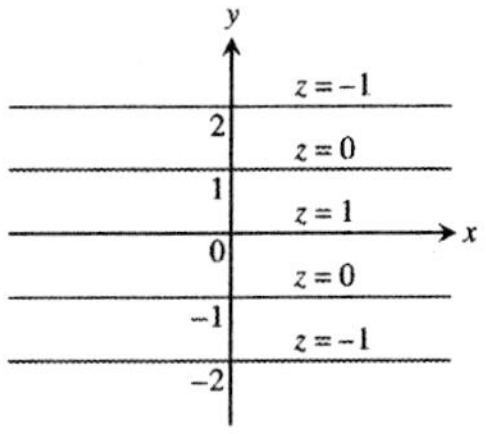

29. $f(x, y) = 16 - x^2 - y^2$ and $\left(2\sqrt{2}, \sqrt{2}\right) \Rightarrow z = 16 - \left(2\sqrt{2}\right)^2 - \left(\sqrt{2}\right)^2 = 6 \Rightarrow 6 = 16 - x^2 - y^2 \Rightarrow x^2 + y^2 = 10$

31. $f(x, y) = \int_x^y \frac{1}{1+t^2}\, dt$ at $\left(-\sqrt{2}, \sqrt{2}\right) \Rightarrow z = \tan^{-1} y - \tan^{-1} x$; at $\left(-\sqrt{2}, \sqrt{2}\right) \Rightarrow z = \tan^{-1} \sqrt{2} - \tan^{-1}\left(-\sqrt{2}\right)$
$= 2 \tan^{-1} \sqrt{2} \Rightarrow \tan^{-1} y - \tan^{-1} x = 2 \tan^{-1} \sqrt{2}$

33.

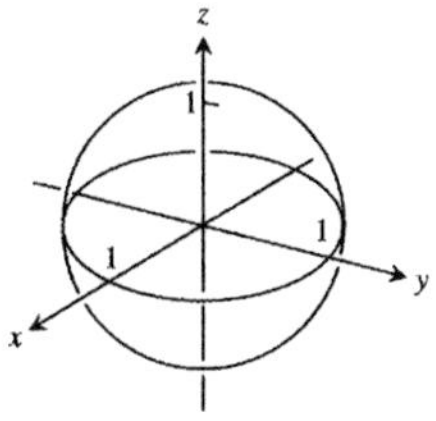

35.

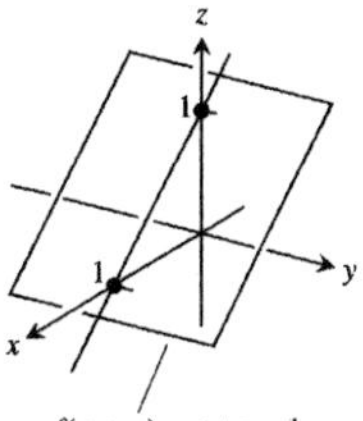

37.

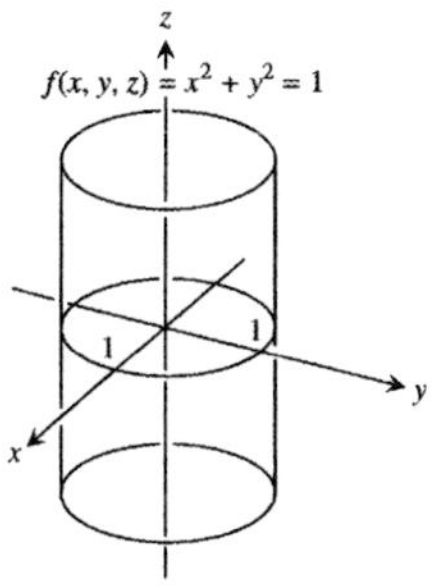

39.

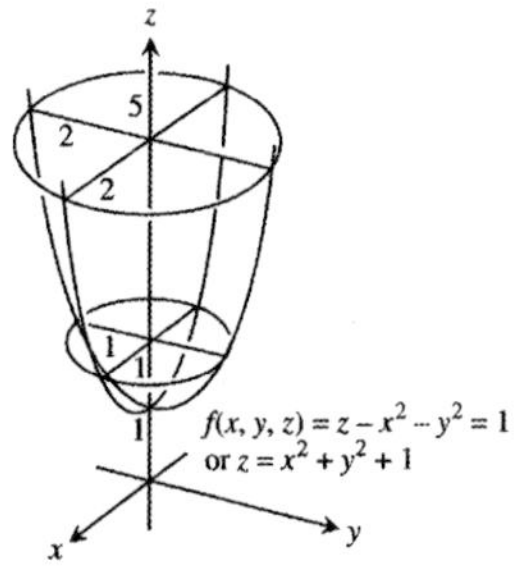

41. $f(x, y, z) = \sqrt{x - y} - \ln z$ at $(3, -1, 1) \Rightarrow w = \sqrt{x - y} - \ln z$; at $(3, -1, 1) \Rightarrow w = \sqrt{3 - (-1)} - \ln 1 = 2$
$\Rightarrow \sqrt{x - y} - \ln z = 2$

43. $g(x, y, z) = \sum_{n=0}^{\infty} \frac{(x+y)^n}{n!\, z^n}$ at $(\ln 2, \ln 4, 3) \Rightarrow w = \sum_{n=0}^{\infty} \frac{(x+y)^n}{n!\, z^n} = e^{(x+y)/z}$; at $(\ln 2, \ln 4, 3) \Rightarrow w = e^{(\ln 2 + \ln 4)/3}$
$= e^{(\ln 8)/3} = e^{\ln 2} = 2 \Rightarrow 2 = e^{(x+y)/z} \Rightarrow \frac{x+y}{z} = \ln 2$

11.2 LIMITS AND CONTINUITY IN HIGHER DIMENSIONS

1. $\lim_{(x, y) \to (0, 0)} \frac{3x^2 - y^2 + 5}{x^2 + y^2 + 2} = \frac{3(0)^2 - 0^2 + 5}{0^2 + 0^2 + 2} = \frac{5}{2}$

3. $\lim_{(x, y) \to (3, 4)} \sqrt{x^2 + y^2 - 1} = \sqrt{3^2 + 4^2 - 1} = \sqrt{24} = 2\sqrt{6}$

5. $\lim_{(x,y)\to(0,\frac{\pi}{4})} \sec x \tan y = (\sec 0)\left(\tan \frac{\pi}{4}\right) = (1)(1) = 1$

7. $\lim_{(x,y)\to(0,\ln 2)} e^{x-y} = e^{0-\ln 2} = e^{\ln\left(\frac{1}{2}\right)} = \frac{1}{2}$

9. $\lim_{(x,y)\to(0,0)} \frac{e^y \sin x}{x} = \lim_{(x,y)\to(0,0)} (e^y)\left(\frac{\sin x}{x}\right) = e^0 \cdot \lim_{x\to 0} \left(\frac{\sin x}{x}\right) = 1 \cdot 1 = 1$

11. $\lim_{(x,y)\to(1,0)} \frac{x \sin y}{x^2+1} = \frac{1\cdot\sin 0}{1^2+1} = \frac{0}{2} = 0$

13. $\lim_{\substack{(x,y)\to(1,1)\\ x\neq y}} \frac{x^2-2xy+y^2}{x-y} = \lim_{(x,y)\to(1,1)} \frac{(x-y)^2}{x-y} = \lim_{(x,y)\to(1,1)} (x-y) = (1-1) = 0$

15. $\lim_{\substack{(x,y)\to(1,1)\\ x\neq 1}} \frac{xy-y-2x+2}{x-1} = \lim_{\substack{(x,y)\to(1,1)\\ x\neq 1}} \frac{(x-1)(y-2)}{x-1} = \lim_{(x,y)\to(1,1)} (y-2) = (1-2) = -1$

17. $\lim_{\substack{(x,y)\to(0,0)\\ x\neq y}} \frac{x-y+2\sqrt{x}-2\sqrt{y}}{\sqrt{x}-\sqrt{y}} = \lim_{\substack{(x,y)\to(0,0)\\ x\neq y}} \frac{\left(\sqrt{x}-\sqrt{y}\right)\left(\sqrt{x}+\sqrt{y}+2\right)}{\sqrt{x}-\sqrt{y}} = \lim_{(x,y)\to(0,0)} \left(\sqrt{x}+\sqrt{y}+2\right)$
$= \left(\sqrt{0}+\sqrt{0}+2\right) = 2$

Note: (x, y) must approach $(0, 0)$ through the first quadrant only with $x \neq y$.

19. $\lim_{\substack{(x,y)\to(2,0)\\ 2x-y\neq 4}} \frac{\sqrt{2x-y}-2}{2x-y-4} = \lim_{\substack{(x,y)\to(2,0)\\ 2x-y\neq 4}} \frac{\sqrt{2x-y}-2}{\left(\sqrt{2x-y}+2\right)\left(\sqrt{2x-y}-2\right)} = \lim_{(x,y)\to(2,0)} \frac{1}{\sqrt{2x-y}+2}$
$= \frac{1}{\sqrt{(2)(2)-0}+2} = \frac{1}{2+2} = \frac{1}{4}$

21. $\lim_{P\to(1,3,4)} \left(\frac{1}{x}+\frac{1}{y}+\frac{1}{z}\right) = \frac{1}{1}+\frac{1}{3}+\frac{1}{4} = \frac{12+4+3}{12} = \frac{19}{12}$

23. $\lim_{P\to(3,3,0)} (\sin^2 x + \cos^2 y + \sec^2 z) = (\sin^2 3 + \cos^2 3) + \sec^2 0 = 1 + 1^2 = 2$

25. $\lim_{P\to(\pi,0,3)} ze^{-2y} \cos 2x = 3e^{-2(0)} \cos 2\pi = (3)(1)(1) = 3$

27. (a) All (x, y)
(b) All (x, y) except $(0, 0)$

29. (a) All (x, y) except where $x = 0$ or $y = 0$
(b) All (x, y)

31. (a) All (x, y, z)
(b) All (x, y, z) except the interior of the cylinder $x^2 + y^2 = 1$

33. (a) All (x, y, z) with $z \neq 0$
(b) All (x, y, z) with $x^2 + z^2 \neq 1$

35. $\lim\limits_{\substack{(x,y)\to(0,0)\\ \text{along } y=x\\ x>0}} -\frac{x}{\sqrt{x^2+y^2}} = \lim\limits_{x\to 0^+} -\frac{x}{\sqrt{x^2+x^2}} = \lim\limits_{x\to 0^+} -\frac{x}{\sqrt{2}\,|x|} = \lim\limits_{x\to 0^+} -\frac{x}{\sqrt{2}\,x} = \lim\limits_{x\to 0^+} -\frac{1}{\sqrt{2}} = -\frac{1}{\sqrt{2}}$;

$\lim\limits_{\substack{(x,y)\to(0,0)\\ \text{along } y=x\\ x<0}} -\frac{x}{\sqrt{x^2+y^2}} = \lim\limits_{x\to 0^-} -\frac{x}{\sqrt{2}\,|x|} = \lim\limits_{x\to 0^-} -\frac{x}{\sqrt{2}(-x)} = \lim\limits_{x\to 0^-} \frac{1}{\sqrt{2}} = \frac{1}{\sqrt{2}}$

37. $\lim\limits_{\substack{(x,y)\to(0,0)\\ \text{along } y=kx^2}} \frac{x^4-y^2}{x^4+y^2} = \lim\limits_{x\to 0} \frac{x^4-(kx^2)^2}{x^4+(kx^2)^2} = \lim\limits_{x\to 0} \frac{x^4-k^2x^4}{x^4+k^2x^4} = \frac{1-k^2}{1+k^2} \Rightarrow$ different limits for different values of k

39. $\lim\limits_{\substack{(x,y)\to(0,0)\\ \text{along } y=kx\\ k\neq -1}} \frac{x-y}{x+y} = \lim\limits_{x\to 0} \frac{x-kx}{x+kx} = \frac{1-k}{1+k} \Rightarrow$ different limits for different values of k, $k \neq -1$

41. $\lim\limits_{\substack{(x,y)\to(0,0)\\ \text{along } y=kx^2\\ k\neq 0}} \frac{x^2+y}{y} = \lim\limits_{x\to 0} \frac{x^2+kx^2}{kx^2} = \frac{1+k}{k} \Rightarrow$ different limits for different values of k, $k \neq 0$

43. First consider the vertical line $x = 0 \Rightarrow \lim\limits_{\substack{(x,y)\to(0,0)\\ \text{along } x=0}} \frac{2x^2y}{x^4+y^2} = \lim\limits_{y\to 0} \frac{2(0)^2y}{(0)^4+y^2} = \lim\limits_{y\to 0} 0 = 0$. Now consider any nonvertical through (0, 0). The equation of any line through (0, 0) is of the form $y = mx \Rightarrow \lim\limits_{\substack{(x,y)\to(0,0)\\ \text{along } y=mx}} f(x,y) = \lim\limits_{\substack{(x,y)\to(0,0)\\ \text{along } y=mx}} \frac{2x^2y}{x^4+y^2}$ $= \lim\limits_{x\to 0} \frac{2x^2(mx)}{x^4+(mx)^2} = \lim\limits_{x\to 0} \frac{2mx^3}{x^4+m^2x^2} = \lim\limits_{x\to 0} \frac{2mx^3}{x^2(x^2+m^2)} = \lim\limits_{x\to 0} \frac{2mx}{(x^2+m^2)} = 0$. Thus $\lim\limits_{\substack{(x,y)\to(0,0)\\ \text{any line though } (0,0)}} \frac{2x^2y}{x^4+y^2} = 0$.

45. $\lim\limits_{(x,y)\to(0,0)} \left(1-\frac{x^2y^2}{3}\right) = 1$ and $\lim\limits_{(x,y)\to(0,0)} 1 = 1 \Rightarrow \lim\limits_{(x,y)\to(0,0)} \frac{\tan^{-1} xy}{xy} = 1$, by the Sandwich Theorem

47. The limit is 0 since $\left|\sin\left(\frac{1}{x}\right)\right| \le 1 \Rightarrow -1 \le \sin\left(\frac{1}{x}\right) \le 1 \Rightarrow -y \le y\sin\left(\frac{1}{x}\right) \le y$ for $y \ge 0$, and $-y \ge y\sin\left(\frac{1}{x}\right) \ge y$ for $y \le 0$. Thus as $(x,y) \to (0,0)$, both $-y$ and y approach $0 \Rightarrow y\sin\left(\frac{1}{x}\right) \to 0$, by the Sandwich Theorem.

49. (a) $f(x,y)\big|_{y=mx} = \frac{2m}{1+m^2} = \frac{2\tan\theta}{1+\tan^2\theta} = \sin 2\theta$. The value of $f(x,y) = \sin 2\theta$ varies with θ, which is the line's angle of inclination.

(b) Since $f(x,y)\big|_{y=mx} = \sin 2\theta$ and since $-1 \le \sin 2\theta \le 1$ for every θ, $\lim\limits_{(x,y)\to(0,0)} f(x,y)$ varies from -1 to 1 along $y = mx$.

51. $\lim\limits_{(x,y)\to(0,0)} \frac{x^3-xy^2}{x^2+y^2} = \lim\limits_{r\to 0} \frac{r^3\cos^3\theta-(r\cos\theta)(r^2\sin^2\theta)}{r^2\cos^2\theta+r^2\sin^2\theta} = \lim\limits_{r\to 0} \frac{r(\cos^3\theta-\cos\theta\sin^2\theta)}{1} = 0$

53. $\lim\limits_{(x,y)\to(0,0)} \frac{y^2}{x^2+y^2} = \lim\limits_{r\to 0} \frac{r^2\sin^2\theta}{r^2} = \lim\limits_{r\to 0} (\sin^2\theta) = \sin^2\theta$; the limit does not exist since $\sin^2\theta$ is between 0 and 1 depending on θ

55. $\lim\limits_{(x,y)\to(0,0)} \tan^{-1}\left[\frac{|x|+|y|}{x^2+y^2}\right] = \lim\limits_{r\to 0} \tan^{-1}\left[\frac{|r\cos\theta|+|r\sin\theta|}{r^2}\right] = \lim\limits_{r\to 0} \tan^{-1}\left[\frac{|r|(|\cos\theta|+|\sin\theta|)}{r^2}\right]$;

if $r \to 0^+$, then $\lim\limits_{r\to 0^+} \tan^{-1}\left[\frac{|r|(|\cos\theta|+|\sin\theta|)}{r^2}\right] = \lim\limits_{r\to 0^+} \tan^{-1}\left[\frac{|\cos\theta|+|\sin\theta|}{r}\right] = \frac{\pi}{2}$; if $r \to 0^-$, then $\lim\limits_{r\to 0^-} \tan^{-1}\left[\frac{|r|(|\cos\theta|+|\sin\theta|)}{r^2}\right] = \lim\limits_{r\to 0^-} \tan^{-1}\left(\frac{|\cos\theta|+|\sin\theta|}{-r}\right) = \frac{\pi}{2} \Rightarrow$ the limit is $\frac{\pi}{2}$

57. Let $\delta = 0.1$. Then $\sqrt{x^2+y^2} < \delta \Rightarrow \sqrt{x^2+y^2} < 0.1 \Rightarrow x^2+y^2 < 0.01 \Rightarrow |x^2+y^2-0| < 0.01$
$\Rightarrow |f(x,y) - f(0,0)| < 0.01 = \epsilon$.

59. Let $\delta = 0.005$. Then $|x| < \delta$ and $|y| < \delta \Rightarrow |f(x,y) - f(0,0)| = \left|\frac{x+y}{x^2+1} - 0\right| = \left|\frac{x+y}{x^2+1}\right| \le |x+y| < |x| + |y|$
$< 0.005 + 0.005 = 0.01 = \epsilon$.

61. Let $\delta = \sqrt{0.015}$. Then $\sqrt{x^2+y^2+z^2} < \delta \Rightarrow |f(x,y,z) - f(0,0,0)| = |x^2+y^2+z^2-0| = |x^2+y^2+z^2|$
$= \left(\sqrt{x^2+t^2+x^2}\right)^2 < \left(\sqrt{0.015}\right)^2 = 0.015 = \epsilon$.

63. Let $\delta = 0.005$. Then $|x| < \delta$, $|y| < \delta$, and $|z| < \delta \Rightarrow |f(x,y,z) - f(0,0,0)| = \left|\frac{x+y+z}{x^2+y^2+z^2+1} - 0\right|$
$= \left|\frac{x+y+z}{x^2+y^2+z^2+1}\right| \le |x+y+z| \le |x| + |y| + |z| < 0.005 + 0.005 + 0.005 = 0.015 = \epsilon$.

65. $\lim\limits_{(x,y,z)\to(x_0,y_0,z_0)} f(x,y,z) = \lim\limits_{(x,y,z)\to(x_0,y_0,z_0)} (x+y-z) = x_0+y_0-z_0 = f(x_0,y_0,z_0) \Rightarrow$ f is continuous at every (x_0,y_0,z_0)

11.3 PARTIAL DERIVATIVES

1. $\frac{\partial f}{\partial x} = 4x$, $\frac{\partial f}{\partial y} = -3$

3. $\frac{\partial f}{\partial x} = 2x(y+2)$, $\frac{\partial f}{\partial y} = x^2 - 1$

5. $\frac{\partial f}{\partial x} = 2y(xy-1)$, $\frac{\partial f}{\partial y} = 2x(xy-1)$

7. $\frac{\partial f}{\partial x} = \frac{x}{\sqrt{x^2+y^2}}$, $\frac{\partial f}{\partial y} = \frac{y}{\sqrt{x^2+y^2}}$

9. $\frac{\partial f}{\partial x} = -\frac{1}{(x+y)^2} \cdot \frac{\partial}{\partial x}(x+y) = -\frac{1}{(x+y)^2}$, $\frac{\partial f}{\partial y} = -\frac{1}{(x+y)^2} \cdot \frac{\partial}{\partial y}(x+y) = -\frac{1}{(x+y)^2}$

11. $\frac{\partial f}{\partial x} = = \frac{(xy-1)(1)-(x+y)(y)}{(xy-1)^2} = \frac{-y^2-1}{(xy-1)^2}$, $\frac{\partial f}{\partial y} = \frac{(xy-1)(1)-(x+y)(x)}{(xy-1)^2} = \frac{-x^2-1}{(xy-1)^2}$

13. $\frac{\partial f}{\partial x} = e^{(x+y+1)} \cdot \frac{\partial}{\partial x}(x+y+1) = e^{(x+y+1)}$, $\frac{\partial f}{\partial y} = e^{(x+y+1)} \cdot \frac{\partial}{\partial y}(x+y+1) = e^{(x+y+1)}$

15. $\frac{\partial f}{\partial x} = \frac{1}{x+y} \cdot \frac{\partial}{\partial x}(x+y) = \frac{1}{x+y}$, $\frac{\partial f}{\partial y} = \frac{1}{x+y} \cdot \frac{\partial}{\partial y}(x+y) = \frac{1}{x+y}$

17. $\frac{\partial f}{\partial x} = 2\sin(x-3y) \cdot \frac{\partial}{\partial x}\sin(x-3y) = 2\sin(x-3y)\cos(x-3y) \cdot \frac{\partial}{\partial x}(x-3y) = 2\sin(x-3y)\cos(x-3y)$,
$\frac{\partial f}{\partial y} = 2\sin(x-3y) \cdot \frac{\partial}{\partial y}\sin(x-3y) = 2\sin(x-3y)\cos(x-3y) \cdot \frac{\partial}{\partial y}(x-3y) = -6\sin(x-3y)\cos(x-3y)$

19. $\frac{\partial f}{\partial x} = yx^{y-1}$, $\frac{\partial f}{\partial y} = x^y \ln x$

21. $\frac{\partial f}{\partial x} = -g(x)$, $\frac{\partial f}{\partial y} = g(y)$

23. $f_x = y^2$, $f_y = 2xy$, $f_z = -4z$

25. $f_x = 1$, $f_y = -\frac{y}{\sqrt{y^2+z^2}}$, $f_z = -\frac{z}{\sqrt{y^2+z^2}}$

27. $f_x = \frac{yz}{\sqrt{1-x^2y^2z^2}}$, $f_y = \frac{xz}{\sqrt{1-x^2y^2z^2}}$, $f_z = \frac{xy}{\sqrt{1-x^2y^2z^2}}$

29. $f_x = \frac{1}{x+2y+3z}$, $f_y = \frac{2}{x+2y+3z}$, $f_z = \frac{3}{x+2y+3z}$

31. $f_x = -2xe^{-(x^2+y^2+z^2)}$, $f_y = -2ye^{-(x^2+y^2+z^2)}$, $f_z = -2ze^{-(x^2+y^2+z^2)}$

33. $f_x = \operatorname{sech}^2(x+2y+3z)$, $f_y = 2\operatorname{sech}^2(x+2y+3z)$, $f_z = 3\operatorname{sech}^2(x+2y+3z)$

35. $\frac{\partial f}{\partial t} = -2\pi \sin(2\pi t - \alpha)$, $\frac{\partial f}{\partial \alpha} = \sin(2\pi t - \alpha)$

37. $\frac{\partial h}{\partial \rho} = \sin\phi \cos\theta$, $\frac{\partial h}{\partial \phi} = \rho \cos\phi \cos\theta$, $\frac{\partial h}{\partial \theta} = -\rho \sin\phi \sin\theta$

39. $W_p = V$, $W_v = P + \frac{\delta v^2}{2g}$, $W_\delta = \frac{Vv^2}{2g}$, $W_v = \frac{2V\delta v}{2g} = \frac{V\delta v}{g}$, $W_g = -\frac{V\delta v^2}{2g^2}$

41. $\frac{\partial f}{\partial x} = 1 + y$, $\frac{\partial f}{\partial y} = 1 + x$, $\frac{\partial^2 f}{\partial x^2} = 0$, $\frac{\partial^2 f}{\partial y^2} = 0$, $\frac{\partial^2 f}{\partial y \partial x} = \frac{\partial^2 f}{\partial x \partial y} = 1$

43. $\frac{\partial g}{\partial x} = 2xy + y\cos x$, $\frac{\partial g}{\partial y} = x^2 - \sin y + \sin x$, $\frac{\partial^2 g}{\partial x^2} = 2y - y\sin x$, $\frac{\partial^2 g}{\partial y^2} = -\cos y$, $\frac{\partial^2 g}{\partial y \partial x} = \frac{\partial^2 g}{\partial x \partial y} = 2x + \cos x$

45. $\frac{\partial r}{\partial x} = \frac{1}{x+y}$, $\frac{\partial r}{\partial y} = \frac{1}{x+y}$, $\frac{\partial^2 r}{\partial x^2} = \frac{-1}{(x+y)^2}$, $\frac{\partial^2 r}{\partial y^2} = \frac{-1}{(x+y)^2}$, $\frac{\partial^2 r}{\partial y \partial x} = \frac{\partial^2 r}{\partial x \partial y} = \frac{-1}{(x+y)^2}$

47. $\frac{\partial w}{\partial x} = \frac{2}{2x+3y}$, $\frac{\partial w}{\partial y} = \frac{3}{2x+3y}$, $\frac{\partial^2 w}{\partial y \partial x} = \frac{-6}{(2x+3y)^2}$, and $\frac{\partial^2 w}{\partial x \partial y} = \frac{-6}{(2x+3y)^2}$

49. $\frac{\partial w}{\partial x} = y^2 + 2xy^3 + 3x^2y^4$, $\frac{\partial w}{\partial y} = 2xy + 3x^2y^2 + 4x^3y^3$, $\frac{\partial^2 w}{\partial y \partial x} = 2y + 6xy^2 + 12x^2y^3$, and $\frac{\partial^2 w}{\partial x \partial y} = 2y + 6xy^2 + 12x^2y^3$

51. (a) x first (b) y first (c) x first (d) x first (e) y first (f) y first

53. $f_x(1,2) = \lim_{h \to 0} \frac{f(1+h,2) - f(1,2)}{h} = \lim_{h \to 0} \frac{[1 - (1+h) + 2 - 6(1+h)^2] - (2-6)}{h} = \lim_{h \to 0} \frac{-h - 6(1 + 2h + h^2) + 6}{h}$
$= \lim_{h \to 0} \frac{-13h - 6h^2}{h} = \lim_{h \to 0} (-13 - 6h) = -13$,
$f_y(1,2) = \lim_{h \to 0} \frac{f(1,2+h) - f(1,2)}{h} = \lim_{h \to 0} \frac{[1 - 1 + (2+h) - 3(2+h)] - (2-6)}{h} = \lim_{h \to 0} \frac{(2 - 6 - 2h) - (2-6)}{h}$
$= \lim_{h \to 0} (-2) = -2$

55. $f_z(x_0, y_0, z_0) = \lim_{h \to 0} \frac{f(x_0, y_0, z_0 + h) - f(x_0, y_0, z_0)}{h}$;
$f_z(1,2,3) = \lim_{h \to 0} \frac{f(1,2,3+h) - f(1,2,3)}{h} = \lim_{h \to 0} \frac{2(3+h)^2 - 2(9)}{h} = \lim_{h \to 0} \frac{12h + 2h^2}{h} = \lim_{h \to 0} (12 + 2h) = 12$

57. $y + \left(3z^2 \frac{\partial z}{\partial x}\right) x + z^3 - 2y \frac{\partial z}{\partial x} = 0 \Rightarrow (3xz^2 - 2y) \frac{\partial z}{\partial x} = -y - z^3 \Rightarrow$ at $(1,1,1)$ we have $(3-2) \frac{\partial z}{\partial x} = -1 - 1$ or $\frac{\partial z}{\partial x} = -2$

59. $a^2 = b^2 + c^2 - 2bc \cos A \Rightarrow 2a = (2bc \sin A) \frac{\partial A}{\partial a} \Rightarrow \frac{\partial A}{\partial a} = \frac{a}{bc \sin A}$; also $0 = 2b - 2c\cos A + (2bc \sin A) \frac{\partial A}{\partial b}$
$\Rightarrow 2c\cos A - 2b = (2bc \sin A) \frac{\partial A}{\partial b} \Rightarrow \frac{\partial A}{\partial b} = \frac{c \cos A - b}{bc \sin A}$

61. Differentiating each equation implicitly gives $1 = v_x \ln u + \left(\frac{v}{u}\right) u_x$ and $0 = u_x \ln v + \left(\frac{u}{v}\right) v_x$ or

$$\left.\begin{array}{l} (\ln u)\, v_x + \left(\frac{v}{u}\right) u_x = 1 \\ \left(\frac{u}{v}\right) v_x + (\ln v)\, u_x = 0 \end{array}\right\} \Rightarrow v_x = \frac{\begin{vmatrix} 1 & \frac{v}{u} \\ 0 & \ln v \end{vmatrix}}{\begin{vmatrix} \ln u & \frac{v}{u} \\ \frac{u}{v} & \ln v \end{vmatrix}} = \frac{\ln v}{(\ln u)(\ln v) - 1}$$

63. $\frac{\partial f}{\partial x} = 2x$, $\frac{\partial f}{\partial y} = 2y$, $\frac{\partial f}{\partial z} = -4z \Rightarrow \frac{\partial^2 f}{\partial x^2} = 2$, $\frac{\partial^2 f}{\partial y^2} = 2$, $\frac{\partial^2 f}{\partial z^2} = -4 \Rightarrow \frac{\partial^2 f}{\partial x^2} + \frac{\partial^2 f}{\partial y^2} + \frac{\partial^2 f}{\partial z^2} = 2 + 2 + (-4) = 0$

65. $\frac{\partial f}{\partial x} = -2e^{-2y} \sin 2x$, $\frac{\partial f}{\partial y} = -2e^{-2y} \cos 2x$, $\frac{\partial^2 f}{\partial x^2} = -4e^{-2y} \cos 2x$, $\frac{\partial^2 f}{\partial y^2} = 4e^{-2y} \cos 2x \Rightarrow \frac{\partial^2 f}{\partial x^2} + \frac{\partial^2 f}{\partial y^2}$
$= -4e^{-2y} \cos 2x + 4e^{-2y} \cos 2x = 0$

67. $\frac{\partial f}{\partial x} = -\frac{1}{2}(x^2+y^2+z^2)^{-3/2}(2x) = -x(x^2+y^2+z^2)^{-3/2}$, $\frac{\partial f}{\partial y} = -\frac{1}{2}(x^2+y^2+z^2)^{-3/2}(2y)$
$= -y(x^2+y^2+z^2)^{-3/2}$, $\frac{\partial f}{\partial z} = -\frac{1}{2}(x^2+y^2+z^2)^{-3/2}(2z) = -z(x^2+y^2+z^2)^{-3/2}$;
$\frac{\partial^2 f}{\partial x^2} = -(x^2+y^2+z^2)^{-3/2} + 3x^2(x^2+y^2+z^2)^{-5/2}$, $\frac{\partial^2 f}{\partial y^2} = -(x^2+y^2+z^2)^{-3/2} + 3y^2(x^2+y^2+z^2)^{-5/2}$,
$\frac{\partial^2 f}{\partial z^2} = -(x^2+y^2+z^2)^{-3/2} + 3z^2(x^2+y^2+z^2)^{-5/2} \Rightarrow \frac{\partial^2 f}{\partial x^2} + \frac{\partial^2 f}{\partial y^2} + \frac{\partial^2 f}{\partial z^2}$
$= \left[-(x^2+y^2+z^2)^{-3/2} + 3x^2(x^2+y^2+z^2)^{-5/2}\right] + \left[-(x^2+y^2+z^2)^{-3/2} + 3y^2(x^2+y^2+z^2)^{-5/2}\right]$
$+ \left[-(x^2+y^2+z^2)^{-3/2} + 3z^2(x^2+y^2+z^2)^{-5/2}\right] = -3(x^2+y^2+z^2)^{-3/2} + (3x^2+3y^2+3z^2)(x^2+y^2+z^2)^{-5/2} = 0$

69. $\frac{\partial w}{\partial x} = \cos(x+ct)$, $\frac{\partial w}{\partial t} = c\cos(x+ct)$; $\frac{\partial^2 w}{\partial x^2} = -\sin(x+ct)$, $\frac{\partial^2 w}{\partial t^2} = -c^2\sin(x+ct) \Rightarrow \frac{\partial^2 w}{\partial t^2} = c^2[-\sin(x+ct)] = c^2\frac{\partial^2 w}{\partial x^2}$

71. $\frac{\partial w}{\partial x} = \cos(x+ct) - 2\sin(2x+2ct)$, $\frac{\partial w}{\partial t} = c\cos(x+ct) - 2c\sin(2x+2ct)$;
$\frac{\partial^2 w}{\partial x^2} = -\sin(x+ct) - 4\cos(2x+2ct)$, $\frac{\partial^2 w}{\partial t^2} = -c^2\sin(x+ct) - 4c^2\cos(2x+2ct)$
$\Rightarrow \frac{\partial^2 w}{\partial t^2} = c^2[-\sin(x+ct) - 4\cos(2x+2ct)] = c^2\frac{\partial^2 w}{\partial x^2}$

73. $\frac{\partial w}{\partial x} = 2\sec^2(2x-2ct)$, $\frac{\partial w}{\partial t} = -2c\sec^2(2x-2ct)$; $\frac{\partial^2 w}{\partial x^2} = 8\sec^2(2x-2ct)\tan(2x-2ct)$,
$\frac{\partial^2 w}{\partial t^2} = 8c^2\sec^2(2x-2ct)\tan(2x-2ct) \Rightarrow \frac{\partial^2 w}{\partial t^2} = c^2[8\sec^2(2x-2ct)\tan(2x-2ct)] = c^2\frac{\partial^2 w}{\partial x^2}$

75. $\frac{\partial w}{\partial t} = \frac{\partial f}{\partial u}\frac{\partial u}{\partial t} = \frac{\partial f}{\partial u}(ac) \Rightarrow \frac{\partial^2 w}{\partial t^2} = (ac)\left(\frac{\partial^2 f}{\partial u^2}\right)(ac) = a^2c^2\frac{\partial^2 f}{\partial u^2}$; $\frac{\partial w}{\partial x} = \frac{\partial f}{\partial u}\frac{\partial u}{\partial x} = \frac{\partial f}{\partial u}\cdot a \Rightarrow \frac{\partial^2 w}{\partial x^2} = \left(a\frac{\partial^2 f}{\partial u^2}\right)\cdot a$
$= a^2\frac{\partial^2 f}{\partial u^2} \Rightarrow \frac{\partial^2 w}{\partial t^2} = a^2c^2\frac{\partial^2 f}{\partial u^2} = c^2\left(a^2\frac{\partial^2 f}{\partial u^2}\right) = c^2\frac{\partial^2 w}{\partial x^2}$

77. Yes, since f_{xx}, f_{yy}, f_{xy}, and f_{yx} are all continuous on R, use the same reasoning as in Exercise 76 with
$f_x(x,y) = f_x(x_0,y_0) + f_{xx}(x_0,y_0)\,\Delta x + f_{xy}(x_0,y_0)\,\Delta y + \epsilon_1\Delta x + \epsilon_2\Delta y$ and
$f_y(x,y) = f_y(x_0,y_0) + f_{yx}(x_0,y_0)\,\Delta x + f_{yy}(x_0,y_0)\,\Delta y + \hat{\epsilon}_1\Delta x + \hat{\epsilon}_2\Delta y$. Then $\lim_{(x,y)\to(x_0,y_0)} f_x(x,y) = f_x(x_0,y_0)$
and $\lim_{(x,y)\to(x_0,y_0)} f_y(x,y) = f_y(x_0,y_0)$.

11.4 THE CHAIN RULE

1. (a) $\frac{\partial w}{\partial x} = 2x$, $\frac{\partial w}{\partial y} = 2y$, $\frac{dx}{dt} = -\sin t$, $\frac{dy}{dt} = \cos t \Rightarrow \frac{dw}{dt} = -2x\sin t + 2y\cos t = -2\cos t\sin t + 2\sin t\cos t$
$= 0$; $w = x^2 + y^2 = \cos^2 t + \sin^2 t = 1 \Rightarrow \frac{dw}{dt} = 0$
(b) $\frac{dw}{dt}(\pi) = 0$

3. (a) $\frac{\partial w}{\partial x} = \frac{1}{z}$, $\frac{\partial w}{\partial y} = \frac{1}{z}$, $\frac{\partial w}{\partial z} = \frac{-(x+y)}{z^2}$, $\frac{dx}{dt} = -2\cos t\sin t$, $\frac{dy}{dt} = 2\sin t\cos t$, $\frac{dz}{dt} = -\frac{1}{t^2}$
$\Rightarrow \frac{dw}{dt} = -\frac{2}{z}\cos t\sin t + \frac{2}{z}\sin t\cos t + \frac{x+y}{z^2t^2} = \frac{\cos^2 t + \sin^2 t}{\left(\frac{1}{t^2}\right)(t^2)} = 1$; $w = \frac{x}{z} + \frac{y}{z} = \frac{\cos^2 t}{\left(\frac{1}{t}\right)} + \frac{\sin^2 t}{\left(\frac{1}{t}\right)} = t \Rightarrow \frac{dw}{dt} = 1$
(b) $\frac{dw}{dt}(3) = 1$

5. (a) $\frac{\partial w}{\partial x} = 2ye^x$, $\frac{\partial w}{\partial y} = 2e^x$, $\frac{\partial w}{\partial z} = -\frac{1}{z}$, $\frac{dx}{dt} = \frac{2t}{t^2+1}$, $\frac{dy}{dt} = \frac{1}{t^2+1}$, $\frac{dz}{dt} = e^t \Rightarrow \frac{dw}{dt} = \frac{4yte^x}{t^2+1} + \frac{2e^x}{t^2+1} - \frac{e^t}{z}$
$= \frac{(4t)(\tan^{-1} t)(t^2+1)}{t^2+1} + \frac{2(t^2+1)}{t^2+1} - \frac{e^t}{e^t} = 4t\tan^{-1} t + 1$; $w = 2ye^x - \ln z = (2\tan^{-1} t)(t^2+1) - t$
$\Rightarrow \frac{dw}{dt} = \left(\frac{2}{t^2+1}\right)(t^2+1) + (2\tan^{-1} t)(2t) - 1 = 4t\tan^{-1} t + 1$
(b) $\frac{dw}{dt}(1) = (4)(1)\left(\frac{\pi}{4}\right) + 1 = \pi + 1$

7. (a) $\frac{\partial z}{\partial u} = \frac{\partial z}{\partial x}\frac{\partial x}{\partial u} + \frac{\partial z}{\partial y}\frac{\partial y}{\partial u} = (4e^x \ln y)\left(\frac{\cos v}{u \cos v}\right) + \left(\frac{4e^x}{y}\right)(\sin v) = \frac{4e^x \ln y}{u} + \frac{4e^x \sin v}{y}$

$= \frac{4(u \cos v) \ln (u \sin v)}{u} + \frac{4(u \cos v)(\sin v)}{u \sin v} = (4 \cos v) \ln (u \sin v) + 4 \cos v;$

$\frac{\partial z}{\partial v} = \frac{\partial z}{\partial x}\frac{\partial x}{\partial v} + \frac{\partial z}{\partial y}\frac{\partial y}{\partial v} = (4e^x \ln y)\left(\frac{-u \sin v}{u \cos v}\right) + \left(\frac{4e^x}{y}\right)(u \cos v) = -(4e^x \ln y)(\tan v) + \frac{4e^x u \cos v}{y}$

$= [-4(u \cos v) \ln (u \sin v)](\tan v) + \frac{4(u \cos v)(u \cos v)}{u \sin v} = (-4u \sin v) \ln (u \sin v) + \frac{4u \cos^2 v}{\sin v};$

$z = 4e^x \ln y = 4(u \cos v) \ln (u \sin v) \Rightarrow \frac{\partial z}{\partial u} = (4 \cos v) \ln (u \sin v) + 4(u \cos v)\left(\frac{\sin v}{u \sin v}\right)$

$= (4 \cos v) \ln (u \sin v) + 4 \cos v$; also $\frac{\partial z}{\partial v} = (-4u \sin v) \ln (u \sin v) + 4(u \cos v)\left(\frac{u \cos v}{u \sin v}\right)$

$= (-4u \sin v) \ln (u \sin v) + \frac{4u \cos^2 v}{\sin v}$

(b) At $\left(2, \frac{\pi}{4}\right)$: $\frac{\partial z}{\partial u} = 4 \cos \frac{\pi}{4} \ln \left(2 \sin \frac{\pi}{4}\right) + 4 \cos \frac{\pi}{4} = 2\sqrt{2} \ln \sqrt{2} + 2\sqrt{2} = \sqrt{2}(\ln 2 + 2);$

$\frac{\partial z}{\partial v} = (-4)(2) \sin \frac{\pi}{4} \ln \left(2 \sin \frac{\pi}{4}\right) + \frac{(4)(2)\left(\cos^2 \frac{\pi}{4}\right)}{\left(\sin \frac{\pi}{4}\right)} = -4\sqrt{2} \ln \sqrt{2} + 4\sqrt{2} = -2\sqrt{2} \ln 2 + 4\sqrt{2}$

9. (a) $\frac{\partial w}{\partial u} = \frac{\partial w}{\partial x}\frac{\partial x}{\partial u} + \frac{\partial w}{\partial y}\frac{\partial y}{\partial u} + \frac{\partial w}{\partial z}\frac{\partial z}{\partial u} = (y+z)(1) + (x+z)(1) + (y+x)(v) = x + y + 2z + v(y+x)$

$= (u+v) + (u-v) + 2uv + v(2u) = 2u + 4uv$; $\frac{\partial w}{\partial v} = \frac{\partial w}{\partial x}\frac{\partial x}{\partial v} + \frac{\partial w}{\partial y}\frac{\partial y}{\partial v} + \frac{\partial w}{\partial z}\frac{\partial z}{\partial v}$

$= (y+z)(1) + (x+z)(-1) + (y+x)(u) = y - x + (y+x)u = -2v + (2u)u = -2v + 2u^2;$

$w = xy + yz + xz = (u^2 - v^2) + (u^2 v - uv^2) + (u^2 v + uv^2) = u^2 - v^2 + 2u^2 v \Rightarrow \frac{\partial w}{\partial u} = 2u + 4uv$ and

$\frac{\partial w}{\partial v} = -2v + 2u^2$

(b) At $\left(\frac{1}{2}, 1\right)$: $\frac{\partial w}{\partial u} = 2\left(\frac{1}{2}\right) + 4\left(\frac{1}{2}\right)(1) = 3$ and $\frac{\partial w}{\partial v} = -2(1) + 2\left(\frac{1}{2}\right)^2 = -\frac{3}{2}$

11. (a) $\frac{\partial u}{\partial x} = \frac{\partial u}{\partial p}\frac{\partial p}{\partial x} + \frac{\partial u}{\partial q}\frac{\partial q}{\partial x} + \frac{\partial u}{\partial r}\frac{\partial r}{\partial x} = \frac{1}{q-r} + \frac{r-p}{(q-r)^2} + \frac{p-q}{(q-r)^2} = \frac{q-r+r-p+p-q}{(q-r)^2} = 0;$

$\frac{\partial u}{\partial y} = \frac{\partial u}{\partial p}\frac{\partial p}{\partial y} + \frac{\partial u}{\partial q}\frac{\partial q}{\partial y} + \frac{\partial u}{\partial r}\frac{\partial r}{\partial y} = \frac{1}{q-r} - \frac{r-p}{(q-r)^2} + \frac{p-q}{(q-r)^2} = \frac{q-r-r+p+p-q}{(q-r)^2} = \frac{2p-2r}{(q-r)^2}$

$= \frac{(2x+2y+2z)-(2x+2y-2z)}{(2z-2y)^2} = \frac{z}{(z-y)^2}$; $\frac{\partial u}{\partial z} = \frac{\partial u}{\partial p}\frac{\partial p}{\partial z} + \frac{\partial u}{\partial q}\frac{\partial q}{\partial z} + \frac{\partial u}{\partial r}\frac{\partial r}{\partial z}$

$= \frac{1}{q-r} + \frac{r-p}{(q-r)^2} - \frac{p-q}{(q-r)^2} = \frac{q-r+r-p-p+q}{(q-r)^2} = \frac{2q-2p}{(q-r)^2} = \frac{-4y}{(2z-2y)^2} = -\frac{y}{(z-y)^2};$

$u = \frac{p-q}{q-r} = \frac{2y}{2z-2y} = \frac{y}{z-y} \Rightarrow \frac{\partial u}{\partial x} = 0$, $\frac{\partial u}{\partial y} = \frac{(z-y)-y(-1)}{(z-y)^2} = \frac{z}{(z-y)^2}$, and $\frac{\partial u}{\partial z} = \frac{(z-y)(0)-y(1)}{(z-y)^2}$

$= -\frac{y}{(z-y)^2}$

(b) At $\left(\sqrt{3}, 2, 1\right)$: $\frac{\partial u}{\partial x} = 0$, $\frac{\partial u}{\partial y} = \frac{1}{(1-2)^2} = 1$, and $\frac{\partial u}{\partial z} = \frac{-2}{(1-2)^2} = -2$

13. $\frac{dz}{dt} = \frac{\partial z}{\partial x}\frac{dx}{dt} + \frac{\partial z}{\partial y}\frac{dy}{dt}$

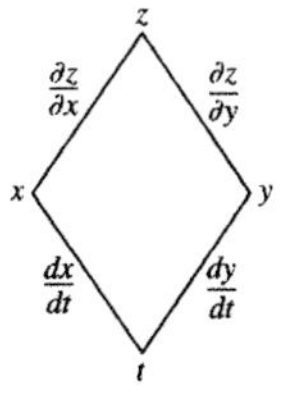

15. $\frac{\partial w}{\partial u} = \frac{\partial w}{\partial x}\frac{\partial x}{\partial u} + \frac{\partial w}{\partial y}\frac{\partial y}{\partial u} + \frac{\partial w}{\partial z}\frac{\partial z}{\partial u}$ $\qquad$ $\frac{\partial w}{\partial v} = \frac{\partial w}{\partial x}\frac{\partial x}{\partial v} + \frac{\partial w}{\partial y}\frac{\partial y}{\partial v} + \frac{\partial w}{\partial z}\frac{\partial z}{\partial v}$

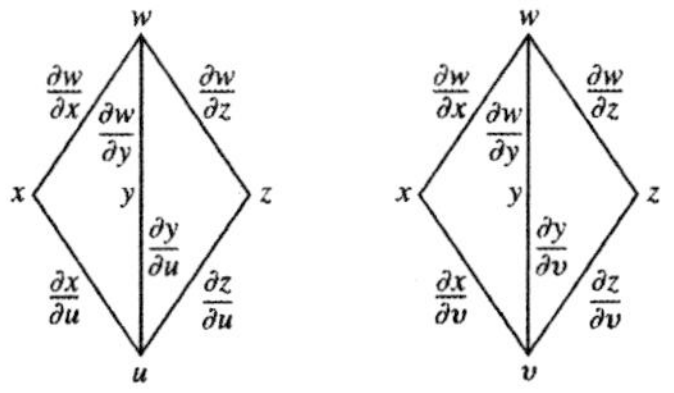

17. $\frac{\partial w}{\partial u} = \frac{\partial w}{\partial x}\frac{\partial x}{\partial u} + \frac{\partial w}{\partial y}\frac{\partial y}{\partial u}$ $\qquad$ $\frac{\partial w}{\partial v} = \frac{\partial w}{\partial x}\frac{\partial x}{\partial v} + \frac{\partial w}{\partial y}\frac{\partial y}{\partial v}$

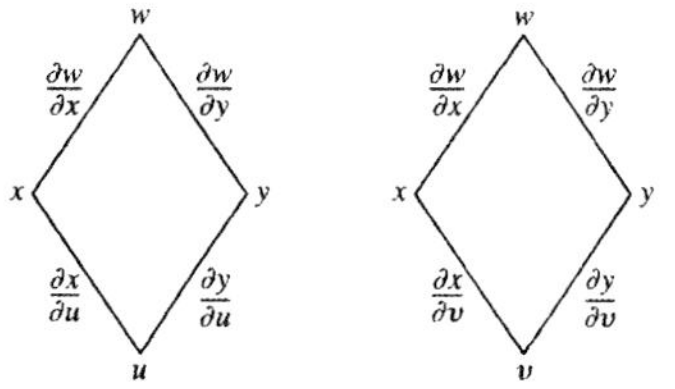

19. $\frac{\partial z}{\partial t} = \frac{\partial z}{\partial x}\frac{\partial x}{\partial t} + \frac{\partial z}{\partial y}\frac{\partial y}{\partial t}$ $\qquad$ $\frac{\partial z}{\partial s} = \frac{\partial z}{\partial x}\frac{\partial x}{\partial s} + \frac{\partial z}{\partial y}\frac{\partial y}{\partial s}$

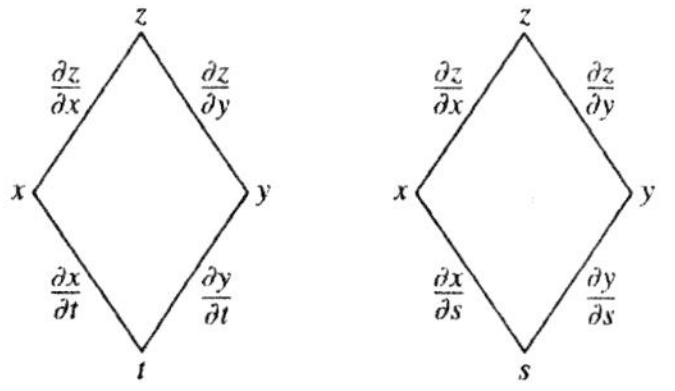

21. $\frac{\partial w}{\partial s} = \frac{dw}{du}\frac{\partial u}{\partial s}$ $\qquad$ $\frac{\partial w}{\partial t} = \frac{dw}{du}\frac{\partial u}{\partial t}$

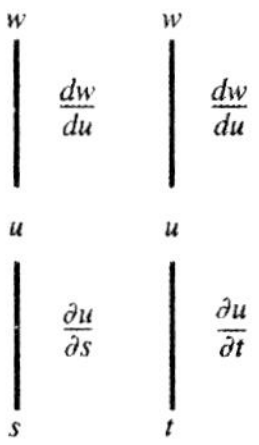

23. $\frac{\partial w}{\partial r} = \frac{\partial w}{\partial x}\frac{dx}{dr} + \frac{\partial w}{\partial y}\frac{dy}{dr} = \frac{\partial w}{\partial x}\frac{dx}{dr}$ since $\frac{dy}{dr} = 0$ $\qquad$ $\frac{\partial w}{\partial s} = \frac{\partial w}{\partial x}\frac{dx}{ds} + \frac{\partial w}{\partial y}\frac{dy}{ds} = \frac{\partial w}{\partial y}\frac{dy}{ds}$ since $\frac{dx}{ds} = 0$

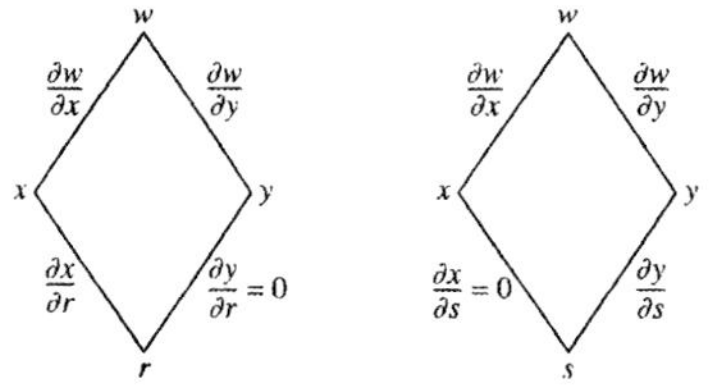

25. Let $F(x, y) = x^3 - 2y^2 + xy = 0 \Rightarrow F_x(x, y) = 3x^2 + y$ and $F_y(x, y) = -4y + x \Rightarrow \frac{dy}{dx} = -\frac{F_x}{F_y} = -\frac{3x^2+y}{(-4y+x)}$ $\Rightarrow \frac{dy}{dx}(1, 1) = \frac{4}{3}$

27. Let $F(x, y) = x^2 + xy + y^2 - 7 = 0 \Rightarrow F_x(x, y) = 2x + y$ and $F_y(x, y) = x + 2y \Rightarrow \frac{dy}{dx} = -\frac{F_x}{F_y} = -\frac{2x+y}{x+2y}$ $\Rightarrow \frac{dy}{dx}(1, 2) = -\frac{4}{5}$

29. Let $F(x, y, z) = z^3 - xy + yz + y^3 - 2 = 0 \Rightarrow F_x(x, y, z) = -y$, $F_y(x, y, z) = -x + z + 3y^2$, $F_z(x, y, z) = 3z^2 + y$ $\Rightarrow \frac{\partial z}{\partial x} = -\frac{F_x}{F_z} = -\frac{-y}{3z^2+y} = \frac{y}{3z^2+y} \Rightarrow \frac{\partial z}{\partial x}(1, 1, 1) = \frac{1}{4}$; $\frac{\partial z}{\partial y} = -\frac{F_y}{F_z} = -\frac{-x+z+3y^2}{3z^2+y} = \frac{x-z-3y^2}{3z^2+y}$ $\Rightarrow \frac{\partial z}{\partial y}(1, 1, 1) = -\frac{3}{4}$

31. Let $F(x,y,z) = \sin(x+y) + \sin(y+z) + \sin(x+z) = 0 \Rightarrow F_x(x,y,z) = \cos(x+y) + \cos(x+z)$, $F_y(x,y,z) = \cos(x+y) + \cos(y+z)$, $F_z(x,y,z) = \cos(y+z) + \cos(x+z) \Rightarrow \frac{\partial z}{\partial x} = -\frac{F_x}{F_z}$ $= -\frac{\cos(x+y)+\cos(x+z)}{\cos(y+z)+\cos(x+z)} \Rightarrow \frac{\partial z}{\partial x}(\pi,\pi,\pi) = -1$; $\frac{\partial z}{\partial y} = -\frac{F_y}{F_z} = -\frac{\cos(x+y)+\cos(y+z)}{\cos(y+z)+\cos(x+z)} \Rightarrow \frac{\partial z}{\partial y}(\pi,\pi,\pi) = -1$

33. $\frac{\partial w}{\partial r} = \frac{\partial w}{\partial x}\frac{\partial x}{\partial r} + \frac{\partial w}{\partial y}\frac{\partial y}{\partial r} + \frac{\partial w}{\partial z}\frac{\partial z}{\partial r} = 2(x+y+z)(1) + 2(x+y+z)[-\sin(r+s)] + 2(x+y+z)[\cos(r+s)]$ $= 2(x+y+z)[1-\sin(r+s)+\cos(r+s)] = 2[r-s+\cos(r+s)+\sin(r+s)][1-\sin(r+s)+\cos(r+s)]$ $\Rightarrow \left.\frac{\partial w}{\partial r}\right|_{r=1,s=-1} = 2(3)(2) = 12$

35. $\frac{\partial w}{\partial v} = \frac{\partial w}{\partial x}\frac{\partial x}{\partial v} + \frac{\partial w}{\partial y}\frac{\partial y}{\partial v} = \left(2x - \frac{y}{x^2}\right)(-2) + \left(\frac{1}{x}\right)(1) = \left[2(u-2v+1) - \frac{2u+v-2}{(u-2v+1)^2}\right](-2) + \frac{1}{u-2v+1}$ $\Rightarrow \left.\frac{\partial w}{\partial v}\right|_{u=0,v=0} = -7$

37. $\frac{\partial z}{\partial u} = \frac{dz}{dx}\frac{\partial x}{\partial u} = \left(\frac{5}{1+x^2}\right)e^u = \left[\frac{5}{1+(e^u+\ln v)^2}\right]e^u \Rightarrow \left.\frac{\partial z}{\partial u}\right|_{u=\ln 2,v=1} = \left[\frac{5}{1+(2)^2}\right](2) = 2$;
$\frac{\partial z}{\partial v} = \frac{dz}{dx}\frac{\partial x}{\partial v} = \left(\frac{5}{1+x^2}\right)\left(\frac{1}{v}\right) = \left[\frac{5}{1+(e^u+\ln v)^2}\right]\left(\frac{1}{v}\right) \Rightarrow \left.\frac{\partial z}{\partial v}\right|_{u=\ln 2,v=1} = \left[\frac{5}{1+(2)^2}\right](1) = 1$

39. $V = IR \Rightarrow \frac{\partial V}{\partial I} = R$ and $\frac{\partial V}{\partial R} = I$; $\frac{dV}{dt} = \frac{\partial V}{\partial I}\frac{dI}{dt} + \frac{\partial V}{\partial R}\frac{dR}{dt} = R\frac{dI}{dt} + I\frac{dR}{dt} \Rightarrow -0.01$ volts/sec $= (600 \text{ ohms})\frac{dI}{dt} + (0.04 \text{ amps})(0.5 \text{ ohms/sec}) \Rightarrow \frac{dI}{dt} = -0.00005$ amps/sec

41. $\frac{\partial f}{\partial x} = \frac{\partial f}{\partial u}\frac{\partial u}{\partial x} + \frac{\partial f}{\partial v}\frac{\partial v}{\partial x} + \frac{\partial f}{\partial w}\frac{\partial w}{\partial x} = \frac{\partial f}{\partial u}(1) + \frac{\partial f}{\partial v}(0) + \frac{\partial f}{\partial w}(-1) = \frac{\partial f}{\partial u} - \frac{\partial f}{\partial w}$,
$\frac{\partial f}{\partial y} = \frac{\partial f}{\partial u}\frac{\partial u}{\partial y} + \frac{\partial f}{\partial v}\frac{\partial v}{\partial y} + \frac{\partial f}{\partial w}\frac{\partial w}{\partial y} = \frac{\partial f}{\partial u}(-1) + \frac{\partial f}{\partial v}(1) + \frac{\partial f}{\partial w}(0) = -\frac{\partial f}{\partial u} + \frac{\partial f}{\partial v}$, and
$\frac{\partial f}{\partial z} = \frac{\partial f}{\partial u}\frac{\partial u}{\partial z} + \frac{\partial f}{\partial v}\frac{\partial v}{\partial z} + \frac{\partial f}{\partial w}\frac{\partial w}{\partial z} = \frac{\partial f}{\partial u}(0) + \frac{\partial f}{\partial v}(-1) + \frac{\partial f}{\partial w}(1) = -\frac{\partial f}{\partial v} + \frac{\partial f}{\partial w} \Rightarrow \frac{\partial f}{\partial x} + \frac{\partial f}{\partial y} + \frac{\partial f}{\partial z} = 0$

43. $w_x = \frac{\partial w}{\partial x} = \frac{\partial w}{\partial u}\frac{\partial u}{\partial x} + \frac{\partial w}{\partial v}\frac{\partial v}{\partial x} = x\frac{\partial w}{\partial u} + y\frac{\partial w}{\partial v} \Rightarrow w_{xx} = \frac{\partial w}{\partial u} + x\frac{\partial}{\partial x}\left(\frac{\partial w}{\partial u}\right) + y\frac{\partial}{\partial x}\left(\frac{\partial w}{\partial v}\right)$
$= \frac{\partial w}{\partial u} + x\left(\frac{\partial^2 w}{\partial u^2}\frac{\partial u}{\partial x} + \frac{\partial^2 w}{\partial v \partial u}\frac{\partial v}{\partial x}\right) + y\left(\frac{\partial^2 w}{\partial u \partial v}\frac{\partial u}{\partial x} + \frac{\partial^2 w}{\partial v^2}\frac{\partial v}{\partial x}\right) = \frac{\partial w}{\partial u} + x\left(x\frac{\partial^2 w}{\partial u^2} + y\frac{\partial^2 w}{\partial v \partial u}\right) + y\left(x\frac{\partial^2 w}{\partial u \partial v} + y\frac{\partial^2 w}{\partial v^2}\right)$
$= \frac{\partial w}{\partial u} + x^2\frac{\partial^2 w}{\partial u^2} + 2xy\frac{\partial^2 w}{\partial v \partial u} + y^2\frac{\partial^2 w}{\partial v^2}$; $w_y = \frac{\partial w}{\partial y} = \frac{\partial w}{\partial u}\frac{\partial u}{\partial y} + \frac{\partial w}{\partial v}\frac{\partial v}{\partial y} = -y\frac{\partial w}{\partial u} + x\frac{\partial w}{\partial v}$
$\Rightarrow w_{yy} = -\frac{\partial w}{\partial u} - y\left(\frac{\partial^2 w}{\partial u^2}\frac{\partial u}{\partial y} + \frac{\partial^2 w}{\partial v \partial u}\frac{\partial v}{\partial y}\right) + x\left(\frac{\partial^2 w}{\partial u \partial v}\frac{\partial u}{\partial y} + \frac{\partial^2 w}{\partial v^2}\frac{\partial v}{\partial y}\right)$
$= -\frac{\partial w}{\partial u} - y\left(-y\frac{\partial^2 w}{\partial u^2} + x\frac{\partial^2 w}{\partial v \partial u}\right) + x\left(-y\frac{\partial^2 w}{\partial u \partial v} + x\frac{\partial^2 w}{\partial v^2}\right) = -\frac{\partial w}{\partial u} + y^2\frac{\partial^2 w}{\partial u^2} - 2xy\frac{\partial^2 w}{\partial v \partial u} + x^2\frac{\partial^2 w}{\partial v^2}$; thus
$w_{xx} + w_{yy} = (x^2+y^2)\frac{\partial^2 w}{\partial u^2} + (x^2+y^2)\frac{\partial^2 w}{\partial v^2} = (x^2+y^2)(w_{uu} + w_{vv}) = 0$, since $w_{uu} + w_{vv} = 0$

45. $f_x(x,y,z) = \cos t$, $f_y(x,y,z) = \sin t$, and $f_z(x,y,z) = t^2 + t - 2 \Rightarrow \frac{df}{dt} = \frac{\partial f}{\partial x}\frac{dx}{dt} + \frac{\partial f}{\partial y}\frac{dy}{dt} + \frac{\partial f}{\partial z}\frac{dz}{dt}$ $= (\cos t)(-\sin t) + (\sin t)(\cos t) + (t^2+t-2)(1) = t^2 + t - 2$; $\frac{df}{dt} = 0 \Rightarrow t^2 + t - 2 = 0 \Rightarrow t = -2$ or $t = 1$; $t = -2 \Rightarrow x = \cos(-2)$, $y = \sin(-2)$, $z = -2$ for the point $(\cos(-2), \sin(-2), -2)$; $t = 1 \Rightarrow x = \cos 1$, $y = \sin 1$, $z = 1$ for the point $(\cos 1, \sin 1, 1)$

47. (a) $\frac{\partial T}{\partial x} = 8x - 4y$ and $\frac{\partial T}{\partial y} = 8y - 4x \Rightarrow \frac{dT}{dt} = \frac{\partial T}{\partial x}\frac{dx}{dt} + \frac{\partial T}{\partial y}\frac{dy}{dt} = (8x-4y)(-\sin t) + (8y-4x)(\cos t)$
$= (8\cos t - 4\sin t)(-\sin t) + (8\sin t - 4\cos t)(\cos t) = 4\sin^2 t - 4\cos^2 t \Rightarrow \frac{d^2T}{dt^2} = 16\sin t\cos t$;
$\frac{dT}{dt} = 0 \Rightarrow 4\sin^2 t - 4\cos^2 t = 0 \Rightarrow \sin^2 t = \cos^2 t \Rightarrow \sin t = \cos t$ or $\sin t = -\cos t \Rightarrow t = \frac{\pi}{4}, \frac{5\pi}{4}, \frac{3\pi}{4}, \frac{7\pi}{4}$ on the interval $0 \le t \le 2\pi$;
$\left.\frac{d^2T}{dt^2}\right|_{t=\frac{\pi}{4}} = 16\sin\frac{\pi}{4}\cos\frac{\pi}{4} > 0 \Rightarrow$ T has a minimum at $(x,y) = \left(\frac{\sqrt{2}}{2}, \frac{\sqrt{2}}{2}\right)$;
$\left.\frac{d^2T}{dt^2}\right|_{t=\frac{3\pi}{4}} = 16\sin\frac{3\pi}{4}\cos\frac{3\pi}{4} < 0 \Rightarrow$ T has a maximum at $(x,y) = \left(-\frac{\sqrt{2}}{2}, \frac{\sqrt{2}}{2}\right)$;

$\left.\frac{d^2T}{dt^2}\right|_{t=\frac{5\pi}{4}} = 16 \sin \frac{5\pi}{4} \cos \frac{5\pi}{4} > 0 \Rightarrow$ T has a minimum at $(x, y) = \left(-\frac{\sqrt{2}}{2}, -\frac{\sqrt{2}}{2}\right)$;

$\left.\frac{d^2T}{dt^2}\right|_{t=\frac{7\pi}{4}} = 16 \sin \frac{7\pi}{4} \cos \frac{7\pi}{4} < 0 \Rightarrow$ T has a maximum at $(x, y) = \left(\frac{\sqrt{2}}{2}, -\frac{\sqrt{2}}{2}\right)$

(b) $T = 4x^2 - 4xy + 4y^2 \Rightarrow \frac{\partial T}{\partial x} = 8x - 4y$, and $\frac{\partial T}{\partial y} = 8y - 4x$ so the extreme values occur at the four points found in part (a): $T\left(-\frac{\sqrt{2}}{2}, \frac{\sqrt{2}}{2}\right) = T\left(\frac{\sqrt{2}}{2}, -\frac{\sqrt{2}}{2}\right) = 4\left(\frac{1}{2}\right) - 4\left(-\frac{1}{2}\right) + 4\left(\frac{1}{2}\right) = 6$, the maximum and $T\left(\frac{\sqrt{2}}{2}, \frac{\sqrt{2}}{2}\right) = T\left(-\frac{\sqrt{2}}{2}, -\frac{\sqrt{2}}{2}\right) = 4\left(\frac{1}{2}\right) - 4\left(\frac{1}{2}\right) + 4\left(\frac{1}{2}\right) = 2$, the minimum

49. $G(u, x) = \int_a^u g(t, x)\, dt$ where $u = f(x) \Rightarrow \frac{dG}{dx} = \frac{\partial G}{\partial u}\frac{du}{dx} + \frac{\partial G}{\partial x}\frac{dx}{dx} = g(u, x)f'(x) + \int_a^u g_x(t, x)\, dt$; thus $F(x) = \int_0^{x^2} \sqrt{t^4 + x^3}\, dt \Rightarrow F'(x) = \sqrt{(x^2)^4 + x^3}\,(2x) + \int_0^{x^2} \frac{\partial}{\partial x}\sqrt{t^4 + x^3}\, dt = 2x\sqrt{x^8 + x^3} + \int_0^{x^2} \frac{3x^2}{2\sqrt{t^4 + x^3}}\, dt$

11.5 DIRECTIONAL DERIVATIVES AND GRADIENT VECTORS

1. $\frac{\partial f}{\partial x} = -1, \frac{\partial f}{\partial y} = 1 \Rightarrow \nabla f = -\mathbf{i} + \mathbf{j}$; $f(2, 1) = -1$
$\Rightarrow -1 = y - x$ is the level curve

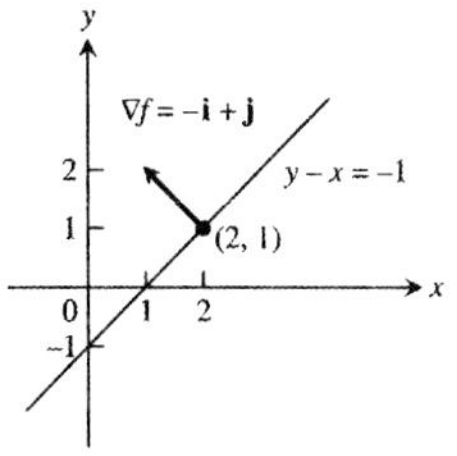

3. $\frac{\partial g}{\partial x} = -2x \Rightarrow \frac{\partial g}{\partial x}(-1, 0) = 2$; $\frac{\partial g}{\partial y} = 1$
$\Rightarrow \nabla g = 2\mathbf{i} + \mathbf{j}$; $g(-1, 0) = -1$
$\Rightarrow -1 = y - x^2$ is the level curve

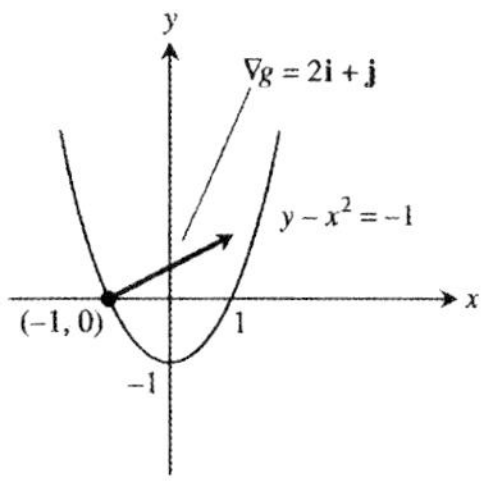

5. $\frac{\partial f}{\partial x} = 2x + \frac{z}{x} \Rightarrow \frac{\partial f}{\partial x}(1, 1, 1) = 3$; $\frac{\partial f}{\partial y} = 2y \Rightarrow \frac{\partial f}{\partial y}(1, 1, 1) = 2$; $\frac{\partial f}{\partial z} = -4z + \ln x \Rightarrow \frac{\partial f}{\partial z}(1, 1, 1) = -4$; thus $\nabla f = 3\mathbf{i} + 2\mathbf{j} - 4\mathbf{k}$

7. $\frac{\partial f}{\partial x} = -\frac{x}{(x^2+y^2+z^2)^{3/2}} + \frac{1}{x} \Rightarrow \frac{\partial f}{\partial x}(-1, 2, -2) = -\frac{26}{27}$; $\frac{\partial f}{\partial y} = -\frac{y}{(x^2+y^2+z^2)^{3/2}} + \frac{1}{y} \Rightarrow \frac{\partial f}{\partial y}(-1, 2, -2) = \frac{23}{54}$; $\frac{\partial f}{\partial z} = -\frac{z}{(x^2+y^2+z^2)^{3/2}} + \frac{1}{z} \Rightarrow \frac{\partial f}{\partial z}(-1, 2, -2) = -\frac{23}{54}$; thus $\nabla f = -\frac{26}{27}\mathbf{i} + \frac{23}{54}\mathbf{j} - \frac{23}{54}\mathbf{k}$

9. $\mathbf{u} = \frac{\mathbf{A}}{|\mathbf{A}|} = \frac{4\mathbf{i} + 3\mathbf{j}}{\sqrt{4^2 + 3^2}} = \frac{4}{5}\mathbf{i} + \frac{3}{5}\mathbf{j}$; $f_x(x, y) = 2y \Rightarrow f_x(5, 5) = 10$; $f_y(x, y) = 2x - 6y \Rightarrow f_y(5, 5) = -20$ $\Rightarrow \nabla f = 10\mathbf{i} - 20\mathbf{j} \Rightarrow (D_{\mathbf{u}}f)_{P_0} = \nabla f \cdot \mathbf{u} = 10\left(\frac{4}{5}\right) - 20\left(\frac{3}{5}\right) = -4$

11. $\mathbf{u} = \frac{\mathbf{A}}{|\mathbf{A}|} = \frac{12\mathbf{i} + 5\mathbf{j}}{\sqrt{12^2 + 5^2}} = \frac{12}{13}\mathbf{i} + \frac{5}{13}\mathbf{j}$; $g_x(x, y) = 1 + \frac{y^2}{x^2} + \frac{2y\sqrt{3}}{2xy\sqrt{4x^2y^2 - 1}} \Rightarrow g_x(1, 1) = 3$; $g_y(x, y) = -\frac{2y}{x} + \frac{2x\sqrt{3}}{2xy\sqrt{4x^2y^2 - 1}} \Rightarrow g_y(1, 1) = -1 \Rightarrow \nabla g = 3\mathbf{i} - \mathbf{j} \Rightarrow (D_{\mathbf{u}}g)_{P_0} = \nabla g \cdot \mathbf{u} = \frac{36}{13} - \frac{5}{13} = \frac{31}{13}$

13. $\mathbf{u} = \frac{\mathbf{A}}{|\mathbf{A}|} = \frac{3\mathbf{i} + 6\mathbf{j} - 2\mathbf{k}}{\sqrt{3^2 + 6^2 + (-2)^2}} = \frac{3}{7}\mathbf{i} + \frac{6}{7}\mathbf{j} - \frac{2}{7}\mathbf{k}$; $f_x(x, y, z) = y + z \Rightarrow f_x(1, -1, 2) = 1$; $f_y(x, y, z) = x + z \Rightarrow f_y(1, -1, 2) = 3$; $f_z(x, y, z) = y + x \Rightarrow f_z(1, -1, 2) = 0 \Rightarrow \nabla f = \mathbf{i} + 3\mathbf{j} \Rightarrow (D_{\mathbf{u}}f)_{P_0} = \nabla f \cdot \mathbf{u} = \frac{3}{7} + \frac{18}{7} = 3$

15. $\mathbf{u} = \frac{\mathbf{A}}{|\mathbf{A}|} = \frac{2\mathbf{i}+\mathbf{j}-2\mathbf{k}}{\sqrt{2^2+1^2+(-2)^2}} = \frac{2}{3}\mathbf{i} + \frac{1}{3}\mathbf{j} - \frac{2}{3}\mathbf{k}$; $g_x(x,y,z) = 3e^x \cos yz \Rightarrow g_x(0,0,0) = 3$; $g_y(x,y,z) = -3ze^x \sin yz$ $\Rightarrow g_y(0,0,0) = 0$; $g_z(x,y,z) = -3ye^x \sin yz \Rightarrow g_z(0,0,0) = 0 \Rightarrow \nabla g = 3\mathbf{i} \Rightarrow (D_{\mathbf{u}}g)_{P_0} = \nabla g \cdot \mathbf{u} = 2$

17. $\nabla f = (2x+y)\mathbf{i} + (x+2y)\mathbf{j} \Rightarrow \nabla f(-1,1) = -\mathbf{i}+\mathbf{j} \Rightarrow \mathbf{u} = \frac{\nabla f}{|\nabla f|} = \frac{-\mathbf{i}+\mathbf{j}}{\sqrt{(-1)^2+1^2}} = -\frac{1}{\sqrt{2}}\mathbf{i} + \frac{1}{\sqrt{2}}\mathbf{j}$; f increases most rapidly in the direction $\mathbf{u} = -\frac{1}{\sqrt{2}}\mathbf{i} + \frac{1}{\sqrt{2}}\mathbf{j}$ and decreases most rapidly in the direction $-\mathbf{u} = \frac{1}{\sqrt{2}}\mathbf{i} - \frac{1}{\sqrt{2}}\mathbf{j}$; $(D_{\mathbf{u}}f)_{P_0} = \nabla f \cdot \mathbf{u} = |\nabla f| = \sqrt{2}$ and $(D_{-\mathbf{u}}f)_{P_0} = -\sqrt{2}$

19. $\nabla f = \frac{1}{y}\mathbf{i} - \left(\frac{x}{y^2}+z\right)\mathbf{j} - y\mathbf{k} \Rightarrow \nabla f(4,1,1) = \mathbf{i} - 5\mathbf{j} - \mathbf{k} \Rightarrow \mathbf{u} = \frac{\nabla f}{|\nabla f|} = \frac{\mathbf{i}-5\mathbf{j}-\mathbf{k}}{\sqrt{1^2+(-5)^2+(-1)^2}}$ $= \frac{1}{3\sqrt{3}}\mathbf{i} - \frac{5}{3\sqrt{3}}\mathbf{j} - \frac{1}{3\sqrt{3}}\mathbf{k}$; f increases most rapidly in the direction of $\mathbf{u} = \frac{1}{3\sqrt{3}}\mathbf{i} - \frac{5}{3\sqrt{3}}\mathbf{j} - \frac{1}{3\sqrt{3}}\mathbf{k}$ and decreases most rapidly in the direction $-\mathbf{u} = -\frac{1}{3\sqrt{3}}\mathbf{i} + \frac{5}{3\sqrt{3}}\mathbf{j} + \frac{1}{3\sqrt{3}}\mathbf{k}$; $(D_{\mathbf{u}}f)_{P_0} = \nabla f \cdot \mathbf{u} = |\nabla f| = 3\sqrt{3}$ and $(D_{-\mathbf{u}}f)_{P_0} = -3\sqrt{3}$

21. $\nabla f = \left(\frac{1}{x}+\frac{1}{x}\right)\mathbf{i} + \left(\frac{1}{y}+\frac{1}{y}\right)\mathbf{j} + \left(\frac{1}{z}+\frac{1}{z}\right)\mathbf{k} \Rightarrow \nabla f(1,1,1) = 2\mathbf{i}+2\mathbf{j}+2\mathbf{k} \Rightarrow \mathbf{u} = \frac{\nabla f}{|\nabla f|} = \frac{1}{\sqrt{3}}\mathbf{i} + \frac{1}{\sqrt{3}}\mathbf{j} + \frac{1}{\sqrt{3}}\mathbf{k}$; f increases most rapidly in the direction $\mathbf{u} = \frac{1}{\sqrt{3}}\mathbf{i} + \frac{1}{\sqrt{3}}\mathbf{j} + \frac{1}{\sqrt{3}}\mathbf{k}$ and decreases most rapidly in the direction $-\mathbf{u} = -\frac{1}{\sqrt{3}}\mathbf{i} - \frac{1}{\sqrt{3}}\mathbf{j} - \frac{1}{\sqrt{3}}\mathbf{k}$; $(D_{\mathbf{u}}f)_{P_0} = \nabla f \cdot \mathbf{u} = |\nabla f| = 2\sqrt{3}$ and $(D_{-\mathbf{u}}f)_{P_0} = -2\sqrt{3}$

23. $\nabla f = 2x\mathbf{i} + 2y\mathbf{j} \Rightarrow \nabla f\left(\sqrt{2},\sqrt{2}\right) = 2\sqrt{2}\mathbf{i} + 2\sqrt{2}\mathbf{j}$
$\Rightarrow$ Tangent line: $2\sqrt{2}\left(x-\sqrt{2}\right) + 2\sqrt{2}\left(y-\sqrt{2}\right) = 0$
$\Rightarrow \sqrt{2}x + \sqrt{2}y = 4$

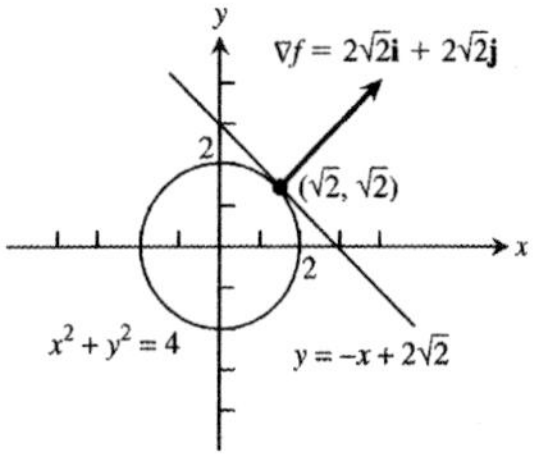

25. $\nabla f = y\mathbf{i} + x\mathbf{j} \Rightarrow \nabla f(2,-2) = -2\mathbf{i} + 2\mathbf{j}$
$\Rightarrow$ Tangent line: $-2(x-2) + 2(y+2) = 0$
$\Rightarrow y = x - 4$

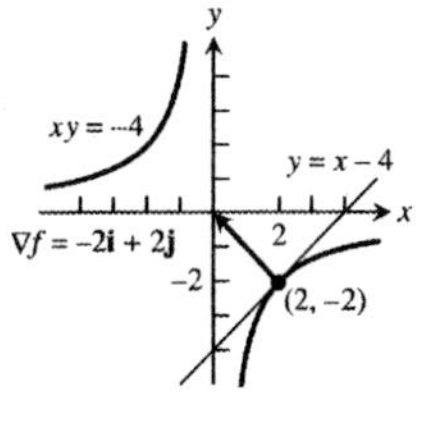

27. $\nabla f = y\mathbf{i} + (x+2y)\mathbf{j} \Rightarrow \nabla f(3,2) = 2\mathbf{i} + 7\mathbf{j}$; a vector orthogonal to ∇f is $\mathbf{v} = 7\mathbf{i} - 2\mathbf{j} \Rightarrow \mathbf{u} = \frac{\mathbf{v}}{|\mathbf{v}|} = \frac{7\mathbf{i}-2\mathbf{j}}{\sqrt{7^2+(-2)^2}}$ $= \frac{7}{\sqrt{53}}\mathbf{i} - \frac{2}{\sqrt{53}}\mathbf{j}$ and $-\mathbf{u} = -\frac{7}{\sqrt{53}}\mathbf{i} + \frac{2}{\sqrt{53}}\mathbf{j}$ are the directions where the derivative is zero

29. $\nabla f = (2x-3y)\mathbf{i} + (-3x+8y)\mathbf{j} \Rightarrow \nabla f(1,2) = -4\mathbf{i} + 13\mathbf{j} \Rightarrow |\nabla f(1,2)| = \sqrt{(-4)^2+(13)^2} = \sqrt{185}$; no, the maximum rate of change is $\sqrt{185} < 14$

31. $\nabla f = f_x(1,2)\mathbf{i} + f_y(1,2)\mathbf{j}$ and $\mathbf{u}_1 = \frac{\mathbf{i}+\mathbf{j}}{\sqrt{1^2+1^2}} = \frac{1}{\sqrt{2}}\mathbf{i} + \frac{1}{\sqrt{2}}\mathbf{j} \Rightarrow (D_{\mathbf{u}_1}f)(1,2) = f_x(1,2)\left(\frac{1}{\sqrt{2}}\right) + f_y(1,2)\left(\frac{1}{\sqrt{2}}\right)$ $= 2\sqrt{2} \Rightarrow f_x(1,2) + f_y(1,2) = 4$; $\mathbf{u}_2 = -\mathbf{j} \Rightarrow (D_{\mathbf{u}_2}f)(1,2) = f_x(1,2)(0) + f_y(1,2)(-1) = -3 \Rightarrow -f_y(1,2) = -3$ $\Rightarrow f_y(1,2) = 3$; then $f_x(1,2) + 3 = 4 \Rightarrow f_x(1,2) = 1$; thus $\nabla f(1,2) = \mathbf{i} + 3\mathbf{j}$ and $\mathbf{u} = \frac{\mathbf{v}}{|\mathbf{v}|} = \frac{-\mathbf{i}-2\mathbf{j}}{\sqrt{(-1)^2+(-2)^2}}$ $= -\frac{1}{\sqrt{5}}\mathbf{i} - \frac{2}{\sqrt{5}}\mathbf{j} \Rightarrow (D_{\mathbf{u}}f)_{P_0} = \nabla f \cdot \mathbf{u} = -\frac{1}{\sqrt{5}} - \frac{6}{\sqrt{5}} = -\frac{7}{\sqrt{5}}$

33. The directional derivative is the scalar component. With ∇f evaluated at P_0, the scalar component of ∇f in the direction of $\mathbf{u}$ is $\nabla f \cdot \mathbf{u} = (D_{\mathbf{u}}f)_{P_0}$.

35. If (x, y) is a point on the line, then $\mathbf{T}(x, y) = (x - x_0)\mathbf{i} + (y - y_0)\mathbf{j}$ is a vector parallel to the line $\Rightarrow \mathbf{T} \cdot \mathbf{N} = 0$ $\Rightarrow A(x - x_0) + B(y - y_0) = 0$, as claimed.

11.6 TANGENT PLANES AND DIFFERENTIALS

1. (a) $\nabla f = 2x\mathbf{i} + 2y\mathbf{j} + 2z\mathbf{k} \Rightarrow \nabla f(1,1,1) = 2\mathbf{i} + 2\mathbf{j} + 2\mathbf{k} \Rightarrow$ Tangent plane: $2(x-1) + 2(y-1) + 2(z-1) = 0$ $\Rightarrow x + y + z = 3$;
   (b) Normal line: $x = 1 + 2t$, $y = 1 + 2t$, $z = 1 + 2t$

3. (a) $\nabla f = -2x\mathbf{i} + 2\mathbf{k} \Rightarrow \nabla f(2,0,2) = -4\mathbf{i} + 2\mathbf{k} \Rightarrow$ Tangent plane: $-4(x-2) + 2(z-2) = 0$ $\Rightarrow -4x + 2z + 4 = 0 \Rightarrow -2x + z + 2 = 0$;
   (b) Normal line: $x = 2 - 4t$, $y = 0$, $z = 2 + 2t$

5. (a) $\nabla f = (-\pi \sin \pi x - 2xy + ze^{xz})\,\mathbf{i} + (-x^2 + z)\,\mathbf{j} + (xe^{xz} + y)\,\mathbf{k} \Rightarrow \nabla f(0,1,2) = 2\mathbf{i} + 2\mathbf{j} + \mathbf{k} \Rightarrow$ Tangent plane: $2(x-0) + 2(y-1) + 1(z-2) = 0 \Rightarrow 2x + 2y + z - 4 = 0$;
   (b) Normal line: $x = 2t$, $y = 1 + 2t$, $z = 2 + t$

7. (a) $\nabla f = \mathbf{i} + \mathbf{j} + \mathbf{k}$ for all points $\Rightarrow \nabla f(0,1,0) = \mathbf{i} + \mathbf{j} + \mathbf{k} \Rightarrow$ Tangent plane: $1(x-0) + 1(y-1) + 1(z-0) = 0$ $\Rightarrow x + y + z - 1 = 0$;
   (b) Normal line: $x = t$, $y = 1 + t$, $z = t$

9. $z = f(x, y) = \ln(x^2 + y^2) \Rightarrow f_x(x, y) = \frac{2x}{x^2+y^2}$ and $f_y(x, y) = \frac{2y}{x^2+y^2} \Rightarrow f_x(1,0) = 2$ and $f_y(1,0) = 0 \Rightarrow$ from Eq. (4) the tangent plane at $(1,0,0)$ is $2(x-1) - z = 0$ or $2x - z - 2 = 0$

11. $z = f(x, y) = \sqrt{y - x} \Rightarrow f_x(x, y) = -\frac{1}{2}(y-x)^{-1/2}$ and $f_y(x, y) = \frac{1}{2}(y-x)^{-1/2} \Rightarrow f_x(1,2) = -\frac{1}{2}$ and $f_y(1,2) = \frac{1}{2}$ $\Rightarrow$ from Eq. (4) the tangent plane at $(1,2,1)$ is $-\frac{1}{2}(x-1) + \frac{1}{2}(y-2) - (z-1) = 0 \Rightarrow x - y + 2z - 1 = 0$

13. $\nabla f = \mathbf{i} + 2y\mathbf{j} + 2\mathbf{k} \Rightarrow \nabla f(1,1,1) = \mathbf{i} + 2\mathbf{j} + 2\mathbf{k}$ and $\nabla g = \mathbf{i}$ for all points; $\mathbf{v} = \nabla f \times \nabla g$

$$\Rightarrow \mathbf{v} = \begin{vmatrix} \mathbf{i} & \mathbf{j} & \mathbf{k} \\ 1 & 2 & 2 \\ 1 & 0 & 0 \end{vmatrix} = 2\mathbf{j} - 2\mathbf{k} \Rightarrow \text{Tangent line: } x = 1,\ y = 1 + 2t,\ z = 1 - 2t$$

15. $\nabla f = 2x\mathbf{i} + 2\mathbf{j} + 2\mathbf{k} \Rightarrow \nabla f\left(1, 1, \frac{1}{2}\right) = 2\mathbf{i} + 2\mathbf{j} + 2\mathbf{k}$ and $\nabla g = \mathbf{j}$ for all points; $\mathbf{v} = \nabla f \times \nabla g$

$$\Rightarrow \mathbf{v} = \begin{vmatrix} \mathbf{i} & \mathbf{j} & \mathbf{k} \\ 2 & 2 & 2 \\ 0 & 1 & 0 \end{vmatrix} = -2\mathbf{i} + 2\mathbf{k} \Rightarrow \text{Tangent line: } x = 1 - 2t,\ y = 1,\ z = \tfrac{1}{2} + 2t$$

17. $\nabla f = (3x^2 + 6xy^2 + 4y)\,\mathbf{i} + (6x^2y + 3y^2 + 4x)\,\mathbf{j} - 2z\mathbf{k} \Rightarrow \nabla f(1,1,3) = 13\mathbf{i} + 13\mathbf{j} - 6\mathbf{k}$; $\nabla g = 2x\mathbf{i} + 2y\mathbf{j} + 2z\mathbf{k}$

$$\Rightarrow \nabla g(1,1,3) = 2\mathbf{i} + 2\mathbf{j} + 6\mathbf{k};\ \mathbf{v} = \nabla f \times \nabla g \Rightarrow \mathbf{v} = \begin{vmatrix} \mathbf{i} & \mathbf{j} & \mathbf{k} \\ 13 & 13 & -6 \\ 2 & 2 & 6 \end{vmatrix} = 90\mathbf{i} - 90\mathbf{j}$$

$\Rightarrow$ Tangent line: $x = 1 + 90t$, $y = 1 - 90t$, $z = 3$

19. $\nabla f = \left(\frac{x}{x^2+y^2+z^2}\right)\mathbf{i} + \left(\frac{y}{x^2+y^2+z^2}\right)\mathbf{j} + \left(\frac{z}{x^2+y^2+z^2}\right)\mathbf{k} \Rightarrow \nabla f(3,4,12) = \frac{3}{169}\mathbf{i} + \frac{4}{169}\mathbf{j} + \frac{12}{169}\mathbf{k}$;
$\mathbf{u} = \frac{\mathbf{v}}{|\mathbf{v}|} = \frac{3\mathbf{i} + 6\mathbf{j} - 2\mathbf{k}}{\sqrt{3^2 + 6^2 + (-2)^2}} = \frac{3}{7}\mathbf{i} + \frac{6}{7}\mathbf{j} - \frac{2}{7}\mathbf{k} \Rightarrow \nabla f \cdot \mathbf{u} = \frac{9}{1183}$ and $df = (\nabla f \cdot \mathbf{u})\,ds = \left(\frac{9}{1183}\right)(0.1) \approx 0.0008$

21. $\nabla g = (1 + \cos z)\mathbf{i} + (1 - \sin z)\mathbf{j} + (-x \sin z - y \cos z)\mathbf{k} \Rightarrow \nabla g(2, -1, 0) = 2\mathbf{i} + \mathbf{j} + \mathbf{k}$; $\mathbf{A} = \overrightarrow{P_0P_1} = -2\mathbf{i} + 2\mathbf{j} + 2\mathbf{k}$
$\Rightarrow \mathbf{u} = \frac{\mathbf{A}}{|\mathbf{A}|} = \frac{-2\mathbf{i} + 2\mathbf{j} + 2\mathbf{k}}{\sqrt{(-2)^2 + 2^2 + 2^2}} = -\frac{1}{\sqrt{3}}\mathbf{i} + \frac{1}{\sqrt{3}}\mathbf{j} + \frac{1}{\sqrt{3}}\mathbf{k} \Rightarrow \nabla g \cdot \mathbf{u} = 0$ and $dg = (\nabla g \cdot \mathbf{u})\, ds = (0)(0.2) = 0$

23. (a) The unit tangent vector at $\left(\frac{1}{2}, \frac{\sqrt{3}}{2}\right)$ in the direction of motion is $\mathbf{u} = \frac{\sqrt{3}}{2}\mathbf{i} - \frac{1}{2}\mathbf{j}$;
$\nabla T = (\sin 2y)\mathbf{i} + (2x \cos 2y)\mathbf{j} \Rightarrow \nabla T\left(\frac{1}{2}, \frac{\sqrt{3}}{2}\right) = \left(\sin \sqrt{3}\right)\mathbf{i} + \left(\cos \sqrt{3}\right)\mathbf{j} \Rightarrow D_{\mathbf{u}}T\left(\frac{1}{2}, \frac{\sqrt{3}}{2}\right) = \nabla T \cdot \mathbf{u}$
$= \frac{\sqrt{3}}{2} \sin \sqrt{3} - \frac{1}{2} \cos \sqrt{3} \approx 0.935°$ C/ft
(b) $\mathbf{r}(t) = (\sin 2t)\mathbf{i} + (\cos 2t)\mathbf{j} \Rightarrow \mathbf{v}(t) = (2 \cos 2t)\mathbf{i} - (2 \sin 2t)\mathbf{j}$ and $|\mathbf{v}| = 2$; $\frac{dT}{dt} = \frac{\partial T}{\partial x}\frac{dx}{dt} + \frac{\partial T}{\partial y}\frac{dy}{dt}$
$= \nabla T \cdot \mathbf{v} = \left(\nabla T \cdot \frac{\mathbf{v}}{|\mathbf{v}|}\right)|\mathbf{v}| = (D_{\mathbf{u}}T)\,|\mathbf{v}|$, where $\mathbf{u} = \frac{\mathbf{v}}{|\mathbf{v}|}$; at $\left(\frac{1}{2}, \frac{\sqrt{3}}{2}\right)$ we have $\mathbf{u} = \frac{\sqrt{3}}{2}\mathbf{i} - \frac{1}{2}\mathbf{j}$ from part (a)
$\Rightarrow \frac{dT}{dt} = \left(\frac{\sqrt{3}}{2} \sin \sqrt{3} - \frac{1}{2} \cos \sqrt{3}\right) \cdot 2 = \sqrt{3} \sin \sqrt{3} - \cos \sqrt{3} \approx 1.87°$ C/sec

25. (a) $f(0,0) = 1$, $f_x(x,y) = 2x \Rightarrow f_x(0,0) = 0$, $f_y(x,y) = 2y \Rightarrow f_y(0,0) = 0 \Rightarrow L(x,y) = 1 + 0(x-0) + 0(y-0) = 1$
(b) $f(1,1) = 3$, $f_x(1,1) = 2$, $f_y(1,1) = 2 \Rightarrow L(x,y) = 3 + 2(x-1) + 2(y-1) = 2x + 2y - 1$

27. (a) $f(0,0) = 5$, $f_x(x,y) = 3$ for all (x,y), $f_y(x,y) = -4$ for all $(x,y) \Rightarrow L(x,y) = 5 + 3(x-0) - 4(y-0) = 3x - 4y + 5$
(b) $f(1,1) = 4$, $f_x(1,1) = 3$, $f_y(1,1) = -4 \Rightarrow L(x,y) = 4 + 3(x-1) - 4(y-1) = 3x - 4y + 5$

29. (a) $f(0,0) = 1$, $f_x(x,y) = e^x \cos y \Rightarrow f_x(0,0) = 1$, $f_y(x,y) = -e^x \sin y \Rightarrow f_y(0,0) = 0$
$\Rightarrow L(x,y) = 1 + 1(x-0) + 0(y-0) = x + 1$
(b) $f\left(0, \frac{\pi}{2}\right) = 0$, $f_x\left(0, \frac{\pi}{2}\right) = 0$, $f_y\left(0, \frac{\pi}{2}\right) = -1 \Rightarrow L(x,y) = 0 + 0(x-0) - 1\left(y - \frac{\pi}{2}\right) = -y + \frac{\pi}{2}$

31. $f(2,1) = 3$, $f_x(x,y) = 2x - 3y \Rightarrow f_x(2,1) = 1$, $f_y(x,y) = -3x \Rightarrow f_y(2,1) = -6 \Rightarrow L(x,y) = 3 + 1(x-2) - 6(y-1)$
$= 7 + x - 6y$; $f_{xx}(x,y) = 2$, $f_{yy}(x,y) = 0$, $f_{xy}(x,y) = -3 \Rightarrow M = 3$; thus $|E(x,y)| \le \left(\frac{1}{2}\right)(3)\left(|x-2| + |y-1|\right)^2$
$\le \left(\frac{3}{2}\right)(0.1 + 0.1)^2 = 0.06$

33. $f(0,0) = 1$, $f_x(x,y) = \cos y \Rightarrow f_x(0,0) = 1$, $f_y(x,y) = 1 - x \sin y \Rightarrow f_y(0,0) = 1$
$\Rightarrow L(x,y) = 1 + 1(x-0) + 1(y-0) = x + y + 1$; $f_{xx}(x,y) = 0$, $f_{yy}(x,y) = -x \cos y$, $f_{xy}(x,y) = -\sin y \Rightarrow M = 1$;
thus $|E(x,y)| \le \left(\frac{1}{2}\right)(1)\left(|x| + |y|\right)^2 \le \left(\frac{1}{2}\right)(0.2 + 0.2)^2 = 0.08$

35. $f(0,0) = 1$, $f_x(x,y) = e^x \cos y \Rightarrow f_x(0,0) = 1$, $f_y(x,y) = -e^x \sin y \Rightarrow f_y(0,0) = 0$
$\Rightarrow L(x,y) = 1 + 1(x-0) + 0(y-0) = 1 + x$; $f_{xx}(x,y) = e^x \cos y$, $f_{yy}(x,y) = -e^x \cos y$, $f_{xy}(x,y) = -e^x \sin y$;
$|x| \le 0.1 \Rightarrow -0.1 \le x \le 0.1$ and $|y| \le 0.1 \Rightarrow -0.1 \le y \le 0.1$; thus the max of $|f_{xx}(x,y)|$ on R is $e^{0.1} \cos(0.1)$
≤ 1.11, the max of $|f_{yy}(x,y)|$ on R is $e^{0.1} \cos(0.1) \le 1.11$, and the max of $|f_{xy}(x,y)|$ on R is $e^{0.1} \sin(0.1)$
$\le 0.12 \Rightarrow M = 1.11$; thus $|E(x,y)| \le \left(\frac{1}{2}\right)(1.11)\left(|x| + |y|\right)^2 \le (0.555)(0.1 + 0.1)^2 = 0.0222$

37. (a) $f(1,1,1) = 3$, $f_x(1,1,1) = y + z\big|_{(1,1,1)} = 2$, $f_y(1,1,1) = x + z\big|_{(1,1,1)} = 2$, $f_z(1,1,1) = y + x\big|_{(1,1,1)} = 2$
$\Rightarrow L(x,y,z) = 3 + 2(x-1) + 2(y-1) + 2(z-1) = 2x + 2y + 2z - 3$
(b) $f(1,0,0) = 0$, $f_x(1,0,0) = 0$, $f_y(1,0,0) = 1$, $f_z(1,0,0) = 1 \Rightarrow L(x,y,z) = 0 + 0(x-1) + (y-0) + (z-0) = y + z$
(c) $f(0,0,0) = 0$, $f_x(0,0,0) = 0$, $f_y(0,0,0) = 0$, $f_z(0,0,0) = 0 \Rightarrow L(x,y,z) = 0$

39. (a) $f(1,0,0) = 1$, $f_x(1,0,0) = \frac{x}{\sqrt{x^2 + y^2 + z^2}}\bigg|_{(1,0,0)} = 1$, $f_y(1,0,0) = \frac{y}{\sqrt{x^2 + y^2 + z^2}}\bigg|_{(1,0,0)} = 0$,
$f_z(1,0,0) = \frac{z}{\sqrt{x^2 + y^2 + z^2}}\bigg|_{(1,0,0)} = 0 \Rightarrow L(x,y,z) = 1 + 1(x-1) + 0(y-0) + 0(z-0) = x$

(b) $f(1,1,0) = \sqrt{2}$, $f_x(1,1,0) = \frac{1}{\sqrt{2}}$, $f_y(1,1,0) = \frac{1}{\sqrt{2}}$, $f_z(1,1,0) = 0$
$\Rightarrow L(x,y,z) = \sqrt{2} + \frac{1}{\sqrt{2}}(x-1) + \frac{1}{\sqrt{2}}(y-1) + 0(z-0) = \frac{1}{\sqrt{2}}x + \frac{1}{\sqrt{2}}y$

(c) $f(1,2,2) = 3$, $f_x(1,2,2) = \frac{1}{3}$, $f_y(1,2,2) = \frac{2}{3}$, $f_z(1,2,2) = \frac{2}{3} \Rightarrow L(x,y,z) = 3 + \frac{1}{3}(x-1) + \frac{2}{3}(y-2) + \frac{2}{3}(z-2)$
$= \frac{1}{3}x + \frac{2}{3}y + \frac{2}{3}z$

41. (a) $f(0,0,0) = 2$, $f_x(0,0,0) = e^x\big|_{(0,0,0)} = 1$, $f_y(0,0,0) = -\sin(y+z)\big|_{(0,0,0)} = 0$,
$f_z(0,0,0) = -\sin(y+z)\big|_{(0,0,0)} = 0 \Rightarrow L(x,y,z) = 2 + 1(x-0) + 0(y-0) + 0(z-0) = 2 + x$

(b) $f\left(0,\frac{\pi}{2},0\right) = 1$, $f_x\left(0,\frac{\pi}{2},0\right) = 1$, $f_y\left(0,\frac{\pi}{2},0\right) = -1$, $f_z\left(0,\frac{\pi}{2},0\right) = -1 \Rightarrow L(x,y,z)$
$= 1 + 1(x-0) - 1\left(y-\frac{\pi}{2}\right) - 1(z-0) = x - y - z + \frac{\pi}{2} + 1$

(c) $f\left(0,\frac{\pi}{4},\frac{\pi}{4}\right) = 1$, $f_x\left(0,\frac{\pi}{4},\frac{\pi}{4}\right) = 1$, $f_y\left(0,\frac{\pi}{4},\frac{\pi}{4}\right) = -1$, $f_z\left(0,\frac{\pi}{4},\frac{\pi}{4}\right) = -1 \Rightarrow L(x,y,z)$
$= 1 + 1(x-0) - 1\left(y-\frac{\pi}{4}\right) - 1\left(z-\frac{\pi}{4}\right) = x - y - z + \frac{\pi}{2} + 1$

43. $f(x,y,z) = xz - 3yz + 2$ at $P_0(1,1,2) \Rightarrow f(1,1,2) = -2$; $f_x = z$, $f_y = -3z$, $f_z = x - 3y \Rightarrow L(x,y,z)$
$= -2 + 2(x-1) - 6(y-1) - 2(z-2) = 2x - 6y - 2z + 6$; $f_{xx} = 0$, $f_{yy} = 0$, $f_{zz} = 0$, $f_{xy} = 0$, $f_{yz} = -3$
$\Rightarrow M = 3$; thus, $|E(x,y,z)| \le \left(\frac{1}{2}\right)(3)(0.01 + 0.01 + 0.02)^2 = 0.0024$

45. $f(x,y,z) = xy + 2yz - 3xz$ at $P_0(1,1,0) \Rightarrow f(1,1,0) = 1$; $f_x = y - 3z$, $f_y = x + 2z$, $f_z = 2y - 3x$
$\Rightarrow L(x,y,z) = 1 + (x-1) + (y-1) - (z-0) = x + y - z - 1$; $f_{xx} = 0$, $f_{yy} = 0$, $f_{zz} = 0$, $f_{xy} = 1$, $f_{xz} = -3$,
$f_{yz} = 2 \Rightarrow M = 3$; thus $|E(x,y,z)| \le \left(\frac{1}{2}\right)(3)(0.01 + 0.01 + 0.01)^2 = 0.00135$

47. $T_x(x,y) = e^y + e^{-y}$ and $T_y(x,y) = x\left(e^y - e^{-y}\right) \Rightarrow dT = T_x(x,y)\,dx + T_y(x,y)\,dy$
$= \left(e^y + e^{-y}\right)dx + x\left(e^y - e^{-y}\right)dy \Rightarrow dT\big|_{(2,\ln 2)} = 2.5\,dx + 3.0\,dy$. If $|dx| \le 0.1$ and $|dy| \le 0.02$, then the maximum possible error in the computed value of T is $(2.5)(0.1) + (3.0)(0.02) = 0.31$ in magnitude.

49. $f(a,b,c,d) = \begin{vmatrix} a & b \\ c & d \end{vmatrix} = ad - bc \Rightarrow f_a = d$, $f_b = -c$, $f_c = -b$, $f_d = a \Rightarrow df = d\,da - c\,db - b\,dc + a\,dd$; since $|a|$ is much greater than $|b|$, $|c|$, and $|d|$, the function f is most sensitive to a change in d.

51. $z = f(x,y) \Rightarrow g(x,y,z) = f(x,y) - z = 0 \Rightarrow g_x(x,y,z) = f_x(x,y)$, $g_y(x,y,z) = f_y(x,y)$ and $g_z(x,y,z) = -1$
$\Rightarrow g_x(x_0,y_0,f(x_0,y_0)) = f_x(x_0,y_0)$, $g_y(x_0,y_0,f(x_0,y_0)) = f_y(x_0,y_0)$ and $g_z(x_0,y_0,f(x_0,y_0)) = -1 \Rightarrow$ the tangent plane at the point P_0 is $f_x(x_0,y_0)(x-x_0) + f_y(x_0,y_0)(y-y_0) - [z - f(x_0,y_0)] = 0$ or
$z = f_x(x_0,y_0)(x-x_0) + f_y(x_0,y_0)(y-y_0) + f(x_0,y_0)$

53. $\nabla f = 2x\mathbf{i} + 2y\mathbf{j} + 2z\mathbf{k} = (2\cos t)\mathbf{i} + (2\sin t)\mathbf{j} + 2t\mathbf{k}$ and $\mathbf{v} = (-\sin t)\mathbf{i} + (\cos t)\mathbf{j} + \mathbf{k} \Rightarrow \mathbf{u} = \frac{\mathbf{v}}{|\mathbf{v}|}$
$= \frac{(-\sin t)\mathbf{i} + (\cos t)\mathbf{j} + \mathbf{k}}{\sqrt{(\sin t)^2 + (\cos t)^2 + 1^2}} = \left(\frac{-\sin t}{\sqrt{2}}\right)\mathbf{i} + \left(\frac{\cos t}{\sqrt{2}}\right)\mathbf{j} + \frac{1}{\sqrt{2}}\mathbf{k} \Rightarrow (D_{\mathbf{u}}f)_{P_0} = \nabla f \cdot \mathbf{u}$
$= (2\cos t)\left(\frac{-\sin t}{\sqrt{2}}\right) + (2\sin t)\left(\frac{\cos t}{\sqrt{2}}\right) + (2t)\left(\frac{1}{\sqrt{2}}\right) = \frac{2t}{\sqrt{2}} \Rightarrow (D_{\mathbf{u}}f)\left(\frac{-\pi}{4}\right) = \frac{-\pi}{2\sqrt{2}}$, $(D_{\mathbf{u}}f)(0) = 0$ and
$(D_{\mathbf{u}}f)\left(\frac{\pi}{4}\right) = \frac{\pi}{2\sqrt{2}}$

55. $r = \sqrt{t}\mathbf{i} + \sqrt{t}\mathbf{j} + (2t-1)\mathbf{k} \Rightarrow \mathbf{v} = \frac{1}{2}t^{-1/2}\mathbf{i} + \frac{1}{2}t^{-1/2}\mathbf{j} + 2\mathbf{k}$; $t = 1 \Rightarrow x = 1$, $y = 1$, $z = 1 \Rightarrow P_0 = (1,1,1)$ and $\mathbf{v}(1) = \frac{1}{2}\mathbf{i} + \frac{1}{2}\mathbf{j} + 2\mathbf{k}$; $f(x,y,z) = x^2 + y^2 - z - 1 = 0 \Rightarrow \nabla f = 2x\mathbf{i} + 2y\mathbf{j} - \mathbf{k} \Rightarrow \nabla f(1,1,1) = 2\mathbf{i} + 2\mathbf{j} - \mathbf{k}$; now $\mathbf{v}(1) \cdot \nabla f(1,1,1) = 0$, thus the curve is tangent to the surface when $t = 1$

11.7 EXTREME VALUES AND SADDLE POINTS

1. $f_x(x, y) = 2x + y + 3 = 0$ and $f_y(x, y) = x + 2y - 3 = 0 \Rightarrow x = -3$ and $y = 3 \Rightarrow$ critical point is $(-3, 3)$; $f_{xx}(-3, 3) = 2$, $f_{yy}(-3, 3) = 2$, $f_{xy}(-3, 3) = 1 \Rightarrow f_{xx}f_{yy} - f_{xy}^2 = 3 > 0$ and $f_{xx} > 0 \Rightarrow$ local minimum of $f(-3, 3) = -5$

3. $f_x(x, y) = 2y - 10x + 4 = 0$ and $f_y(x, y) = 2x - 4y + 4 = 0 \Rightarrow x = \frac{2}{3}$ and $y = \frac{4}{3} \Rightarrow$ critical point is $\left(\frac{2}{3}, \frac{4}{3}\right)$; $f_{xx}\left(\frac{2}{3}, \frac{4}{3}\right) = -10$, $f_{yy}\left(\frac{2}{3}, \frac{4}{3}\right) = -4$, $f_{xy}\left(\frac{2}{3}, \frac{4}{3}\right) = 2 \Rightarrow f_{xx}f_{yy} - f_{xy}^2 = 36 > 0$ and $f_{xx} < 0 \Rightarrow$ local maximum of $f\left(\frac{2}{3}, \frac{4}{3}\right) = 0$

5. $f_x(x, y) = 2x + y + 3 = 0$ and $f_y(x, y) = x + 2 = 0 \Rightarrow x = -2$ and $y = 1 \Rightarrow$ critical point is $(-2, 1)$; $f_{xx}(-2, 1) = 2$, $f_{yy}(-2, 1) = 0$, $f_{xy}(-2, 1) = 1 \Rightarrow f_{xx}f_{yy} - f_{xy}^2 = -1 < 0 \Rightarrow$ saddle point

7. $f_x(x, y) = 5y - 14x + 3 = 0$ and $f_y(x, y) = 5x - 6 = 0 \Rightarrow x = \frac{6}{5}$ and $y = \frac{69}{25} \Rightarrow$ critical point is $\left(\frac{6}{5}, \frac{69}{25}\right)$; $f_{xx}\left(\frac{6}{5}, \frac{69}{25}\right) = -14$, $f_{yy}\left(\frac{6}{5}, \frac{69}{25}\right) = 0$, $f_{xy}\left(\frac{6}{5}, \frac{69}{25}\right) = 5 \Rightarrow f_{xx}f_{yy} - f_{xy}^2 = -25 < 0 \Rightarrow$ saddle point

9. $f_x(x, y) = 2x - 4y = 0$ and $f_y(x, y) = -4x + 2y + 6 = 0 \Rightarrow x = 2$ and $y = 1 \Rightarrow$ critical point is $(2, 1)$; $f_{xx}(2, 1) = 2$, $f_{yy}(2, 1) = 2$, $f_{xy}(2, 1) = -4 \Rightarrow f_{xx}f_{yy} - f_{xy}^2 = -12 < 0 \Rightarrow$ saddle point

11. $f_x(x, y) = 4x + 3y - 5 = 0$ and $f_y(x, y) = 3x + 8y + 2 = 0 \Rightarrow x = 2$ and $y = -1 \Rightarrow$ critical point is $(2, -1)$; $f_{xx}(2, -1) = 4$, $f_{yy}(2, -1) = 8$, $f_{xy}(2, -1) = 3 \Rightarrow f_{xx}f_{yy} - f_{xy}^2 = 23 > 0$ and $f_{xx} > 0 \Rightarrow$ local minimum of $f(2, -1) = -6$

13. $f_x(x, y) = 2x - 2 = 0$ and $f_y(x, y) = -2y + 4 = 0 \Rightarrow x = 1$ and $y = 2 \Rightarrow$ critical point is $(1, 2)$; $f_{xx}(1, 2) = 2$, $f_{yy}(1, 2) = -2$, $f_{xy}(1, 2) = 0 \Rightarrow f_{xx}f_{yy} - f_{xy}^2 = -4 < 0 \Rightarrow$ saddle point

15. $f_x(x, y) = 2x + 2y = 0$ and $f_y(x, y) = 2x = 0 \Rightarrow x = 0$ and $y = 0 \Rightarrow$ critical point is $(0, 0)$; $f_{xx}(0, 0) = 2$, $f_{yy}(0, 0) = 0$, $f_{xy}(0, 0) = 2 \Rightarrow f_{xx}f_{yy} - f_{xy}^2 = -4 < 0 \Rightarrow$ saddle point

17. $f_x(x, y) = 3x^2 - 2y = 0$ and $f_y(x, y) = -3y^2 - 2x = 0 \Rightarrow x = 0$ and $y = 0$, or $x = -\frac{2}{3}$ and $y = \frac{2}{3} \Rightarrow$ critical points are $(0, 0)$ and $\left(-\frac{2}{3}, \frac{2}{3}\right)$; for $(0, 0)$: $f_{xx}(0, 0) = 6x|_{(0,0)} = 0$, $f_{yy}(0, 0) = -6y|_{(0,0)} = 0$, $f_{xy}(0, 0) = -2$ $\Rightarrow f_{xx}f_{yy} - f_{xy}^2 = -4 < 0 \Rightarrow$ saddle point; for $\left(-\frac{2}{3}, \frac{2}{3}\right)$: $f_{xx}\left(-\frac{2}{3}, \frac{2}{3}\right) = -4$, $f_{yy}\left(-\frac{2}{3}, \frac{2}{3}\right) = -4$, $f_{xy}\left(-\frac{2}{3}, \frac{2}{3}\right) = -2$ $\Rightarrow f_{xx}f_{yy} - f_{xy}^2 = 12 > 0$ and $f_{xx} < 0 \Rightarrow$ local maximum of $f\left(-\frac{2}{3}, \frac{2}{3}\right) = \frac{170}{27}$

19. $f_x(x, y) = 12x - 6x^2 + 6y = 0$ and $f_y(x, y) = 6y + 6x = 0 \Rightarrow x = 0$ and $y = 0$, or $x = 1$ and $y = -1 \Rightarrow$ critical points are $(0, 0)$ and $(1, -1)$; for $(0, 0)$: $f_{xx}(0, 0) = 12 - 12x|_{(0,0)} = 12$, $f_{yy}(0, 0) = 6$, $f_{xy}(0, 0) = 6 \Rightarrow f_{xx}f_{yy} - f_{xy}^2 = 36 > 0$ and $f_{xx} > 0 \Rightarrow$ local minimum of $f(0, 0) = 0$; for $(1, -1)$: $f_{xx}(1, -1) = 0$, $f_{yy}(1, -1) = 6$, $f_{xy}(1, -1) = 6 \Rightarrow f_{xx}f_{yy} - f_{xy}^2 = -36 < 0 \Rightarrow$ saddle point

21. $f_x(x, y) = 27x^2 - 4y = 0$ and $f_y(x, y) = y^2 - 4x = 0 \Rightarrow x = 0$ and $y = 0$, or $x = \frac{4}{9}$ and $y = \frac{4}{3} \Rightarrow$ critical points are $(0, 0)$ and $\left(\frac{4}{9}, \frac{4}{3}\right)$; for $(0, 0)$: $f_{xx}(0, 0) = 54x|_{(0,0)} = 0$, $f_{yy}(0, 0) = 2y|_{(0,0)} = 0$, $f_{xy}(0, 0) = -4 \Rightarrow f_{xx}f_{yy} - f_{xy}^2 = -16 < 0 \Rightarrow$ saddle point; for $\left(\frac{4}{9}, \frac{4}{3}\right)$: $f_{xx}\left(\frac{4}{9}, \frac{4}{3}\right) = 24$, $f_{yy}\left(\frac{4}{9}, \frac{4}{3}\right) = \frac{8}{3}$, $f_{xy}\left(\frac{4}{9}, \frac{4}{3}\right) = -4 \Rightarrow f_{xx}f_{yy} - f_{xy}^2 = 48 > 0$ and $f_{xx} > 0 \Rightarrow$ local minimum of $f\left(\frac{4}{9}, \frac{4}{3}\right) = -\frac{64}{81}$

23. $f_x(x, y) = 3x^2 + 6x = 0 \Rightarrow x = 0$ or $x = -2$; $f_y(x, y) = 3y^2 - 6y = 0 \Rightarrow y = 0$ or $y = 2 \Rightarrow$ the critical points are $(0, 0)$, $(0, 2)$, $(-2, 0)$, and $(-2, 2)$; for $(0, 0)$: $f_{xx}(0, 0) = 6x + 6|_{(0,0)} = 6$, $f_{yy}(0, 0) = 6y - 6|_{(0,0)} = -6$, $f_{xy}(0, 0) = 0 \Rightarrow f_{xx}f_{yy} - f_{xy}^2 = -36 < 0 \Rightarrow$ saddle point; for $(0, 2)$: $f_{xx}(0, 2) = 6$, $f_{yy}(0, 2) = 6$, $f_{xy}(0, 2) = 0$

$\Rightarrow f_{xx}f_{yy} - f_{xy}^2 = 36 > 0$ and $f_{xx} > 0 \Rightarrow$ local minimum of $f(0,2) = -12$; for $(-2,0)$: $f_{xx}(-2,0) = -6$, $f_{yy}(-2,0) = -6$, $f_{xy}(-2,0) = 0 \Rightarrow f_{xx}f_{yy} - f_{xy}^2 = 36 > 0$ and $f_{xx} < 0 \Rightarrow$ local maximum of $f(-2,0) = -4$; for $(-2,2)$: $f_{xx}(-2,2) = -6$, $f_{yy}(-2,2) = 6$, $f_{xy}(-2,2) = 0 \Rightarrow f_{xx}f_{yy} - f_{xy}^2 = -36 < 0 \Rightarrow$ saddle point

25. $f_x(x,y) = 4y - 4x^3 = 0$ and $f_y(x,y) = 4x - 4y^3 = 0 \Rightarrow x = y \Rightarrow x(1-x^2) = 0 \Rightarrow x = 0, 1, -1 \Rightarrow$ the critical points are $(0,0)$, $(1,1)$, and $(-1,-1)$; for $(0,0)$: $f_{xx}(0,0) = -12x^2|_{(0,0)} = 0$, $f_{yy}(0,0) = -12y^2|_{(0,0)} = 0$, $f_{xy}(0,0) = 4 \Rightarrow f_{xx}f_{yy} - f_{xy}^2 = -16 < 0 \Rightarrow$ saddle point; for $(1,1)$: $f_{xx}(1,1) = -12$, $f_{yy}(1,1) = -12$, $f_{xy}(1,1) = 4$ $\Rightarrow f_{xx}f_{yy} - f_{xy}^2 = 128 > 0$ and $f_{xx} < 0 \Rightarrow$ local maximum of $f(1,1) = 2$; for $(-1,-1)$: $f_{xx}(-1,-1) = -12$, $f_{yy}(-1,-1) = -12$, $f_{xy}(-1,-1) = 4 \Rightarrow f_{xx}f_{yy} - f_{xy}^2 = 128 > 0$ and $f_{xx} < 0 \Rightarrow$ local maximum of $f(-1,-1) = 2$

27. $f_x(x,y) = \frac{-2x}{(x^2+y^2-1)^2} = 0$ and $f_y(x,y) = \frac{-2y}{(x^2+y^2-1)^2} = 0 \Rightarrow x = 0$ and $y = 0 \Rightarrow$ the critical point is $(0,0)$; $f_{xx} = \frac{4x^2-2y^2+2}{(x^2+y^2-1)^3}$, $f_{yy} = \frac{-2x^2+4y^2+2}{(x^2+y^2-1)^3}$, $f_{xy} = \frac{8xy}{(x^2+y^2-1)^3}$; $f_{xx}(0,0) = -2$, $f_{yy}(0,0) = -2$, $f_{xy}(0,0) = 0$ $\Rightarrow f_{xx}f_{yy} - f_{xy}^2 = 4 > 0$ and $f_{xx} < 0 \Rightarrow$ local maximum of $f(0,0) = -1$

29. $f_x(x,y) = y\cos x = 0$ and $f_y(x,y) = \sin x = 0 \Rightarrow x = n\pi$, n an integer, and $y = 0 \Rightarrow$ the critical points are $(n\pi, 0)$, n an integer (Note: $\cos x$ and $\sin x$ cannot both be 0 for the same x, so $\sin x$ must be 0 and $y = 0$); $f_{xx} = -y\sin x$, $f_{yy} = 0$, $f_{xy} = \cos x$; $f_{xx}(n\pi,0) = 0$, $f_{yy}(n\pi,0) = 0$, $f_{xy}(n\pi,0) = 1$ if n is even and $f_{xy}(n\pi,0) = -1$ if n is odd $\Rightarrow f_{xx}f_{yy} - f_{xy}^2 = -1 < 0 \Rightarrow$ saddle point.

31. (i) On OA, $f(x,y) = f(0,y) = y^2 - 4y + 1$ on $0 \le y \le 2$; $f'(0,y) = 2y - 4 = 0 \Rightarrow y = 2$; $f(0,0) = 1$ and $f(0,2) = -3$

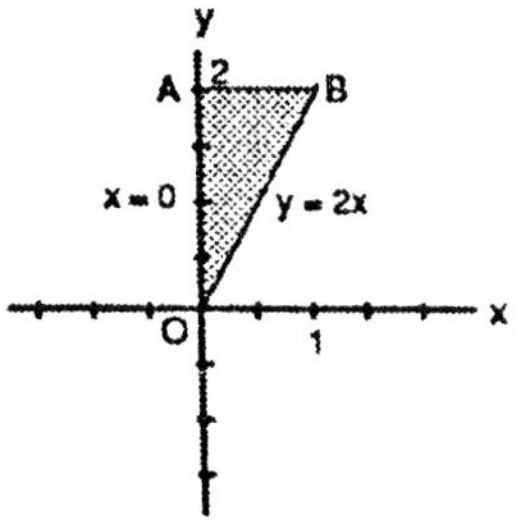

(ii) On AB, $f(x,y) = f(x,2) = 2x^2 - 4x - 3$ on $0 \le x \le 1$; $f'(x,2) = 4x - 4 = 0 \Rightarrow x = 1$; $f(0,2) = -3$ and $f(1,2) = -5$

(iii) On OB, $f(x,y) = f(x,2x) = 6x^2 - 12x + 1$ on $0 \le x \le 1$; endpoint values have been found above; $f'(x,2x) = 12x - 12 = 0 \Rightarrow x = 1$ and $y = 2$, but $(1,2)$ is not an interior point of OB

(iv) For interior points of the triangular region, $f_x(x,y) = 4x - 4 = 0$ and $f_y(x,y) = 2y - 4 = 0$ $\Rightarrow x = 1$ and $y = 2$, but $(1,2)$ is not an interior point of the region. Therefore, the absolute maximum is 1 at $(0,0)$ and the absolute minimum is -5 at $(1,2)$.

33. (i) On OA, $f(x,y) = f(0,y) = y^2$ on $0 \le y \le 2$; $f'(0,y) = 2y = 0 \Rightarrow y = 0$ and $x = 0$; $f(0,0) = 0$ and $f(0,2) = 4$

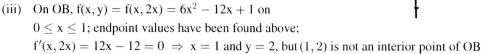

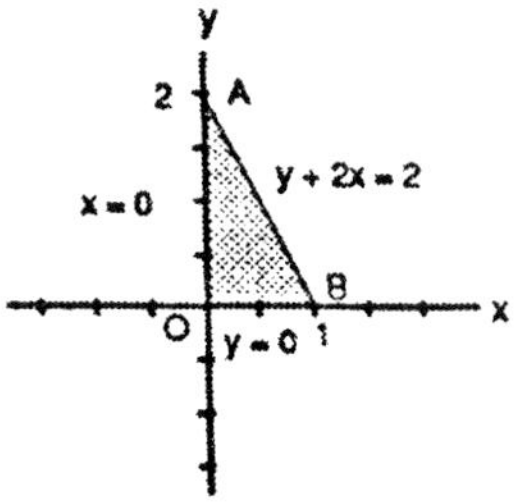

(ii) On OB, $f(x,y) = f(x,0) = x^2$ on $0 \le x \le 1$; $f'(x,0) = 2x = 0 \Rightarrow x = 0$ and $y = 0$; $f(0,0) = 0$ and $f(1,0) = 1$

(iii) On AB, $f(x,y) = f(x,-2x+2) = 5x^2 - 8x + 4$ on $0 \le x \le 1$; $f'(x,-2x+2) = 10x - 8 = 0 \Rightarrow x = \frac{4}{5}$ and $y = \frac{2}{5}$; $f\left(\frac{4}{5},\frac{2}{5}\right) = \frac{4}{5}$; endpoint values have been found above.

(iv) For interior points of the triangular region, $f_x(x,y) = 2x = 0$ and $f_y(x,y) = 2y = 0 \Rightarrow x = 0$ and $y = 0$, but $(0,0)$ is not an interior point of the region. Therefore the absolute maximum is 4 at $(0,2)$ and the absolute minimum is 0 at $(0,0)$.

35. (i) On OC, $T(x,y) = T(x,0) = x^2 - 6x + 2$ on $0 \le x \le 5$; $T'(x,0) = 2x - 6 = 0 \Rightarrow x = 3$ and $y = 0$; $T(3,0) = -7$, $T(0,0) = 2$, and $T(5,0) = -3$

(ii) On CB, $T(x,y) = T(5,y) = y^2 + 5y - 3$ on $-3 \le y \le 0$; $T'(5,y) = 2y + 5 = 0 \Rightarrow y = -\frac{5}{2}$ and $x = 5$; $T\left(5, -\frac{5}{2}\right) = -\frac{37}{4}$ and $T(5,-3) = -9$

(iii) On AB, $T(x,y) = T(x,-3) = x^2 - 9x + 11$ on $0 \le x \le 5$; $T'(x,-3) = 2x - 9 = 0 \Rightarrow x = \frac{9}{2}$ and $y = -3$; $T\left(\frac{9}{2}, -3\right) = -\frac{37}{4}$ and $T(0,-3) = 11$

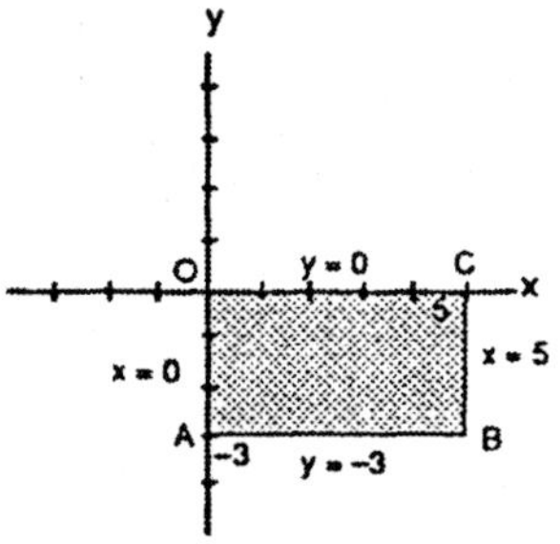

(iv) On AO, $T(x,y) = T(0,y) = y^2 + 2$ on $-3 \le y \le 0$; $T'(0,y) = 2y = 0 \Rightarrow y = 0$ and $x = 0$, but $(0,0)$ is not an interior point of AO

(v) For interior points of the rectangular region, $T_x(x,y) = 2x + y - 6 = 0$ and $T_y(x,y) = x + 2y = 0 \Rightarrow x = 4$ and $y = -2$, an interior critical point with $T(4,-2) = -10$. Therefore the absolute maximum is 11 at $(0,-3)$ and the absolute minimum is -10 at $(4,-2)$.

37. (i) On AB, $f(x,y) = f(1,y) = 3\cos y$ on $-\frac{\pi}{4} \le y \le \frac{\pi}{4}$; $f'(1,y) = -3\sin y = 0 \Rightarrow y = 0$ and $x = 1$; $f(1,0) = 3$, $f\left(1, -\frac{\pi}{4}\right) = \frac{3\sqrt{2}}{2}$, and $f\left(1, \frac{\pi}{4}\right) = \frac{3\sqrt{2}}{2}$

(ii) On CD, $f(x,y) = f(3,y) = 3\cos y$ on $-\frac{\pi}{4} \le y \le \frac{\pi}{4}$; $f'(3,y) = -3\sin y = 0 \Rightarrow y = 0$ and $x = 3$; $f(3,0) = 3$, $f\left(3, -\frac{\pi}{4}\right) = \frac{3\sqrt{2}}{2}$ and $f\left(3, \frac{\pi}{4}\right) = \frac{3\sqrt{2}}{2}$

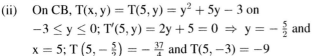

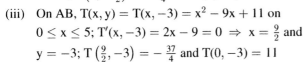

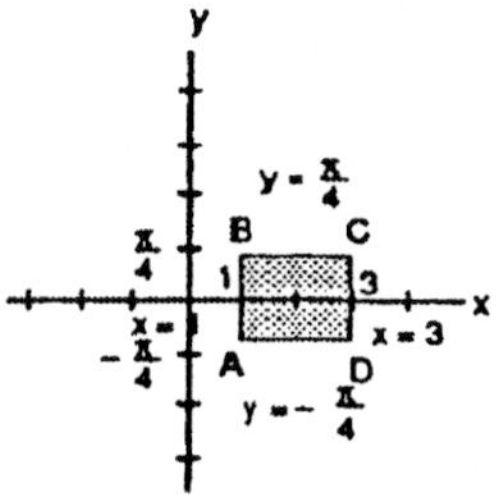

(iii) On BC, $f(x,y) = f\left(x, \frac{\pi}{4}\right) = \frac{\sqrt{2}}{2}(4x - x^2)$ on $1 \le x \le 3$; $f'\left(x, \frac{\pi}{4}\right) = \sqrt{2}(2 - x) = 0 \Rightarrow x = 2$ and $y = \frac{\pi}{4}$; $f\left(2, \frac{\pi}{4}\right) = 2\sqrt{2}$, $f\left(1, \frac{\pi}{4}\right) = \frac{3\sqrt{2}}{2}$, and $f\left(3, \frac{\pi}{4}\right) = \frac{3\sqrt{2}}{2}$

(iv) On AD, $f(x,y) = f\left(x, -\frac{\pi}{4}\right) = \frac{\sqrt{2}}{2}(4x - x^2)$ on $1 \le x \le 3$; $f'\left(x, -\frac{\pi}{4}\right) = \sqrt{2}(2 - x) = 0 \Rightarrow x = 2$ and $y = -\frac{\pi}{4}$; $f\left(2, -\frac{\pi}{4}\right) = 2\sqrt{2}$, $f\left(1, -\frac{\pi}{4}\right) = \frac{3\sqrt{2}}{2}$, and $f\left(3, -\frac{\pi}{4}\right) = \frac{3\sqrt{2}}{2}$

(v) For interior points of the region, $f_x(x,y) = (4 - 2x)\cos y = 0$ and $f_y(x,y) = -(4x - x^2)\sin y = 0 \Rightarrow x = 2$ and $y = 0$, which is an interior critical point with $f(2,0) = 4$. Therefore the absolute maximum is 4 at $(2,0)$ and the absolute minimum is $\frac{3\sqrt{2}}{2}$ at $\left(3, -\frac{\pi}{4}\right)$, $\left(3, \frac{\pi}{4}\right)$, $\left(1, -\frac{\pi}{4}\right)$, and $\left(1, \frac{\pi}{4}\right)$.

39. Let $F(a,b) = \int_a^b (6 - x - x^2)\,dx$ where $a \le b$. The boundary of the domain of F is the line $a = b$ in the ab-plane, and $F(a,a) = 0$, so F is identically 0 on the boundary of its domain. For interior critical points we have: $\frac{\partial F}{\partial a} = -(6 - a - a^2) = 0 \Rightarrow a = -3, 2$ and $\frac{\partial F}{\partial b} = (6 - b - b^2) = 0 \Rightarrow b = -3, 2$. Since $a \le b$, there is only one interior critical point $(-3, 2)$ and $F(-3,2) = \int_{-3}^{2} (6 - x - x^2)\,dx$ gives the area under the parabola $y = 6 - x - x^2$ that is above the x-axis. Therefore, $a = -3$ and $b = 2$.

41. (a) $f_x(x,y) = 2x - 4y = 0$ and $f_y(x,y) = 2y - 4x = 0 \Rightarrow x = 0$ and $y = 0$; $f_{xx}(0,0) = 2$, $f_{yy}(0,0) = 2$, $f_{xy}(0,0) = -4 \Rightarrow f_{xx}f_{yy} - f_{xy}^2 = -12 < 0 \Rightarrow$ saddle point at $(0,0)$

(b) $f_x(x,y) = 2x - 2 = 0$ and $f_y(x,y) = 2y - 4 = 0 \Rightarrow x = 1$ and $y = 2$; $f_{xx}(1,2) = 2$, $f_{yy}(1,2) = 2$, $f_{xy}(1,2) = 0 \Rightarrow f_{xx}f_{yy} - f_{xy}^2 = 4 > 0$ and $f_{xx} > 0 \Rightarrow$ local minimum at $(1,2)$

(c) $f_x(x,y) = 9x^2 - 9 = 0$ and $f_y(x,y) = 2y + 4 = 0 \Rightarrow x = \pm 1$ and $y = -2$; $f_{xx}(1,-2) = 18x|_{(1,-2)} = 18$, $f_{yy}(1,-2) = 2$, $f_{xy}(1,-2) = 0 \Rightarrow f_{xx}f_{yy} - f_{xy}^2 = 36 > 0$ and $f_{xx} > 0 \Rightarrow$ local minimum at $(1,-2)$; $f_{xx}(-1,-2) = -18$, $f_{yy}(-1,-2) = 2$, $f_{xy}(-1,-2) = 0 \Rightarrow f_{xx}f_{yy} - f_{xy}^2 = -36 < 0 \Rightarrow$ saddle point at $(-1,-2)$

43. If $k = 0$, then $f(x, y) = x^2 + y^2 \Rightarrow f_x(x, y) = 2x = 0$ and $f_y(x, y) = 2y = 0 \Rightarrow x = 0$ and $y = 0 \Rightarrow (0, 0)$ is the only critical point. If $k \neq 0$, $f_x(x, y) = 2x + ky = 0 \Rightarrow y = -\frac{2}{k}x$; $f_y(x, y) = kx + 2y = 0 \Rightarrow kx + 2\left(-\frac{2}{k}x\right) = 0$ $\Rightarrow kx - \frac{4x}{k} = 0 \Rightarrow \left(k - \frac{4}{k}\right)x = 0 \Rightarrow x = 0$ or $k = \pm 2 \Rightarrow y = \left(-\frac{2}{k}\right)(0) = 0$ or $y = \pm x$; in any case $(0, 0)$ is a critical point.

45. No; for example $f(x, y) = xy$ has a saddle point at $(a, b) = (0, 0)$ where $f_x = f_y = 0$.

47. We want the point on $z = 10 - x^2 - y^2$ where the tangent plane is parallel to the plane $x + 2y + 3z = 0$. To find a normal vector to $z = 10 - x^2 - y^2$ let $w = z + x^2 + y^2 - 10$. Then $\nabla w = 2x\mathbf{i} + 2y\mathbf{j} + \mathbf{k}$ is normal to $z = 10 - x^2 - y^2$ at (x, y). The vector ∇w is parallel to $\mathbf{i} + 2\mathbf{j} + 3\mathbf{k}$ which is normal to the plane $x + 2y + 3z = 0$ if $6x\mathbf{i} + 6y\mathbf{j} + 3\mathbf{k} = \mathbf{i} + 2\mathbf{j} + 3\mathbf{k}$ or $x = \frac{1}{6}$ and $y = \frac{1}{3}$. Thus the point is $\left(\frac{1}{6}, \frac{1}{3}, 10 - \frac{1}{36} - \frac{1}{9}\right)$ or $\left(\frac{1}{6}, \frac{1}{3}, \frac{355}{36}\right)$.

49. No, because the domain $x \geq 0$ and $y \geq 0$ is unbounded since x and y can be as large as we please. Absolute extrema are guaranteed for continuous functions defined over closed <u>and</u> <u>bounded</u> domains in the plane. Since the domain is unbounded, the continuous function $f(x, y) = x + y$ need not have an absolute maximum (although, in this case, it does have an absolute minimum value of $f(0, 0) = 0$).

51. (a) $\frac{df}{dt} = \frac{\partial f}{\partial x}\frac{dx}{dt} + \frac{\partial f}{\partial y}\frac{dy}{dt} = \frac{dx}{dt} + \frac{dy}{dt} = -2\sin t + 2\cos t = 0 \Rightarrow \cos t = \sin t \Rightarrow x = y$

(i) On the semicircle $x^2 + y^2 = 4$, $y \geq 0$, we have $t = \frac{\pi}{4}$ and $x = y = \sqrt{2} \Rightarrow f\left(\sqrt{2}, \sqrt{2}\right) = 2\sqrt{2}$. At the endpoints, $f(-2, 0) = -2$ and $f(2, 0) = 2$. Therefore the absolute minimum is $f(-2, 0) = -2$ when $t = \pi$; the absolute maximum is $f\left(\sqrt{2}, \sqrt{2}\right) = 2\sqrt{2}$ when $t = \frac{\pi}{4}$.

(ii) On the quartercircle $x^2 + y^2 = 4$, $x \geq 0$ and $y \geq 0$, the endpoints give $f(0, 2) = 2$ and $f(2, 0) = 2$. Therefore the absolute minimum is $f(2, 0) = 2$ and $f(0, 2) = 2$ when $t = 0, \frac{\pi}{2}$ respectively; the absolute maximum is $f\left(\sqrt{2}, \sqrt{2}\right) = 2\sqrt{2}$ when $t = \frac{\pi}{4}$.

(b) $\frac{dg}{dt} = \frac{\partial g}{\partial x}\frac{dx}{dt} + \frac{\partial g}{\partial y}\frac{dy}{dt} = y\frac{dx}{dt} + x\frac{dy}{dt} = -4\sin^2 t + 4\cos^2 t = 0 \Rightarrow \cos t = \pm\sin t \Rightarrow x = \pm y$.

(i) On the semicircle $x^2 + y^2 = 4$, $y \geq 0$, we obtain $x = y = \sqrt{2}$ at $t = \frac{\pi}{4}$ and $x = -\sqrt{2}$, $y = \sqrt{2}$ at $t = \frac{3\pi}{4}$. Then $g\left(\sqrt{2}, \sqrt{2}\right) = 2$ and $g\left(-\sqrt{2}, \sqrt{2}\right) = -2$. At the endpoints, $g(-2, 0) = g(2, 0) = 0$. Therefore the absolute minimum is $g\left(-\sqrt{2}, \sqrt{2}\right) = -2$ when $t = \frac{3\pi}{4}$; the absolute maximum is $g\left(\sqrt{2}, \sqrt{2}\right) = 2$ when $t = \frac{\pi}{4}$.

(ii) On the quartercircle $x^2 + y^2 = 4$, $x \geq 0$ and $y \geq 0$, the endpoints give $g(0, 2) = 0$ and $g(2, 0) = 0$. Therefore the absolute minimum is $g(2, 0) = 0$ and $g(0, 2) = 0$ when $t = 0, \frac{\pi}{2}$ respectively; the absolute maximum is $g\left(\sqrt{2}, \sqrt{2}\right) = 2$ when $t = \frac{\pi}{4}$.

(c) $\frac{dh}{dt} = \frac{\partial h}{\partial x}\frac{dx}{dt} + \frac{\partial h}{\partial y}\frac{dy}{dt} = 4x\frac{dx}{dt} + 2y\frac{dy}{dt} = (8\cos t)(-2\sin t) + (4\sin t)(2\cos t) = -8\cos t\sin t = 0$ $\Rightarrow t = 0, \frac{\pi}{2}, \pi$ yielding the points $(2, 0)$, $(0, 2)$ for $0 \leq t \leq \pi$.

(i) On the semicircle $x^2 + y^2 = 4$, $y \geq 0$ we have $h(2, 0) = 8$, $h(0, 2) = 4$, and $h(-2, 0) = 8$. Therefore, the absolute minimum is $h(0, 2) = 4$ when $t = \frac{\pi}{2}$; the absolute maximum is $h(2, 0) = 8$ and $h(-2, 0) = 8$ when $t = 0, \pi$ respectively.

(ii) On the quartercircle $x^2 + y^2 = 4$, $x \geq 0$ and $y \geq 0$ the absolute minimum is $h(0, 2) = 4$ when $t = \frac{\pi}{2}$; the absolute maximum is $h(2, 0) = 8$ when $t = 0$.

53. $\frac{df}{dt} = \frac{\partial f}{\partial x}\frac{dx}{dt} + \frac{\partial f}{\partial y}\frac{dy}{dt} = y\frac{dx}{dt} + x\frac{dy}{dt}$

(i) $x = 2t$ and $y = t + 1 \Rightarrow \frac{df}{dt} = (t + 1)(2) + (2t)(1) = 4t + 2 = 0 \Rightarrow t = -\frac{1}{2} \Rightarrow x = -1$ and $y = \frac{1}{2}$ with $f\left(-1, \frac{1}{2}\right) = -\frac{1}{2}$. The absolute minimum is $f\left(-1, \frac{1}{2}\right) = -\frac{1}{2}$ when $t = -\frac{1}{2}$; there is no absolute maximum.

(ii) For the endpoints: $t=-1 \Rightarrow x=-2$ and $y=0$ with $f(-2,0)=0$; $t=0 \Rightarrow x=0$ and $y=1$ with $f(0,1)=0$. The absolute minimum is $f\left(-1,\frac{1}{2}\right)=-\frac{1}{2}$ when $t=-\frac{1}{2}$; the absolute maximum is $f(0,1)=0$ and $f(-2,0)=0$ when $t=-1, 0$ respectively.

(iii) There are no interior critical points. For the endpoints: $t=0 \Rightarrow x=0$ and $y=1$ with $f(0,1)=0$; $t=1 \Rightarrow x=2$ and $y=2$ with $f(2,2)=4$. The absolute minimum is $f(0,1)=0$ when $t=0$; the absolute maximum is $f(2,2)=4$ when $t=1$.

55. $w=(m x_1+b-y_1)^2+(m x_2+b-y_2)^2+\cdots+(m x_n+b-y_n)^2$

$\Rightarrow \frac{\partial w}{\partial m}=2(m x_1+b-y_1)(x_1)+2(m x_2+b-y_2)(x_2)+\cdots+2(m x_n+b-y_n)(x_n)$

$\Rightarrow \frac{\partial w}{\partial b}=2(m x_1+b-y_1)(1)+2(m x_2+b-y_2)(1)+\cdots+2(m x_n+b-y_n)(1)$

$\frac{\partial w}{\partial m}=0 \Rightarrow 2\left[(m x_1+b-y_1)(x_1)+(m x_2+b-y_2)(x_2)+\cdots+(m x_n+b-y_n)(x_n)\right]=0$

$\Rightarrow m x_1^2+b x_1-x_1 y_1+m x_2^2+b x_2-x_2 y_2+\cdots+m x_n^2+b x_n-x_n y_n=0$

$\Rightarrow m(x_1^2+x_2^2+\cdots+x_n^2)+b(x_1+x_2+\cdots+x_n)-(x_1 y_1+x_2 y_2+\cdots+x_n y_n)=0$

$\Rightarrow m\sum_{k=1}^{n}(x_k^2)+b\sum_{k=1}^{n}x_k-\sum_{k=1}^{n}(x_k y_k)=0$

$\frac{\partial w}{\partial b}=0 \Rightarrow 2\left[(m x_1+b-y_1)+(m x_2+b-y_2)+\cdots+(m x_n+b-y_n)\right]=0$

$\Rightarrow m x_1+b-y_1+m x_2+b-y_2+\cdots+m x_n+b-y_n=0$

$\Rightarrow m(x_1+x_2+\cdots+x_n)+(b+b+\cdots+b)-(y_1+y_2+\cdots+y_n)=0$

$\Rightarrow m\sum_{k=1}^{n}x_k+b\sum_{k=1}^{n}1-\sum_{k=1}^{n}y_k=0 \Rightarrow m\sum_{k=1}^{n}x_k+bn-\sum_{k=1}^{n}y_k=0 \Rightarrow b=\frac{1}{n}\left(\sum_{k=1}^{n}y_k-m\sum_{k=1}^{n}x_k\right).$

Substituting for b in the equation obtained for $\frac{\partial w}{\partial m}$ we get $m\sum_{k=1}^{n}(x_k^2)+\frac{1}{n}\left(\sum_{k=1}^{n}y_k-m\sum_{k=1}^{n}x_k\right)\sum_{k=1}^{n}x_k-\sum_{k=1}^{n}(x_k y_k)=0.$

Multiply both sides by n to obtain $m n\sum_{k=1}^{n}(x_k^2)+\left(\sum_{k=1}^{n}y_k-m\sum_{k=1}^{n}x_k\right)\sum_{k=1}^{n}x_k-n\sum_{k=1}^{n}(x_k y_k)=0$

$\Rightarrow m n\sum_{k=1}^{n}(x_k^2)+\left(\sum_{k=1}^{n}x_k\right)\left(\sum_{k=1}^{n}y_k\right)-m\left(\sum_{k=1}^{n}x_k\right)^2-n\sum_{k=1}^{n}(x_k y_k)=0$

$\Rightarrow m n\sum_{k=1}^{n}(x_k^2)-m\left(\sum_{k=1}^{n}x_k\right)^2=n\sum_{k=1}^{n}(x_k y_k)-\left(\sum_{k=1}^{n}x_k\right)\left(\sum_{k=1}^{n}y_k\right)$

$\Rightarrow m\left[n\sum_{k=1}^{n}(x_k^2)-\left(\sum_{k=1}^{n}x_k\right)^2\right]=n\sum_{k=1}^{n}(x_k y_k)-\left(\sum_{k=1}^{n}x_k\right)\left(\sum_{k=1}^{n}y_k\right)$

$\Rightarrow m=\dfrac{n\sum_{k=1}^{n}(x_k y_k)-\left(\sum_{k=1}^{n}x_k\right)\left(\sum_{k=1}^{n}y_k\right)}{n\sum_{k=1}^{n}(x_k^2)-\left(\sum_{k=1}^{n}x_k\right)^2}=\dfrac{\left(\sum_{k=1}^{n}x_k\right)\left(\sum_{k=1}^{n}y_k\right)-n\sum_{k=1}^{n}(x_k y_k)}{\left(\sum_{k=1}^{n}x_k\right)^2-n\sum_{k=1}^{n}(x_k^2)}$

To show that these values for m and b minimize the sum of the squares of the distances, use second derivative test.

$\frac{\partial^2 w}{\partial m^2}=2x_1^2+2x_2^2+\cdots+2x_n^2=2\sum_{k=1}^{n}(x_k^2)$; $\frac{\partial^2 w}{\partial m\,\partial b}=2x_1+2x_2+\cdots+2x_n=2\sum_{k=1}^{n}x_k$; $\frac{\partial^2 w}{\partial b^2}=2+2+\cdots+2=2n$

The discriminant is: $\left(\frac{\partial^2 w}{\partial m^2}\right)\left(\frac{\partial^2 w}{\partial b^2}\right)-\left(\frac{\partial^2 w}{\partial m\,\partial b}\right)^2=\left[2\sum_{k=1}^{n}(x_k^2)\right](2n)-\left[2\sum_{k=1}^{n}x_k\right]^2=4\left[n\sum_{k=1}^{n}(x_k^2)-\left(\sum_{k=1}^{n}x_k\right)^2\right].$

Now, $n\sum_{k=1}^{n}(x_k^2)-\left(\sum_{k=1}^{n}x_k\right)^2=n(x_1^2+x_2^2+\cdots+x_n^2)-(x_1+x_2+\cdots+x_n)(x_1+x_2+\cdots+x_n)$

$=n x_1^2+n x_2^2+\cdots+n x_n^2-x_1^2-x_1x_2-\cdots-x_1x_n-x_2x_1-x_2^2-\cdots-x_2x_n-x_nx_1-x_nx_2-\cdots-x_n^2$

$=(n-1)x_1^2+(n-1)x_2^2+\cdots+(n-1)x_n^2-2x_1x_2-2x_1x_3-\cdots-2x_1x_n-2x_2x_3-\cdots-2x_2x_n-\cdots-2x_{n-1}x_n$

$=(x_1^2-2x_1x_2+x_2^2)+(x_1^2-2x_1x_3+x_3^2)+\cdots+(x_1^2-2x_1x_n+x_n^2)+(x_2^2-2x_2x_3+x_3^2)+\cdots+(x_2^2-2x_2x_n+x_n^2)$
$+\cdots+(x_{n-1}^2-2x_{n-1}x_n+x_n^2)$

$=(x_1-x_2)^2+(x_1-x_3)^2+\cdots+(x_1-x_n)^2+(x_2-x_3)^2+\cdots+(x_2-x_n)^2+\cdots+(x_{n-1}-x_n)^2\geq 0.$

Thus we have : $\left(\frac{\partial^2 w}{\partial m^2}\right)\left(\frac{\partial^2 w}{\partial b^2}\right)-\left(\frac{\partial^2 w}{\partial m\,\partial b}\right)^2=4\left[n\sum_{k=1}^{n}(x_k^2)-\left(\sum_{k=1}^{n}x_k\right)^2\right]\geq 4(0)=0$. If $x_1=x_2=\cdots=x_n$ then

$\left(\frac{\partial^2 w}{\partial m^2}\right)\left(\frac{\partial^2 w}{\partial b^2}\right) - \left(\frac{\partial^2 w}{\partial m\,\partial b}\right)^2 = 0$. Also, $\frac{\partial^2 w}{\partial m^2} = 2\sum_{k=1}^{n}(x_k^2) \ge 0$. If $x_1 = x_2 = \cdots = x_n = 0$, then $\frac{\partial^2 w}{\partial m^2} = 0$.

Provided that at least one x_i is nonzero and different from the rest of x_j, $j \ne i$, then $\left(\frac{\partial^2 w}{\partial m^2}\right)\left(\frac{\partial^2 w}{\partial b^2}\right) - \left(\frac{\partial^2 w}{\partial m\,\partial b}\right)^2 > 0$ and $\frac{\partial^2 w}{\partial m^2} > 0 \Rightarrow$ the values given above for m and b minimize w.

57. $m = \frac{(2)(-1) - 3(-14)}{(2)^2 - 3(10)} = -\frac{20}{13}$ and

$b = \frac{1}{3}\left[-1 - \left(-\frac{20}{13}\right)(2)\right] = \frac{9}{13}$

$\Rightarrow y = -\frac{20}{13}x + \frac{9}{13}$; $y\big|_{x=4} = -\frac{71}{13}$

k	x_k	y_k	x_k^2	$x_k y_k$
1	−1	2	1	−2
2	0	1	0	0
3	3	−4	9	−12
Σ	2	−1	10	−14

11.8 LAGRANGE MULTIPLIERS

1. $\nabla f = y\mathbf{i} + x\mathbf{j}$ and $\nabla g = 2x\mathbf{i} + 4y\mathbf{j}$ so that $\nabla f = \lambda \nabla g \Rightarrow y\mathbf{i} + x\mathbf{j} = \lambda(2x\mathbf{i} + 4y\mathbf{j}) \Rightarrow y = 2x\lambda$ and $x = 4y\lambda$ $\Rightarrow x = 8x\lambda^2 \Rightarrow \lambda = \pm\frac{\sqrt{2}}{4}$ or $x = 0$.

   CASE 1: If $x = 0$, then $y = 0$. But $(0, 0)$ is not on the ellipse so $x \ne 0$.

   CASE 2: $x \ne 0 \Rightarrow \lambda = \pm\frac{\sqrt{2}}{4} \Rightarrow x = \pm\sqrt{2}y \Rightarrow \left(\pm\sqrt{2}y\right)^2 + 2y^2 = 1 \Rightarrow y = \pm\frac{1}{2}$.

   Therefore f takes on its extreme values at $\left(\pm\frac{\sqrt{2}}{2}, \frac{1}{2}\right)$ and $\left(\pm\frac{\sqrt{2}}{2}, -\frac{1}{2}\right)$. The extreme values of f on the ellipse are $\pm\frac{\sqrt{2}}{2}$.

3. $\nabla f = -2x\mathbf{i} - 2y\mathbf{j}$ and $\nabla g = \mathbf{i} + 3\mathbf{j}$ so that $\nabla f = \lambda \nabla g \Rightarrow -2x\mathbf{i} - 2y\mathbf{j} = \lambda(\mathbf{i} + 3\mathbf{j}) \Rightarrow x = -\frac{\lambda}{2}$ and $y = -\frac{3\lambda}{2}$ $\Rightarrow \left(-\frac{\lambda}{2}\right) + 3\left(-\frac{3\lambda}{2}\right) = 10 \Rightarrow \lambda = -2 \Rightarrow x = 1$ and $y = 3 \Rightarrow$ f takes on its extreme value at $(1, 3)$ on the line. The extreme value is $f(1, 3) = 49 - 1 - 9 = 39$.

5. We optimize $f(x, y) = x^2 + y^2$, the square of the distance to the origin, subject to the constraint $g(x, y) = xy^2 - 54 = 0$. Thus $\nabla f = 2x\mathbf{i} + 2y\mathbf{j}$ and $\nabla g = y^2\mathbf{i} + 2xy\mathbf{j}$ so that $\nabla f = \lambda \nabla g \Rightarrow 2x\mathbf{i} + 2y\mathbf{j}$ $= \lambda\left(y^2\mathbf{i} + 2xy\mathbf{j}\right) \Rightarrow 2x = \lambda y^2$ and $2y = 2\lambda xy$.

   CASE 1: If $y = 0$, then $x = 0$. But $(0, 0)$ does not satisfy the constraint $xy^2 = 54$ so $y \ne 0$.

   CASE 2: If $y \ne 0$, then $2 = 2\lambda x \Rightarrow x = \frac{1}{\lambda} \Rightarrow 2\left(\frac{1}{\lambda}\right) = \lambda y^2 \Rightarrow y^2 = \frac{2}{\lambda^2}$. Then $xy^2 = 54 \Rightarrow \left(\frac{1}{\lambda}\right)\left(\frac{2}{\lambda^2}\right) = 54$ $\Rightarrow \lambda^3 = \frac{1}{27} \Rightarrow \lambda = \frac{1}{3} \Rightarrow x = 3$ and $y^2 = 18 \Rightarrow x = 3$ and $y = \pm 3\sqrt{2}$.

   Therefore $\left(3, \pm 3\sqrt{2}\right)$ are the points on the curve $xy^2 = 54$ nearest the origin (since $xy^2 = 54$ has points increasingly far away as y gets close to 0, no points are farthest away).

7. (a) $\nabla f = \mathbf{i} + \mathbf{j}$ and $\nabla g = y\mathbf{i} + x\mathbf{j}$ so that $\nabla f = \lambda \nabla g \Rightarrow \mathbf{i} + \mathbf{j} = \lambda(y\mathbf{i} + x\mathbf{j}) \Rightarrow 1 = \lambda y$ and $1 = \lambda x \Rightarrow y = \frac{1}{\lambda}$ and $x = \frac{1}{\lambda} \Rightarrow \frac{1}{\lambda^2} = 16 \Rightarrow \lambda = \pm\frac{1}{4}$. Use $\lambda = \frac{1}{4}$ since $x > 0$ and $y > 0$. Then $x = 4$ and $y = 4 \Rightarrow$ the minimum value is 8 at the point $(4, 4)$. Now, $xy = 16$, $x > 0$, $y > 0$ is a branch of a hyperbola in the first quadrant with the x-and y-axes as asymptotes. The equations $x + y = c$ give a family of parallel lines with $m = -1$. As these lines move away from the origin, the number c increases. Thus the minimum value of c occurs where $x + y = c$ is tangent to the hyperbola's branch.

   (b) $\nabla f = y\mathbf{i} + x\mathbf{j}$ and $\nabla g = \mathbf{i} + \mathbf{j}$ so that $\nabla f = \lambda \nabla g \Rightarrow y\mathbf{i} + x\mathbf{j} = \lambda(\mathbf{i} + \mathbf{j}) \Rightarrow y = \lambda = x\; y + y = 16 \Rightarrow y = 8$ $\Rightarrow x = 8 \Rightarrow f(8, 8) = 64$ is the maximum value. The equations $xy = c$ ($x > 0$ and $y > 0$ or $x < 0$ and $y < 0$ to get a maximum value) give a family of hyperbolas in the first and third quadrants with the x- and y-axes as asymptotes. The maximum value of c occurs where the hyperbola $xy = c$ is tangent to the line $x + y = 16$.

9. $V = \pi r^2 h \Rightarrow 16\pi = \pi r^2 h \Rightarrow 16 = r^2 h \Rightarrow g(r,h) = r^2 h - 16$; $S = 2\pi rh + 2\pi r^2 \Rightarrow \nabla S = (2\pi h + 4\pi r)\mathbf{i} + 2\pi r\mathbf{j}$ and $\nabla g = 2rh\mathbf{i} + r^2\mathbf{j}$ so that $\nabla S = \lambda \nabla g \Rightarrow (2\pi rh + 4\pi r)\mathbf{i} + 2\pi r\mathbf{j} = \lambda(2rh\mathbf{i} + r^2\mathbf{j}) \Rightarrow 2\pi rh + 4\pi r = 2rh\lambda$ and $2\pi r = \lambda r^2$ $\Rightarrow r = 0$ or $\lambda = \frac{2\pi}{r}$. But $r = 0$ gives no physical can, so $r \neq 0 \Rightarrow \lambda = \frac{2\pi}{r} \Rightarrow 2\pi h + 4\pi r = 2rh\left(\frac{2\pi}{r}\right) \Rightarrow 2r = h$ $\Rightarrow 16 = r^2(2r) \Rightarrow r = 2 \Rightarrow h = 4$; thus $r = 2$ cm and $h = 4$ cm give the only extreme surface area of 24π cm^2. Since $r = 4$ cm and $h = 1$ cm $\Rightarrow V = 16\pi$ cm^3 and $S = 40\pi$ cm^2, which is a larger surface area, then 24π cm^2 must be the minimum surface area.

11. $A = (2x)(2y) = 4xy$ subject to $g(x,y) = \frac{x^2}{16} + \frac{y^2}{9} - 1 = 0$; $\nabla A = 4y\mathbf{i} + 4x\mathbf{j}$ and $\nabla g = \frac{x}{8}\mathbf{i} + \frac{2y}{9}\mathbf{j}$ so that $\nabla A = \lambda \nabla g \Rightarrow 4y\mathbf{i} + 4x\mathbf{j} = \lambda\left(\frac{x}{8}\mathbf{i} + \frac{2y}{9}\mathbf{j}\right) \Rightarrow 4y = \left(\frac{x}{8}\right)\lambda$ and $4x = \left(\frac{2y}{9}\right)\lambda \Rightarrow \lambda = \frac{32y}{x}$ and $4x = \left(\frac{2y}{9}\right)\left(\frac{32y}{x}\right)$ $\Rightarrow y = \pm\frac{3}{4}x \Rightarrow \frac{x^2}{16} + \frac{\left(\frac{\pm 3}{4}x\right)^2}{9} = 1 \Rightarrow x^2 = 8 \Rightarrow x = \pm 2\sqrt{2}$. We use $x = 2\sqrt{2}$ since x represents distance. Then $y = \frac{3}{4}\left(2\sqrt{2}\right) = \frac{3\sqrt{2}}{2}$, so the length is $2x = 4\sqrt{2}$ and the width is $2y = 3\sqrt{2}$.

13. $\nabla f = 2x\mathbf{i} + 2y\mathbf{j}$ and $\nabla g = (2x-2)\mathbf{i} + (2y-4)\mathbf{j}$ so that $\nabla f = \lambda \nabla g = 2x\mathbf{i} + 2y\mathbf{j} = \lambda[(2x-2)\mathbf{i} + (2y-4)\mathbf{j}]$ $\Rightarrow 2x = \lambda(2x-2)$ and $2y = \lambda(2y-4) \Rightarrow x = \frac{\lambda}{\lambda-1}$ and $y = \frac{2\lambda}{\lambda-1}$, $\lambda \neq 1 \Rightarrow y = 2x \Rightarrow x^2 - 2x + (2x)^2 - 4(2x) = 0$ $\Rightarrow x = 0$ and $y = 0$, or $x = 2$ and $y = 4$. Therefore $f(0,0) = 0$ is the minimum value and $f(2,4) = 20$ is the maximum value. (Note that $\lambda = 1$ gives $2x = 2x - 2$ or $0 = -2$, which is impossible.)

15. $\nabla T = (8x - 4y)\mathbf{i} + (-4x + 2y)\mathbf{j}$ and $g(x,y) = x^2 + y^2 - 25 = 0 \Rightarrow \nabla g = 2x\mathbf{i} + 2y\mathbf{j}$ so that $\nabla T = \lambda \nabla g$ $\Rightarrow (8x - 4y)\mathbf{i} + (-4x + 2y)\mathbf{j} = \lambda(2x\mathbf{i} + 2y\mathbf{j}) \Rightarrow 8x - 4y = 2\lambda x$ and $-4x + 2y = 2\lambda y \Rightarrow y = \frac{-2x}{\lambda-1}$, $\lambda \neq 1$ $\Rightarrow 8x - 4\left(\frac{-2x}{\lambda-1}\right) = 2\lambda x \Rightarrow x = 0$, or $\lambda = 0$, or $\lambda = 5$.

CASE 1: $x = 0 \Rightarrow y = 0$; but $(0,0)$ is not on $x^2 + y^2 = 25$ so $x \neq 0$.

CASE 2: $\lambda = 0 \Rightarrow y = 2x \Rightarrow x^2 + (2x)^2 = 25 \Rightarrow x = \pm\sqrt{5}$ and $y = 2x$.

CASE 3: $\lambda = 5 \Rightarrow y = \frac{-2x}{4} = -\frac{x}{2} \Rightarrow x^2 + \left(-\frac{x}{2}\right)^2 = 25 \Rightarrow x = \pm 2\sqrt{5} \Rightarrow x = 2\sqrt{5}$ and $y = -\sqrt{5}$, or $x = -2\sqrt{5}$ and $y = \sqrt{5}$.

Therefore $T\left(\sqrt{5}, 2\sqrt{5}\right) = 0° = T\left(-\sqrt{5}, -2\sqrt{5}\right)$ is the minimum value and $T\left(2\sqrt{5}, -\sqrt{5}\right) = 125°$ $= T\left(-2\sqrt{5}, \sqrt{5}\right)$ is the maximum value. (Note: $\lambda = 1 \Rightarrow x = 0$ from the equation $-4x + 2y = 2\lambda y$; but we found $x \neq 0$ in CASE 1.)

17. Let $f(x,y,z) = (x-1)^2 + (y-1)^2 + (z-1)^2$ be the square of the distance from $(1,1,1)$. Then $\nabla f = 2(x-1)\mathbf{i} + 2(y-1)\mathbf{j} + 2(z-1)\mathbf{k}$ and $\nabla g = \mathbf{i} + 2\mathbf{j} + 3\mathbf{k}$ so that $\nabla f = \lambda \nabla g$ $\Rightarrow 2(x-1)\mathbf{i} + 2(y-1)\mathbf{j} + 2(z-1)\mathbf{k} = \lambda(\mathbf{i} + 2\mathbf{j} + 3\mathbf{k}) \Rightarrow 2(x-1) = \lambda$, $2(y-1) = 2\lambda$, $2(z-1) = 3\lambda$ $\Rightarrow 2(y-1) = 2[2(x-1)]$ and $2(z-1) = 3[2(x-1)] \Rightarrow x = \frac{y+1}{2} \Rightarrow z + 2 = 3\left(\frac{y+1}{2}\right)$ or $z = \frac{3y-1}{2}$; thus $\frac{y+1}{2} + 2y + 3\left(\frac{3y-1}{2}\right) - 13 = 0 \Rightarrow y = 2 \Rightarrow x = \frac{3}{2}$ and $z = \frac{5}{2}$. Therefore the point $\left(\frac{3}{2}, 2, \frac{5}{2}\right)$ is closest (since no point on the plane is farthest from the point $(1,1,1)$).

19. Let $f(x,y,z) = x^2 + y^2 + z^2$ be the square of the distance from the origin. Then $\nabla f = 2x\mathbf{i} + 2y\mathbf{j} + 2z\mathbf{k}$ and $\nabla g = 2x\mathbf{i} - 2y\mathbf{j} - 2z\mathbf{k}$ so that $\nabla f = \lambda \nabla g \Rightarrow 2x\mathbf{i} + 2y\mathbf{j} + 2z\mathbf{k} = \lambda(2x\mathbf{i} - 2y\mathbf{j} - 2z\mathbf{k}) \Rightarrow 2x = 2x\lambda$, $2y = 2y\lambda$, and $2z = -2z\lambda \Rightarrow x = 0$ or $\lambda = 1$.

CASE 1: $\lambda = 1 \Rightarrow 2y = -2y \Rightarrow y = 0$; $2z = -2z \Rightarrow z = 0 \Rightarrow x^2 - 1 = 0 \Rightarrow x^2 - 1 = 0 \Rightarrow x = \pm 1$ and $y = z = 0$.

CASE 2: $x = 0 \Rightarrow y^2 - z^2 = 1$, which has no solution.

Therefore the points on the unit circle $x^2 + y^2 = 1$, are the points on the surface $x^2 + y^2 - z^2 = 1$ closest to the origin. The minimum distance is 1.

21. Let $f(x,y,z) = x^2 + y^2 + z^2$ be the square of the distance to the origin. Then $\nabla f = 2x\mathbf{i} + 2y\mathbf{j} + 2z\mathbf{k}$ and $\nabla g = -y\mathbf{i} - x\mathbf{j} + 2z\mathbf{k}$ so that $\nabla f = \lambda \nabla g \Rightarrow 2x\mathbf{i} + 2y\mathbf{j} + 2z\mathbf{k} = \lambda(-y\mathbf{i} - x\mathbf{j} + 2z\mathbf{k}) \Rightarrow 2x = -y\lambda$, $2y = -x\lambda$, and $2z = 2z\lambda \Rightarrow \lambda = 1$ or $z = 0$.

CASE 1: $\lambda = 1 \Rightarrow 2x = -y$ and $2y = -x \Rightarrow y = 0$ and $x = 0 \Rightarrow z^2 - 4 = 0 \Rightarrow z = \pm 2$ and $x = y = 0$.

CASE 2: $z = 0 \Rightarrow -xy - 4 = 0 \Rightarrow y = -\frac{4}{x}$. Then $2x = \frac{4}{x}\lambda \Rightarrow \lambda = \frac{x^2}{2}$, and $-\frac{8}{x} = -x\lambda \Rightarrow -\frac{8}{x} = -x\left(\frac{x^2}{2}\right)$

$\Rightarrow x^4 = 16 \Rightarrow x = \pm 2$. Thus, $x = 2$ and $y = -2$, or $x = -2$ and $y = 2$.

Therefore we get four points: $(2, -2, 0)$, $(-2, 2, 0)$, $(0, 0, 2)$ and $(0, 0, -2)$. But the points $(0, 0, 2)$ and $(0, 0, -2)$ are closest to the origin since they are 2 units away and the others are $2\sqrt{2}$ units away.

23. $\nabla f = \mathbf{i} - 2\mathbf{j} + 5\mathbf{k}$ and $\nabla g = 2x\mathbf{i} + 2y\mathbf{j} + 2z\mathbf{k}$ so that $\nabla f = \lambda \nabla g \Rightarrow \mathbf{i} - 2\mathbf{j} + 5\mathbf{k} = \lambda(2x\mathbf{i} + 2y\mathbf{j} + 2z\mathbf{k}) \Rightarrow 1 = 2x\lambda$, $-2 = 2y\lambda$, and $5 = 2z\lambda \Rightarrow x = \frac{1}{2\lambda}$, $y = -\frac{1}{\lambda} = -2x$, and $z = \frac{5}{2\lambda} = 5x \Rightarrow x^2 + (-2x)^2 + (5x)^2 = 30 \Rightarrow x = \pm 1$. Thus, $x = 1, y = -2, z = 5$ or $x = -1, y = 2, z = -5$. Therefore $f(1, -2, 5) = 30$ is the maximum value and $f(-1, 2, -5) = -30$ is the minimum value.

25. $f(x, y, z) = x^2 + y^2 + z^2$ and $g(x, y, z) = x + y + z - 9 = 0 \Rightarrow \nabla f = 2x\mathbf{i} + 2y\mathbf{j} + 2z\mathbf{k}$ and $\nabla g = \mathbf{i} + \mathbf{j} + \mathbf{k}$ so that $\nabla f = \lambda \nabla g \Rightarrow 2x\mathbf{i} + 2y\mathbf{j} + 2z\mathbf{k} = \lambda(\mathbf{i} + \mathbf{j} + \mathbf{k}) \Rightarrow 2x = \lambda$, $2y = \lambda$, and $2z = \lambda \Rightarrow x = y = z \Rightarrow x + x + x - 9 = 0$ $\Rightarrow x = 3$, $y = 3$, and $z = 3$.

27. $V = xyz$ and $g(x, y, z) = x^2 + y^2 + z^2 - 1 = 0 \Rightarrow \nabla V = yz\mathbf{i} + xz\mathbf{j} + xy\mathbf{k}$ and $\nabla g = 2x\mathbf{i} + 2y\mathbf{j} + 2z\mathbf{k}$ so that $\nabla V = \lambda \nabla g \Rightarrow yz = \lambda x$, $xz = \lambda y$, and $xy = \lambda z \Rightarrow xyz = \lambda x^2$ and $xyz = \lambda y^2 \Rightarrow y = \pm x \Rightarrow z = \pm x$ $\Rightarrow x^2 + x^2 + x^2 = 1 \Rightarrow x = \frac{1}{\sqrt{3}}$ since $x > 0 \Rightarrow$ the dimensions of the box are $\frac{1}{\sqrt{3}}$ by $\frac{1}{\sqrt{3}}$ by $\frac{1}{\sqrt{3}}$ for maximum volume. (Note that there is no minimum volume since the box could be made arbitrarily thin.)

29. $\nabla T = 16x\mathbf{i} + 4z\mathbf{j} + (4y - 16)\mathbf{k}$ and $\nabla g = 8x\mathbf{i} + 2y\mathbf{j} + 8z\mathbf{k}$ so that $\nabla T = \lambda \nabla g \Rightarrow 16x\mathbf{i} + 4z\mathbf{j} + (4y - 16)\mathbf{k}$ $= \lambda(8x\mathbf{i} + 2y\mathbf{j} + 8z\mathbf{k}) \Rightarrow 16x = 8x\lambda$, $4z = 2y\lambda$, and $4y - 16 = 8z\lambda \Rightarrow \lambda = 2$ or $x = 0$.

CASE 1: $\lambda = 2 \Rightarrow 4z = 2y(2) \Rightarrow z = y$. Then $4z - 16 = 16z \Rightarrow z = -\frac{4}{3} \Rightarrow y = -\frac{4}{3}$. Then

$4x^2 + \left(-\frac{4}{3}\right)^2 + 4\left(-\frac{4}{3}\right)^2 = 16 \Rightarrow x = \pm\frac{4}{3}$.

CASE 2: $x = 0 \Rightarrow \lambda = \frac{2z}{y} \Rightarrow 4y - 16 = 8z\left(\frac{2z}{y}\right) \Rightarrow y^2 - 4y = 4z^2 \Rightarrow 4(0)^2 + y^2 + (y^2 - 4y) - 16 = 0$

$\Rightarrow y^2 - 2y - 8 = 0 \Rightarrow (y - 4)(y + 2) = 0 \Rightarrow y = 4$ or $y = -2$. Now $y = 4 \Rightarrow 4z^2 = 4^2 - 4(4)$

$\Rightarrow z = 0$ and $y = -2 \Rightarrow 4z^2 = (-2)^2 - 4(-2) \Rightarrow z = \pm\sqrt{3}$.

The temperatures are $T\left(\pm\frac{4}{3}, -\frac{4}{3}, -\frac{4}{3}\right) = 642\frac{2}{3}^{\circ}$, $T(0, 4, 0) = 600^{\circ}$, $T\left(0, -2, \sqrt{3}\right) = \left(600 - 24\sqrt{3}\right)^{\circ}$, and $T\left(0, -2, -\sqrt{3}\right) = \left(600 + 24\sqrt{3}\right)^{\circ} \approx 641.6^{\circ}$. Therefore $\left(\pm\frac{4}{3}, -\frac{4}{3}, -\frac{4}{3}\right)$ are the hottest points on the space probe.

31. Let $g_1(x, y, z) = 2x - y = 0$ and $g_2(x, y, z) = y + z = 0 \Rightarrow \nabla g_1 = 2\mathbf{i} - \mathbf{j}$, $\nabla g_2 = \mathbf{j} + \mathbf{k}$, and $\nabla f = 2x\mathbf{i} + 2\mathbf{j} - 2z\mathbf{k}$ so that $\nabla f = \lambda \nabla g_1 + \mu \nabla g_2 \Rightarrow 2x\mathbf{i} + 2\mathbf{j} - 2z\mathbf{k} = \lambda(2\mathbf{i} - \mathbf{j}) + \mu(\mathbf{j} + \mathbf{k}) \Rightarrow 2x\mathbf{i} + 2\mathbf{j} - 2z\mathbf{k} = 2\lambda\mathbf{i} + (\mu - \lambda)\mathbf{j} + \mu\mathbf{k}$ $\Rightarrow 2x = 2\lambda$, $2 = \mu - \lambda$, and $-2z = \mu \Rightarrow x = \lambda$. Then $2 = -2z - x \Rightarrow x = -2z - 2$ so that $2x - y = 0$ $\Rightarrow 2(-2z - 2) - y = 0 \Rightarrow -4z - 4 - y = 0$. This equation coupled with $y + z = 0$ implies $z = -\frac{4}{3}$ and $y = \frac{4}{3}$. Then $x = \frac{2}{3}$ so that $\left(\frac{2}{3}, \frac{4}{3}, -\frac{4}{3}\right)$ is the point that gives the maximum value $f\left(\frac{2}{3}, \frac{4}{3}, -\frac{4}{3}\right) = \left(\frac{2}{3}\right)^2 + 2\left(\frac{4}{3}\right) - \left(-\frac{4}{3}\right)^2 = \frac{4}{3}$.

33. Let $f(x, y, z) = x^2 + y^2 + z^2$ be the square of the distance from the origin. We want to minimize $f(x, y, z)$ subject to the constraints $g_1(x, y, z) = y + 2z - 12 = 0$ and $g_2(x, y, z) = x + y - 6 = 0$. Thus $\nabla f = 2x\mathbf{i} + 2y\mathbf{j} + 2z\mathbf{k}$, $\nabla g_1 = \mathbf{j} + 2\mathbf{k}$, and $\nabla g_2 = \mathbf{i} + \mathbf{j}$ so that $\nabla f = \lambda \nabla g_1 + \mu \nabla g_2 \Rightarrow 2x = \mu$, $2y = \lambda + \mu$, and $2z = 2\lambda$. Then $0 = y + 2z - 12 = \left(\frac{\lambda}{2} + \frac{\mu}{2}\right) + 2\lambda - 12 \Rightarrow \frac{5}{2}\lambda + \frac{1}{2}\mu = 12 \Rightarrow 5\lambda + \mu = 24$; $0 = x + y - 6 = \frac{\mu}{2} + \left(\frac{\lambda}{2} + \frac{\mu}{2}\right) - 6$ $\Rightarrow \frac{1}{2}\lambda + \mu = 6 \Rightarrow \lambda + 2\mu = 12$. Solving these two equations for λ and μ gives $\lambda = 4$ and $\mu = 4 \Rightarrow x = \frac{\mu}{2} = 2$, $y = \frac{\lambda + \mu}{2} = 4$, and $z = \lambda = 4$. The point $(2, 4, 4)$ on the line of intersection is closest to the origin. (There is no maximum distance from the origin since points on the line can be arbitrarily far away.)

35. Let $g_1(x,y,z) = z - 1 = 0$ and $g_2(x,y,z) = x^2 + y^2 + z^2 - 10 = 0 \Rightarrow \nabla g_1 = \mathbf{k}$, $\nabla g_2 = 2x\mathbf{i} + 2y\mathbf{j} + 2z\mathbf{k}$, and $\nabla f = 2xyz\mathbf{i} + x^2z\mathbf{j} + x^2y\mathbf{k}$ so that $\nabla f = \lambda \nabla g_1 + \mu \nabla g_2 \Rightarrow 2xyz\mathbf{i} + x^2z\mathbf{j} + x^2y\mathbf{k} = \lambda(\mathbf{k}) + \mu(2x\mathbf{i} + 2y\mathbf{j} + 2z\mathbf{k})$ $\Rightarrow 2xyz = 2x\mu$, $x^2z = 2y\mu$, and $x^2y = 2z\mu + \lambda \Rightarrow xyz = x\mu \Rightarrow x = 0$ or $yz = \mu \Rightarrow \mu = y$ since $z = 1$.

CASE 1: $x = 0$ and $z = 1 \Rightarrow y^2 - 9 = 0$ (from g_2) $\Rightarrow y = \pm 3$ yielding the points $(0, \pm 3, 1)$.

CASE 2: $\mu = y \Rightarrow x^2z = 2y^2 \Rightarrow x^2 = 2y^2$ (since $z = 1$) $\Rightarrow 2y^2 + y^2 + 1 - 10 = 0$ (from g_2) $\Rightarrow 3y^2 - 9 = 0$ $\Rightarrow y = \pm\sqrt{3} \Rightarrow x^2 = 2\left(\pm\sqrt{3}\right)^2 \Rightarrow x = \pm\sqrt{6}$ yielding the points $\left(\pm\sqrt{6}, \pm\sqrt{3}, 1\right)$.

Now $f(0, \pm 3, 1) = 1$ and $f\left(\pm\sqrt{6}, \pm\sqrt{3}, 1\right) = 6\left(\pm\sqrt{3}\right) + 1 = 1 \pm 6\sqrt{3}$. Therefore the maximum of f is $1 + 6\sqrt{3}$ at $\left(\pm\sqrt{6}, \sqrt{3}, 1\right)$, and the minimum of f is $1 - 6\sqrt{3}$ at $\left(\pm\sqrt{6}, -\sqrt{3}, 1\right)$.

37. Let $g_1(x,y,z) = y - x = 0$ and $g_2(x,y,z) = x^2 + y^2 + z^2 - 4 = 0$. Then $\nabla f = y\mathbf{i} + x\mathbf{j} + 2z\mathbf{k}$, $\nabla g_1 = -\mathbf{i} + \mathbf{j}$, and $\nabla g_2 = 2x\mathbf{i} + 2y\mathbf{j} + 2z\mathbf{k}$ so that $\nabla f = \lambda \nabla g_1 + \mu \nabla g_2 \Rightarrow y\mathbf{i} + x\mathbf{j} + 2z\mathbf{k} = \lambda(-\mathbf{i} + \mathbf{j}) + \mu(2x\mathbf{i} + 2y\mathbf{j} + 2z\mathbf{k})$ $\Rightarrow y = -\lambda + 2x\mu$, $x = \lambda + 2y\mu$, and $2z = 2z\mu \Rightarrow z = 0$ or $\mu = 1$.

CASE 1: $z = 0 \Rightarrow x^2 + y^2 - 4 = 0 \Rightarrow 2x^2 - 4 = 0$ (since $x = y$) $\Rightarrow x = \pm\sqrt{2}$ and $y = \pm\sqrt{2}$ yielding the points $\left(\pm\sqrt{2}, \pm\sqrt{2}, 0\right)$.

CASE 2: $\mu = 1 \Rightarrow y = -\lambda + 2x$ and $x = \lambda + 2y \Rightarrow x + y = 2(x + y) \Rightarrow 2x = 2(2x)$ since $x = y \Rightarrow x = 0 \Rightarrow y = 0$ $\Rightarrow z^2 - 4 = 0 \Rightarrow z = \pm 2$ yielding the points $(0, 0, \pm 2)$.

Now, $f(0, 0, \pm 2) = 4$ and $f\left(\pm\sqrt{2}, \pm\sqrt{2}, 0\right) = 2$. Therefore the maximum value of f is 4 at $(0, 0, \pm 2)$ and the minimum value of f is 2 at $\left(\pm\sqrt{2}, \pm\sqrt{2}, 0\right)$.

39. $\nabla f = \mathbf{i} + \mathbf{j}$ and $\nabla g = y\mathbf{i} + x\mathbf{j}$ so that $\nabla f = \lambda \nabla g \Rightarrow \mathbf{i} + \mathbf{j} = \lambda(y\mathbf{i} + x\mathbf{j}) \Rightarrow 1 = y\lambda$ and $1 = x\lambda \Rightarrow y = x$ $\Rightarrow y^2 = 16 \Rightarrow y = \pm 4 \Rightarrow (4, 4)$ and $(-4, -4)$ are candidates for the location of extreme values. But as $x \to \infty$, $y \to \infty$ and $f(x,y) \to \infty$; as $x \to -\infty$, $y \to 0$ and $f(x,y) \to -\infty$. Therefore no maximum or minimum value exists subject to the constraint.

41. (a) Maximize $f(a,b,c) = a^2b^2c^2$ subject to $a^2 + b^2 + c^2 = r^2$. Thus $\nabla f = 2ab^2c^2\mathbf{i} + 2a^2bc^2\mathbf{j} + 2a^2b^2c\mathbf{k}$ and $\nabla g = 2a\mathbf{i} + 2b\mathbf{j} + 2c\mathbf{k}$ so that $\nabla f = \lambda \nabla g \Rightarrow 2ab^2c^2 = 2a\lambda$, $2a^2bc^2 = 2b\lambda$, and $2a^2b^2c = 2c\lambda$ $\Rightarrow 2a^2b^2c^2 = 2a^2\lambda = 2b^2\lambda = 2c^2\lambda \Rightarrow \lambda = 0$ or $a^2 = b^2 = c^2$.

CASE 1: $\lambda = 0 \Rightarrow a^2b^2c^2 = 0$.

CASE 2: $a^2 = b^2 = c^2 \Rightarrow f(a,b,c) = a^2a^2a^2$ and $3a^2 = r^2 \Rightarrow f(a,b,c) = \left(\frac{r^2}{3}\right)^3$ is the maximum value.

(b) The point $\left(\sqrt{a}, \sqrt{b}, \sqrt{c}\right)$ is on the sphere if $a + b + c = r^2$. Moreover, by part (a), $abc = f\left(\sqrt{a}, \sqrt{b}, \sqrt{c}\right)$ $\le \left(\frac{r^2}{3}\right)^3 \Rightarrow (abc)^{1/3} \le \frac{r^2}{3} = \frac{a+b+c}{3}$, as claimed.

CHAPTER 11 PRACTICE AND ADDITIONAL EXERCISES

1. Domain: All points in the xy-plane

Range: $z \ge 0$

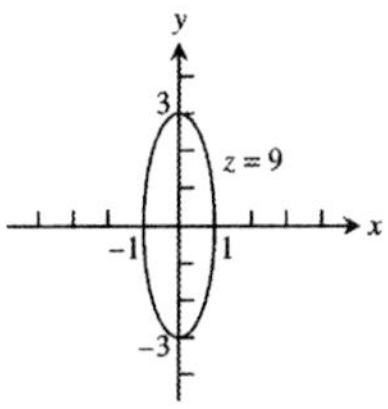

Level curves are ellipses with major axis along the y-axis and minor axis along the x-axis.

3. Domain: All (x, y) such that $x \neq 0$ and $y \neq 0$
Range: $z \neq 0$

Level curves are hyperbolas with the x- and y-axes as asymptotes.

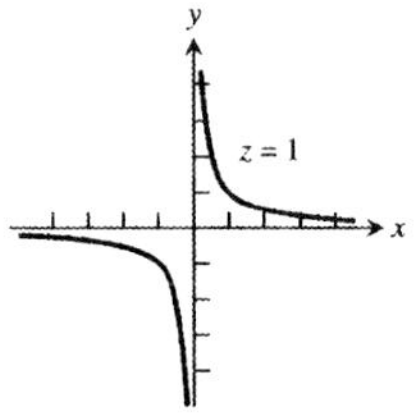

5. Domain: All points (x, y, z) in space
Range: All real numbers

Level surfaces are paraboloids of revolution with the z-axis as axis.

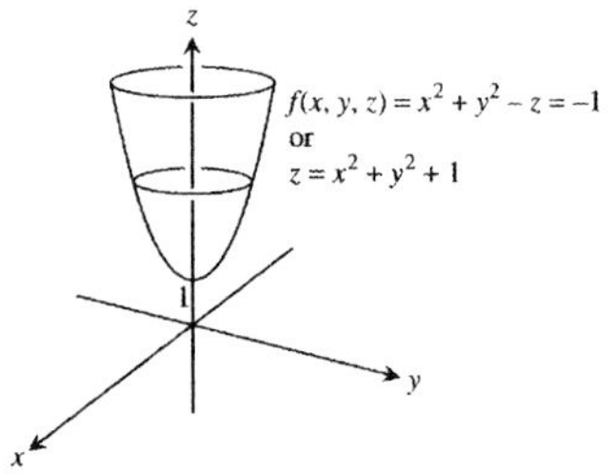

7. Domain: All (x, y, z) such that $(x, y, z) \neq (0, 0, 0)$
Range: Positive real numbers

Level surfaces are spheres with center $(0, 0, 0)$ and radius $r > 0$.

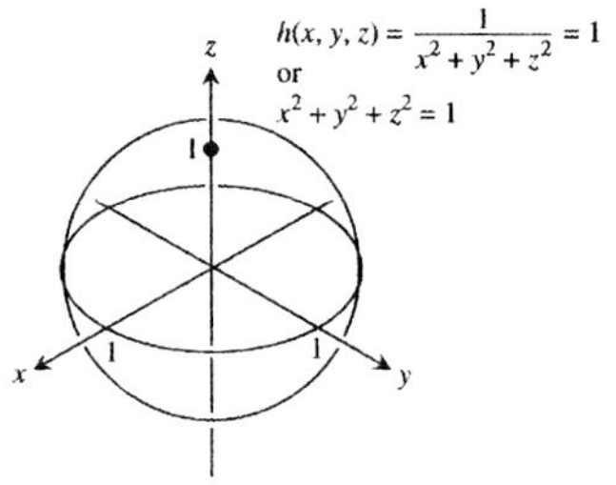

9. $\lim\limits_{(x,y) \to (\pi, \ln 2)} e^y \cos x = e^{\ln 2} \cos \pi = (2)(-1) = -2$

11. $\lim\limits_{\substack{(x,y) \to (1,1) \\ x \neq \pm y}} \frac{x-y}{x^2-y^2} = \lim\limits_{\substack{(x,y) \to (1,1) \\ x \neq \pm y}} \frac{x-y}{(x-y)(x+y)} = \lim\limits_{(x,y) \to (1,1)} \frac{1}{x+y} = \frac{1}{1+1} = \frac{1}{2}$

13. $\lim\limits_{P \to (1,-1,e)} \ln|x+y+z| = \ln|1+(-1)+e| = \ln e = 1$

15. Let $y = kx^2$, $k \neq 1$. Then $\lim\limits_{\substack{(x,y) \to (0,0) \\ y \neq x^2}} \frac{y}{x^2-y} = \lim\limits_{(x, kx^2) \to (0,0)} \frac{kx^2}{x^2 - kx^2} = \frac{k}{1-k^2}$ which gives different limits for different values of k $\Rightarrow$ the limit does not exist.

17. $\lim\limits_{(x,y) \to (0,0)} \frac{x^3-y^3}{x^2+y^2} = \lim\limits_{r \to 0} \frac{r^3 \cos^3 \theta - r^3 \sin^3 \theta}{r^2 \cos^2 \theta + r^2 \sin^2 \theta} = \lim\limits_{r \to 0} \frac{r(\cos^3 \theta - \sin^3 \theta)}{1} = 0$

19. $\frac{\partial g}{\partial r} = \cos\theta + \sin\theta$, $\frac{\partial g}{\partial \theta} = -r\sin\theta + r\cos\theta$

21. $\frac{\partial f}{\partial R_1} = -\frac{1}{R_1^2}$, $\frac{\partial f}{\partial R_2} = -\frac{1}{R_2^2}$, $\frac{\partial f}{\partial R_3} = -\frac{1}{R_3^2}$

23. $\frac{\partial P}{\partial n} = \frac{RT}{V}$, $\frac{\partial P}{\partial R} = \frac{nT}{V}$, $\frac{\partial P}{\partial T} = \frac{nR}{V}$, $\frac{\partial P}{\partial V} = -\frac{nRT}{V^2}$

25. $\frac{\partial g}{\partial x} = \frac{1}{y}$, $\frac{\partial g}{\partial y} = 1 - \frac{x}{y^2} \Rightarrow \frac{\partial^2 g}{\partial x^2} = 0$, $\frac{\partial^2 g}{\partial y^2} = \frac{2x}{y^3}$, $\frac{\partial^2 g}{\partial y \partial x} = \frac{\partial^2 g}{\partial x \partial y} = -\frac{1}{y^2}$

27. $\frac{\partial f}{\partial x} = 1 + y - 15x^2 + \frac{2x}{x^2+1}$, $\frac{\partial f}{\partial y} = x \Rightarrow \frac{\partial^2 f}{\partial x^2} = -30x + \frac{2-2x^2}{(x^2+1)^2}$, $\frac{\partial^2 f}{\partial y^2} = 0$, $\frac{\partial^2 f}{\partial y \partial x} = \frac{\partial^2 f}{\partial x \partial y} = 1$

29. $\frac{\partial w}{\partial x} = y\cos(xy+\pi)$, $\frac{\partial w}{\partial y} = x\cos(xy+\pi)$, $\frac{dx}{dt} = e^t$, $\frac{dy}{dt} = \frac{1}{t+1}$
$\Rightarrow \frac{dw}{dt} = [y\cos(xy+\pi)]e^t + [x\cos(xy+\pi)]\left(\frac{1}{t+1}\right)$; $t = 0 \Rightarrow x = 1$ and $y = 0$
$\Rightarrow \left.\frac{dw}{dt}\right|_{t=0} = 0\cdot 1 + [1\cdot(-1)]\left(\frac{1}{0+1}\right) = -1$

31. $\frac{\partial w}{\partial x} = 2\cos(2x-y)$, $\frac{\partial w}{\partial y} = -\cos(2x-y)$, $\frac{\partial x}{\partial r} = 1$, $\frac{\partial x}{\partial s} = \cos s$, $\frac{\partial y}{\partial r} = s$, $\frac{\partial y}{\partial s} = r$
$\Rightarrow \frac{\partial w}{\partial r} = [2\cos(2x-y)](1) + [-\cos(2x-y)](s)$; $r = \pi$ and $s = 0 \Rightarrow x = \pi$ and $y = 0$
$\Rightarrow \left.\frac{\partial w}{\partial r}\right|_{(\pi,0)} = (2\cos 2\pi) - (\cos 2\pi)(0) = 2$; $\frac{\partial w}{\partial s} = [2\cos(2x-y)](\cos s) + [-\cos(2x-y)](r)$
$\Rightarrow \left.\frac{\partial w}{\partial s}\right|_{(\pi,0)} = (2\cos 2\pi)(\cos 0) - (\cos 2\pi)(\pi) = 2 - \pi$

33. $\frac{\partial f}{\partial x} = y + z$, $\frac{\partial f}{\partial y} = x + z$, $\frac{\partial f}{\partial z} = y + x$, $\frac{dx}{dt} = -\sin t$, $\frac{dy}{dt} = \cos t$, $\frac{dz}{dt} = -2\sin 2t$
$\Rightarrow \frac{df}{dt} = -(y+z)(\sin t) + (x+z)(\cos t) - 2(y+x)(\sin 2t)$; $t = 1 \Rightarrow x = \cos 1$, $y = \sin 1$, and $z = \cos 2$
$\Rightarrow \left.\frac{df}{dt}\right|_{t=1} = -(\sin 1 + \cos 2)(\sin 1) + (\cos 1 + \cos 2)(\cos 1) - 2(\sin 1 + \cos 1)(\sin 2)$

35. $F(x,y) = 1 - x - y^2 - \sin xy \Rightarrow F_x = -1 - y\cos xy$ and $F_y = -2y - x\cos xy \Rightarrow \frac{dy}{dx} = -\frac{F_x}{F_y} = -\frac{-1-y\cos xy}{-2y - x\cos xy}$
$= \frac{1 + y\cos xy}{-2y - x\cos xy} \Rightarrow$ at $(x,y) = (0,1)$ we have $\left.\frac{dy}{dx}\right|_{(0,1)} = \frac{1+1}{-2} = -1$

37. $\nabla f = (-\sin x\cos y)\mathbf{i} - (\cos x\sin y)\mathbf{j} \Rightarrow \nabla f|_{(\frac{\pi}{4},\frac{\pi}{4})} = -\frac{1}{2}\mathbf{i} - \frac{1}{2}\mathbf{j} \Rightarrow |\nabla f| = \sqrt{\left(-\frac{1}{2}\right)^2 + \left(-\frac{1}{2}\right)^2} = \frac{1}{\sqrt{2}} = \frac{\sqrt{2}}{2}$;
$\mathbf{u} = \frac{\nabla f}{|\nabla f|} = -\frac{\sqrt{2}}{2}\mathbf{i} - \frac{\sqrt{2}}{2}\mathbf{j} \Rightarrow$ f increases most rapidly in the direction $\mathbf{u} = -\frac{\sqrt{2}}{2}\mathbf{i} - \frac{\sqrt{2}}{2}\mathbf{j}$ and decreases most rapidly in the direction $-\mathbf{u} = \frac{\sqrt{2}}{2}\mathbf{i} + \frac{\sqrt{2}}{2}\mathbf{j}$; $(D_{\mathbf{u}}f)_{P_0} = |\nabla f| = \frac{\sqrt{2}}{2}$ and $(D_{-\mathbf{u}}f)_{P_0} = -\frac{\sqrt{2}}{2}$;
$\mathbf{u}_1 = \frac{\mathbf{v}}{|\mathbf{v}|} = \frac{3\mathbf{i}+4\mathbf{j}}{\sqrt{3^2+4^2}} = \frac{3}{5}\mathbf{i} + \frac{4}{5}\mathbf{j} \Rightarrow (D_{\mathbf{u}_1}f)_{P_0} = \nabla f\cdot\mathbf{u}_1 = \left(-\frac{1}{2}\right)\left(\frac{3}{5}\right) + \left(-\frac{1}{2}\right)\left(\frac{4}{5}\right) = -\frac{7}{10}$

39. $\nabla f = \left(\frac{2}{2x+3y+6z}\right)\mathbf{i} + \left(\frac{3}{2x+3y+6z}\right)\mathbf{j} + \left(\frac{6}{2x+3y+6z}\right)\mathbf{k} \Rightarrow \nabla f|_{(-1,-1,1)} = 2\mathbf{i} + 3\mathbf{j} + 6\mathbf{k}$;
$\mathbf{u} = \frac{\nabla f}{|\nabla f|} = \frac{2\mathbf{i}+3\mathbf{j}+6\mathbf{k}}{\sqrt{2^2+3^2+6^2}} = \frac{2}{7}\mathbf{i} + \frac{3}{7}\mathbf{j} + \frac{6}{7}\mathbf{k} \Rightarrow$ f increases most rapidly in the direction $\mathbf{u} = \frac{2}{7}\mathbf{i} + \frac{3}{7}\mathbf{j} + \frac{6}{7}\mathbf{k}$ and decreases most rapidly in the direction $-\mathbf{u} = -\frac{2}{7}\mathbf{i} - \frac{3}{7}\mathbf{j} - \frac{6}{7}\mathbf{k}$; $(D_{\mathbf{u}}f)_{P_0} = |\nabla f| = 7$, $(D_{-\mathbf{u}}f)_{P_0} = -7$;
$\mathbf{u}_1 = \frac{\mathbf{v}}{|\mathbf{v}|} = \frac{2}{7}\mathbf{i} + \frac{3}{7}\mathbf{j} + \frac{6}{7}\mathbf{k} \Rightarrow (D_{\mathbf{u}_1}f)_{P_0} = (D_{\mathbf{u}}f)_{P_0} = 7$

41. $\mathbf{r} = (\cos 3t)\mathbf{i} + (\sin 3t)\mathbf{j} + 3t\mathbf{k} \Rightarrow \mathbf{v}(t) = (-3\sin 3t)\mathbf{i} + (3\cos 3t)\mathbf{j} + 3\mathbf{k} \Rightarrow \mathbf{v}\left(\frac{\pi}{3}\right) = -3\mathbf{j} + 3\mathbf{k}$
$\Rightarrow \mathbf{u} = -\frac{1}{\sqrt{2}}\mathbf{j} + \frac{1}{\sqrt{2}}\mathbf{k}$; $f(x,y,z) = xyz \Rightarrow \nabla f = yz\mathbf{i} + xz\mathbf{j} + xy\mathbf{k}$; $t = \frac{\pi}{3}$ yields the point on the helix $(-1,0,\pi)$
$\Rightarrow \nabla f|_{(-1,0,\pi)} = -\pi\mathbf{j} \Rightarrow \nabla f\cdot\mathbf{u} = (-\pi\mathbf{j})\cdot\left(-\frac{1}{\sqrt{2}}\mathbf{j} + \frac{1}{\sqrt{2}}\mathbf{k}\right) = \frac{\pi}{\sqrt{2}}$

43. (a) Let $\nabla f = a\mathbf{i} + b\mathbf{j}$ at $(1,2)$. The direction toward $(2,2)$ is determined by $\mathbf{v}_1 = (2-1)\mathbf{i} + (2-2)\mathbf{j} = \mathbf{i} = \mathbf{u}$ so that $\nabla f\cdot\mathbf{u} = 2 \Rightarrow a = 2$. The direction toward $(1,1)$ is determined by $\mathbf{v}_2 = (1-1)\mathbf{i} + (1-2)\mathbf{j} = -\mathbf{j} = \mathbf{u}$ so that $\nabla f\cdot\mathbf{u} = -2 \Rightarrow -b = -2 \Rightarrow b = 2$. Therefore $\nabla f = 2\mathbf{i} + 2\mathbf{j}$; $f_x(1,2) = f_y(1,2) = 2$.
(b) The direction toward $(4,6)$ is determined by $\mathbf{v}_3 = (4-1)\mathbf{i} + (6-2)\mathbf{j} = 3\mathbf{i} + 4\mathbf{j} \Rightarrow \mathbf{u} = \frac{3}{5}\mathbf{i} + \frac{4}{5}\mathbf{j}$
$\Rightarrow \nabla f\cdot\mathbf{u} = \frac{14}{5}$.

45. $\nabla f = 2x\mathbf{i} + \mathbf{j} + 2z\mathbf{k} \Rightarrow$
$\nabla f|_{(0,-1,-1)} = \mathbf{j} - 2\mathbf{k}$,
$\nabla f|_{(0,0,0)} = \mathbf{j}$,
$\nabla f|_{(0,-1,1)} = \mathbf{j} + 2\mathbf{k}$

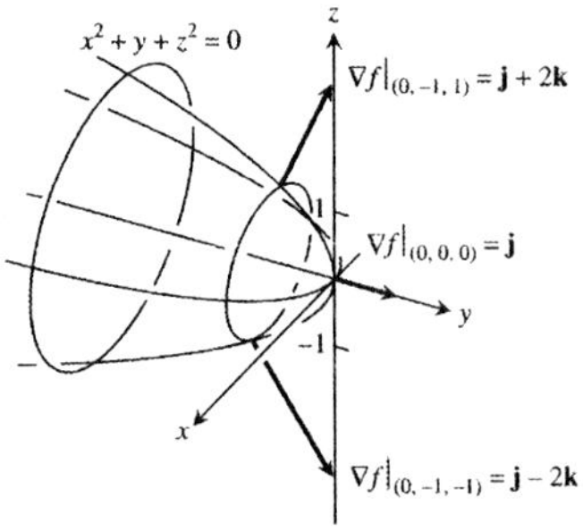

47. $\nabla f = 2x\mathbf{i} - \mathbf{j} - 5\mathbf{k} \Rightarrow \nabla f|_{(2,-1,1)} = 4\mathbf{i} - \mathbf{j} - 5\mathbf{k} \Rightarrow$ Tangent Plane: $4(x-2) - (y+1) - 5(z-1) = 0$ $\Rightarrow 4x - y - 5z = 4$; Normal Line: $x = 2 + 4t$, $y = -1 - t$, $z = 1 - 5t$

49. $\frac{\partial z}{\partial x} = \frac{2x}{x^2+y^2} \Rightarrow \left.\frac{\partial z}{\partial x}\right|_{(0,1,0)} = 0$ and $\frac{\partial z}{\partial y} = \frac{2y}{x^2+y^2} \Rightarrow \left.\frac{\partial z}{\partial y}\right|_{(0,1,0)} = 2$; thus the tangent plane is $2(y-1) - (z-0) = 0$ or $2y - z - 2 = 0$

51. $\nabla f = (-\cos x)\mathbf{i} + \mathbf{j} \Rightarrow \nabla f|_{(\pi,1)} = \mathbf{i} + \mathbf{j} \Rightarrow$ the tangent line is $(x - \pi) + (y - 1) = 0 \Rightarrow x + y = \pi + 1$; the normal line is $y - 1 = 1(x - \pi) \Rightarrow y = x - \pi + 1$

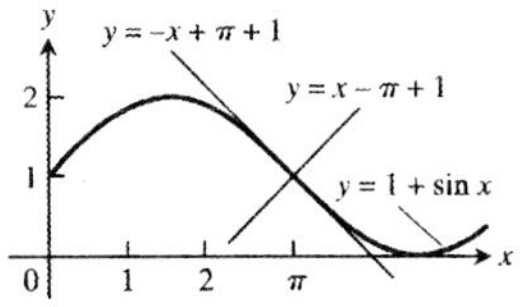

53. Let $f(x,y,z) = x^2 + 2y + 2z - 4$ and $g(x,y,z) = y - 1$. Then $\nabla f = 2x\mathbf{i} + 2\mathbf{j} + 2\mathbf{k}|_{(1,1,\frac{1}{2})} = 2\mathbf{i} + 2\mathbf{j} + 2\mathbf{k}$ and $\nabla g = \mathbf{j} \Rightarrow \nabla f \times \nabla g = \begin{vmatrix} \mathbf{i} & \mathbf{j} & \mathbf{k} \\ 2 & 2 & 2 \\ 0 & 1 & 0 \end{vmatrix} = -2\mathbf{i} + 2\mathbf{k} \Rightarrow$ the line is $x = 1 - 2t$, $y = 1$, $z = \frac{1}{2} + 2t$

55. $f\left(\frac{\pi}{4}, \frac{\pi}{4}\right) = \frac{1}{2}$, $f_x\left(\frac{\pi}{4}, \frac{\pi}{4}\right) = \cos x \cos y|_{(\pi/4,\pi/4)} = \frac{1}{2}$, $f_y\left(\frac{\pi}{4}, \frac{\pi}{4}\right) = -\sin x \sin y|_{(\pi/4,\pi/4)} = -\frac{1}{2}$ $\Rightarrow L(x,y) = \frac{1}{2} + \frac{1}{2}\left(x - \frac{\pi}{4}\right) - \frac{1}{2}\left(y - \frac{\pi}{4}\right) = \frac{1}{2} + \frac{1}{2}x - \frac{1}{2}y$; $f_{xx}(x,y) = -\sin x \cos y$, $f_{yy}(x,y) = -\sin x \cos y$, and $f_{xy}(x,y) = -\cos x \sin y$. Thus an upper bound for E depends on the bound M used for $|f_{xx}|$, $|f_{xy}|$, and $|f_{yy}|$. With $M = \frac{\sqrt{2}}{2}$ we have $|E(x,y)| \le \frac{1}{2}\left(\frac{\sqrt{2}}{2}\right)\left(\left|x - \frac{\pi}{4}\right| + \left|y - \frac{\pi}{4}\right|\right)^2 \le \frac{\sqrt{2}}{4}(0.2)^2 \le 0.0142$; with $M = 1$, $|E(x,y)| \le \frac{1}{2}(1)\left(\left|x - \frac{\pi}{4}\right| + \left|y - \frac{\pi}{4}\right|\right)^2 = \frac{1}{2}(0.2)^2 = 0.02$.

57. $f(1,0,0) = 0$, $f_x(1,0,0) = y - 3z|_{(1,0,0)} = 0$, $f_y(1,0,0) = x + 2z|_{(1,0,0)} = 1$, $f_z(1,0,0) = 2y - 3x|_{(1,0,0)} = -3$ $\Rightarrow L(x,y,z) = 0(x-1) + (y-0) - 3(z-0) = y - 3z$; $f(1,1,0) = 1$, $f_x(1,1,0) = 1$, $f_y(1,1,0) = 1$, $f_z(1,1,0) = -1$ $\Rightarrow L(x,y,z) = 1 + (x-1) + (y-1) - 1(z-0) = x + y - z - 1$

59. $f_x(x,y) = 2x - y + 2 = 0$ and $f_y(x,y) = -x + 2y + 2 = 0 \Rightarrow x = -2$ and $y = -2 \Rightarrow (-2,-2)$ is the critical point; $f_{xx}(-2,-2) = 2$, $f_{yy}(-2,-2) = 2$, $f_{xy}(-2,-2) = -1 \Rightarrow f_{xx}f_{yy} - f_{xy}^2 = 3 > 0$ and $f_{xx} > 0 \Rightarrow$ local minimum value of $f(-2,-2) = -8$

61. $f_x(x,y) = 6x^2 + 3y = 0$ and $f_y(x,y) = 3x + 6y^2 = 0 \Rightarrow y = -2x^2$ and $3x + 6(4x^4) = 0 \Rightarrow x(1 + 8x^3) = 0$ $\Rightarrow x = 0$ and $y = 0$, or $x = -\frac{1}{2}$ and $y = -\frac{1}{2} \Rightarrow$ the critical points are $(0,0)$ and $\left(-\frac{1}{2}, -\frac{1}{2}\right)$. For $(0,0)$: $f_{xx}(0,0) = 12x|_{(0,0)} = 0$, $f_{yy}(0,0) = 12y|_{(0,0)} = 0$, $f_{xy}(0,0) = 3 \Rightarrow f_{xx}f_{yy} - f_{xy}^2 = -9 < 0 \Rightarrow$ saddle point with $f(0,0) = 0$. For $\left(-\frac{1}{2}, -\frac{1}{2}\right)$: $f_{xx} = -6$, $f_{yy} = -6$, $f_{xy} = 3 \Rightarrow f_{xx}f_{yy} - f_{xy}^2 = 27 > 0$ and $f_{xx} < 0 \Rightarrow$ local maximum value of $f\left(-\frac{1}{2}, -\frac{1}{2}\right) = \frac{1}{4}$

63. $f_x(x,y) = 3x^2 + 6x = 0$ and $f_y(x,y) = 3y^2 - 6y = 0 \Rightarrow x(x+2) = 0$ and $y(y-2) = 0 \Rightarrow x = 0$ or $x = -2$ and $y = 0$ or $y = 2 \Rightarrow$ the critical points are $(0,0)$, $(0,2)$, $(-2,0)$, and $(-2,2)$. For $(0,0)$: $f_{xx}(0,0) = 6x + 6|_{(0,0)} = 6$, $f_{yy}(0,0) = 6y - 6|_{(0,0)} = -6$, $f_{xy}(0,0) = 0 \Rightarrow f_{xx}f_{yy} - f_{xy}^2 = -36 < 0 \Rightarrow$ saddle point with $f(0,0) = 0$. For $(0,2)$: $f_{xx}(0,2) = 6$, $f_{yy}(0,2) = 6$, $f_{xy}(0,2) = 0 \Rightarrow f_{xx}f_{yy} - f_{xy}^2 = 36 > 0$ and $f_{xx} > 0 \Rightarrow$ local minimum value of $f(0,2) = -4$. For $(-2,0)$: $f_{xx}(-2,0) = -6$, $f_{yy}(-2,0) = -6$, $f_{xy}(-2,0) = 0 \Rightarrow f_{xx}f_{yy} - f_{xy}^2 = 36 > 0$ and $f_{xx} < 0 \Rightarrow$ local maximum value of $f(-2,0) = 4$. For $(-2,2)$: $f_{xx}(-2,2) = -6$, $f_{yy}(-2,2) = 6$, $f_{xy}(-2,2) = 0 \Rightarrow f_{xx}f_{yy} - f_{xy}^2 = -36 < 0 \Rightarrow$ saddle point with $f(-2,2) = 0$.

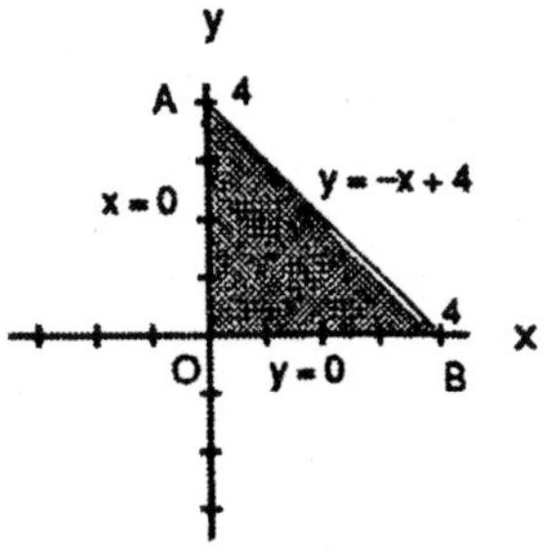

65. (i) On OA, $f(x,y) = f(0,y) = y^2 + 3y$ for $0 \le y \le 4 \Rightarrow f'(0,y) = 2y + 3 = 0 \Rightarrow y = -\frac{3}{2}$. But $\left(0, -\frac{3}{2}\right)$ is not in the region.
Endpoints: $f(0,0) = 0$ and $f(0,4) = 28$.
(ii) On AB, $f(x,y) = f(x, -x+4) = x^2 - 10x + 28$ for $0 \le x \le 4 \Rightarrow f'(x, -x+4) = 2x - 10 = 0 \Rightarrow x = 5, y = -1$. But $(5,-1)$ is not in the region.
Endpoints: $f(4,0) = 4$ and $f(0,4) = 28$.
(iii) On OB, $f(x,y) = f(x,0) = x^2 - 3x$ for $0 \le x \le 4 \Rightarrow f'(x,0) = 2x - 3 \Rightarrow x = \frac{3}{2}$ and $y = 0 \Rightarrow \left(\frac{3}{2}, 0\right)$ is a critical point with $f\left(\frac{3}{2}, 0\right) = -\frac{9}{4}$.
Endpoints: $f(0,0) = 0$ and $f(4,0) = 4$.
(iv) For the interior of the triangular region, $f_x(x,y) = 2x + y - 3 = 0$ and $f_y(x,y) = x + 2y + 3 = 0 \Rightarrow x = 3$ and $y = -3$. But $(3,-3)$ is not in the region. Therefore the absolute maximum is 28 at $(0,4)$ and the absolute minimum is $-\frac{9}{4}$ at $\left(\frac{3}{2}, 0\right)$.

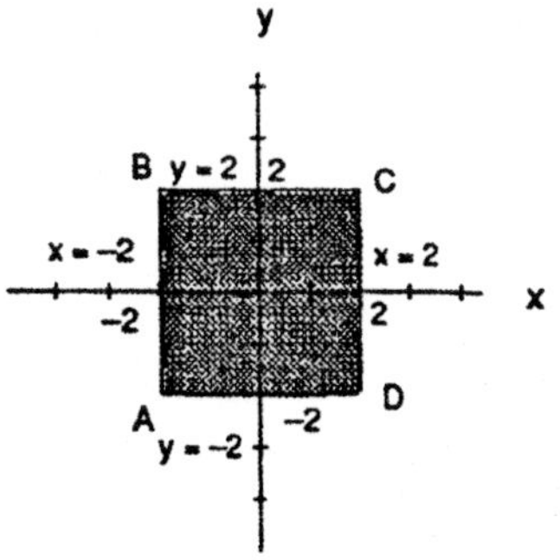

67. (i) On AB, $f(x,y) = f(-2,y) = y^2 - y - 4$ for $-2 \le y \le 2 \Rightarrow f'(-2,y) = 2y - 1 \Rightarrow y = \frac{1}{2}$ and $x = -2 \Rightarrow \left(-2, \frac{1}{2}\right)$ is an interior critical point in AB with $f\left(-2, \frac{1}{2}\right) = -\frac{17}{4}$. Endpoints: $f(-2,-2) = 2$ and $f(2,2) = -2$.
(ii) On BC, $f(x,y) = f(x,2) = -2$ for $-2 \le x \le 2 \Rightarrow f'(x,2) = 0 \Rightarrow$ no critical points in the interior of BC. Endpoints: $f(-2,2) = -2$ and $f(2,2) = -2$.
(iii) On CD, $f(x,y) = f(2,y) = y^2 - 5y + 4$ for $-2 \le y \le 2 \Rightarrow f'(2,y) = 2y - 5 = 0 \Rightarrow y = \frac{5}{2}$ and $x = 2$. But $\left(2, \frac{5}{2}\right)$ is not in the region.
Endpoints: $f(2,-2) = 18$ and $f(2,2) = -2$.
(iv) On AD, $f(x,y) = f(x,-2) = 4x + 10$ for $-2 \le x \le 2 \Rightarrow f'(x,-2) = 4 \Rightarrow$ no critical points in the interior of AD. Endpoints: $f(-2,-2) = 2$ and $f(2,-2) = 18$.
(v) For the interior of the square, $f_x(x,y) = -y + 2 = 0$ and $f_y(x,y) = 2y - x - 3 = 0 \Rightarrow y = 2$ and $x = 1 \Rightarrow (1,2)$ is an interior critical point of the square with $f(1,2) = -2$. Therefore the absolute maximum is 18 at $(2,-2)$ and the absolute minimum is $-\frac{17}{4}$ at $\left(-2, \frac{1}{2}\right)$.

69. (i) On AB, $f(x, y) = f(x, x + 2) = -2x + 4$ for $-2 \le x \le 2 \Rightarrow f'(x, x + 2) = -2 = 0 \Rightarrow$ no critical points in the interior of AB. Endpoints: $f(-2, 0) = 8$ and $f(2, 4) = 0$.

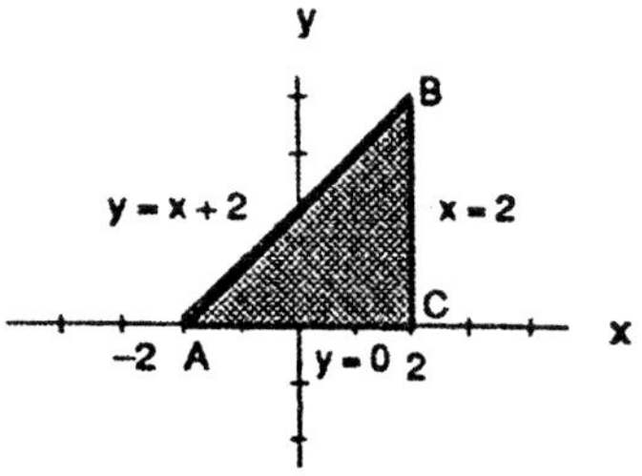

(ii) On BC, $f(x, y) = f(2, y) = -y^2 + 4y$ for $0 \le y \le 4$ $\Rightarrow f'(2, y) = -2y + 4 = 0 \Rightarrow y = 2$ and $x = 2$ $\Rightarrow (2, 2)$ is an interior critical point of BC with $f(2, 2) = 4$. Endpoints: $f(2, 0) = 0$ and $f(2, 4) = 0$.

(iii) On AC, $f(x, y) = f(x, 0) = x^2 - 2x$ for $-2 \le x \le 2$ $\Rightarrow f'(x, 0) = 2x - 2 \Rightarrow x = 1$ and $y = 0 \Rightarrow (1, 0)$ is an interior critical point of AC with $f(1, 0) = -1$. Endpoints: $f(-2, 0) = 8$ and $f(2, 0) = 0$.

(iv) For the interior of the triangular region, $f_x(x, y) = 2x - 2 = 0$ and $f_y(x, y) = -2y + 4 = 0 \Rightarrow x = 1$ and $y = 2 \Rightarrow (1, 2)$ is an interior critical point of the region with $f(1, 2) = 3$. Therefore the absolute maximum is 8 at $(-2, 0)$ and the absolute minimum is -1 at $(1, 0)$.

71. (i) On AB, $f(x, y) = f(-1, y) = y^3 - 3y^2 + 2$ for $-1 \le y \le 1 \Rightarrow f'(-1, y) = 3y^2 - 6y = 0 \Rightarrow y = 0$ and $x = -1$, or $y = 2$ and $x = -1 \Rightarrow (-1, 0)$ is an interior critical point of AB with $f(-1, 0) = 2$; $(-1, 2)$ is outside the boundary. Endpoints: $f(-1, -1) = -2$ and $f(-1, 1) = 0$.

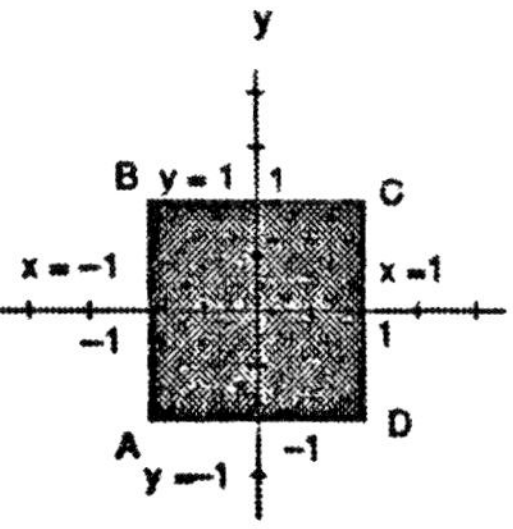

(ii) On BC, $f(x, y) = f(x, 1) = x^3 + 3x^2 - 2$ for $-1 \le x \le 1 \Rightarrow f'(x, 1) = 3x^2 + 6x = 0 \Rightarrow x = 0$ and $y = 1$, or $x = -2$ and $y = 1 \Rightarrow (0, 1)$ is an interior critical point of BC with $f(0, 1) = -2$; $(-2, 1)$ is outside the boundary. Endpoints: $f(-1, 1) = 0$ and $f(1, 1) = 2$.

(iii) On CD, $f(x, y) = f(1, y) = y^3 - 3y^2 + 4$ for $-1 \le y \le 1 \Rightarrow f'(1, y) = 3y^2 - 6y = 0 \Rightarrow y = 0$ and $x = 1$, or $y = 2$ and $x = 1 \Rightarrow (1, 0)$ is an interior critical point of CD with $f(1, 0) = 4$; $(1, 2)$ is outside the boundary. Endpoints: $f(1, 1) = 2$ and $f(1, -1) = 0$.

(iv) On AD, $f(x, y) = f(x, -1) = x^3 + 3x^2 - 4$ for $-1 \le x \le 1 \Rightarrow f'(x, -1) = 3x^2 + 6x = 0 \Rightarrow x = 0$ and $y = -1$, or $x = -2$ and $y = -1 \Rightarrow (0, -1)$ is an interior point of AD with $f(0, -1) = -4$; $(-2, -1)$ is outside the boundary. Endpoints: $f(-1, -1) = -2$ and $f(1, -1) = 0$.

(v) For the interior of the square, $f_x(x, y) = 3x^2 + 6x = 0$ and $f_y(x, y) = 3y^2 - 6y = 0 \Rightarrow x = 0$ or $x = -2$, and $y = 0$ or $y = 2 \Rightarrow (0, 0)$ is an interior critical point of the square region with $f(0, 0) = 0$; the points $(0, 2)$, $(-2, 0)$, and $(-2, 2)$ are outside the region. Therefore the absolute maximum is 4 at $(1, 0)$ and the absolute minimum is -4 at $(0, -1)$.

73. $\nabla f = 3x^2\mathbf{i} + 2y\mathbf{j}$ and $\nabla g = 2x\mathbf{i} + 2y\mathbf{j}$ so that $\nabla f = \lambda \nabla g \Rightarrow 3x^2\mathbf{i} + 2y\mathbf{j} = \lambda(2x\mathbf{i} + 2y\mathbf{j}) \Rightarrow 3x^2 = 2x\lambda$ and $2y = 2y\lambda \Rightarrow \lambda = 1$ or $y = 0$.

CASE 1: $\lambda = 1 \Rightarrow 3x^2 = 2x \Rightarrow x = 0$ or $x = \frac{2}{3}$; $x = 0 \Rightarrow y = \pm 1$ yielding the points $(0, 1)$ and $(0, -1)$; $x = \frac{2}{3}$ $\Rightarrow y = \pm \frac{\sqrt{5}}{3}$ yielding the points $\left(\frac{2}{3}, \frac{\sqrt{5}}{3}\right)$ and $\left(\frac{2}{3}, -\frac{\sqrt{5}}{3}\right)$.

CASE 2: $y = 0 \Rightarrow x^2 - 1 = 0 \Rightarrow x = \pm 1$ yielding the points $(1, 0)$ and $(-1, 0)$.

Evaluations give $f(0, \pm 1) = 1$, $f\left(\frac{2}{3}, \pm \frac{\sqrt{5}}{3}\right) = \frac{23}{27}$, $f(1, 0) = 1$, and $f(-1, 0) = -1$. Therefore the absolute maximum is 1 at $(0, \pm 1)$ and $(1, 0)$, and the absolute minimum is -1 at $(-1, 0)$.

75. (i) $f(x, y) = x^2 + 3y^2 + 2y$ on $x^2 + y^2 = 1 \Rightarrow \nabla f = 2x\mathbf{i} + (6y + 2)\mathbf{j}$ and $\nabla g = 2x\mathbf{i} + 2y\mathbf{j}$ so that $\nabla f = \lambda \nabla g$ $\Rightarrow 2x\mathbf{i} + (6y + 2)\mathbf{j} = \lambda(2x\mathbf{i} + 2y\mathbf{j}) \Rightarrow 2x = 2x\lambda$ and $6y + 2 = 2y\lambda \Rightarrow \lambda = 1$ or $x = 0$.

CASE 1: $\lambda = 1 \Rightarrow 6y + 2 = 2y \Rightarrow y = -\frac{1}{2}$ and $x = \pm\frac{\sqrt{3}}{2}$ yielding the points $\left(\pm\frac{\sqrt{3}}{2}, -\frac{1}{2}\right)$.

CASE 2: $x = 0 \Rightarrow y^2 = 1 \Rightarrow y = \pm 1$ yielding the points $(0, \pm 1)$.

Evaluations give $f\left(\pm\frac{\sqrt{3}}{2}, -\frac{1}{2}\right) = \frac{1}{2}$, $f(0, 1) = 5$, and $f(0, -1) = 1$. Therefore $\frac{1}{2}$ and 5 are the extreme values on the boundary of the disk.

(ii) For the interior of the disk, $f_x(x, y) = 2x = 0$ and $f_y(x, y) = 6y + 2 = 0 \Rightarrow x = 0$ and $y = -\frac{1}{3}$ $\Rightarrow \left(0, -\frac{1}{3}\right)$ is an interior critical point with $f\left(0, -\frac{1}{3}\right) = -\frac{1}{3}$. Therefore the absolute maximum of f on the disk is 5 at (0, 1) and the absolute minimum of f on the disk is $-\frac{1}{3}$ at $\left(0, -\frac{1}{3}\right)$.

77. $\nabla f = \mathbf{i} - \mathbf{j} + \mathbf{k}$ and $\nabla g = 2x\mathbf{i} + 2y\mathbf{j} + 2z\mathbf{k}$ so that $\nabla f = \lambda \nabla g \Rightarrow \mathbf{i} - \mathbf{j} + \mathbf{k} = \lambda(2x\mathbf{i} + 2y\mathbf{j} + 2z\mathbf{k}) \Rightarrow 1 = 2x\lambda$, $-1 = 2y\lambda$, $1 = 2z\lambda \Rightarrow x = -y = z = \frac{1}{\lambda}$. Thus $x^2 + y^2 + z^2 = 1 \Rightarrow 3x^2 = 1 \Rightarrow x = \pm\frac{1}{\sqrt{3}}$ yielding the points $\left(\frac{1}{\sqrt{3}}, -\frac{1}{\sqrt{3}}, \frac{1}{\sqrt{3}}\right)$ and $\left(-\frac{1}{\sqrt{3}}, \frac{1}{\sqrt{3}}, -\frac{1}{\sqrt{3}}\right)$. Evaluations give the absolute maximum value of $f\left(\frac{1}{\sqrt{3}}, -\frac{1}{\sqrt{3}}, \frac{1}{\sqrt{3}}\right) = \frac{3}{\sqrt{3}} = \sqrt{3}$ and the absolute minimum value of $f\left(-\frac{1}{\sqrt{3}}, \frac{1}{\sqrt{3}}, -\frac{1}{\sqrt{3}}\right) = -\sqrt{3}$.

79. The cost is $f(x, y, z) = 2axy + 2bxz + 2cyz$ subject to the constraint $xyz = V$. Then $\nabla f = \lambda \nabla g$ $\Rightarrow 2ay + 2bz = \lambda yz$, $2ax + 2cz = \lambda xz$, and $2bx + 2cy = \lambda xy \Rightarrow 2axy + 2bxz = \lambda xyz$, $2axy + 2cyz = \lambda xyz$, and $2bxz + 2cyz = \lambda xyz \Rightarrow 2axy + 2bxz = 2axy + 2cyz \Rightarrow y = \left(\frac{b}{c}\right)x$. Also $2axy + 2bxz = 2bxz + 2cyz \Rightarrow z = \left(\frac{a}{c}\right)x$. Then $x\left(\frac{b}{c}x\right)\left(\frac{a}{c}x\right) = V \Rightarrow x^3 = \frac{c^2V}{ab} \Rightarrow$ width $= x = \left(\frac{c^2V}{ab}\right)^{1/3}$, Depth $= y = \left(\frac{b}{c}\right)\left(\frac{c^2V}{ab}\right)^{1/3} = \left(\frac{b^2V}{ac}\right)^{1/3}$, and Height $= z = \left(\frac{a}{c}\right)\left(\frac{c^2V}{ab}\right)^{1/3} = \left(\frac{a^2V}{bc}\right)^{1/3}$.

81. $\nabla f = (y + z)\mathbf{i} + x\mathbf{j} + x\mathbf{k}$, $\nabla g = 2x\mathbf{i} + 2y\mathbf{j}$, and $\nabla h = z\mathbf{i} + x\mathbf{k}$ so that $\nabla f = \lambda \nabla g + \mu \nabla h$ $\Rightarrow (y + z)\mathbf{i} + x\mathbf{j} + x\mathbf{k} = \lambda(2x\mathbf{i} + 2y\mathbf{j}) + \mu(z\mathbf{i} + x\mathbf{k}) \Rightarrow y + z = 2\lambda x + \mu z$, $x = 2\lambda y$, $x = \mu x \Rightarrow x = 0$ or $\mu = 1$.

CASE 1: $x = 0$ which is impossible since $xz = 1$.

CASE 2: $\mu = 1 \Rightarrow y + z = 2\lambda x + z \Rightarrow y = 2\lambda x$ and $x = 2\lambda y \Rightarrow y = (2\lambda)(2\lambda y) \Rightarrow y = 0$ or $4\lambda^2 = 1$. If $y = 0$, then $x^2 = 1 \Rightarrow x = \pm 1$ so with $xz = 1$ we obtain the points $(1, 0, 1)$ and $(-1, 0, -1)$. If $4\lambda^2 = 1$, then $\lambda = \pm\frac{1}{2}$. For $\lambda = -\frac{1}{2}$, $y = -x$ so $x^2 + y^2 = 1 \Rightarrow x^2 = \frac{1}{2}$ $\Rightarrow x = \pm\frac{1}{\sqrt{2}}$ with $xz = 1 \Rightarrow z = \pm\sqrt{2}$, and we obtain the points $\left(\frac{1}{\sqrt{2}}, -\frac{1}{\sqrt{2}}, \sqrt{2}\right)$ and $\left(-\frac{1}{\sqrt{2}}, \frac{1}{\sqrt{2}}, -\sqrt{2}\right)$. For $\lambda = \frac{1}{2}$, $y = x \Rightarrow x^2 = \frac{1}{2} \Rightarrow x = \pm\frac{1}{\sqrt{2}}$ with $xz = 1 \Rightarrow z = \pm\sqrt{2}$, and we obtain the points $\left(\frac{1}{\sqrt{2}}, \frac{1}{\sqrt{2}}, \sqrt{2}\right)$ and $\left(-\frac{1}{\sqrt{2}}, -\frac{1}{\sqrt{2}}, -\sqrt{2}\right)$.

Evaluations give $f(1, 0, 1) = 1$, $f(-1, 0, -1) = 1$, $f\left(\frac{1}{\sqrt{2}}, -\frac{1}{\sqrt{2}}, \sqrt{2}\right) = \frac{1}{2}$, $f\left(-\frac{1}{\sqrt{2}}, \frac{1}{\sqrt{2}}, -\sqrt{2}\right) = \frac{1}{2}$, $f\left(\frac{1}{\sqrt{2}}, \frac{1}{\sqrt{2}}, \sqrt{2}\right) = \frac{3}{2}$, and $f\left(-\frac{1}{\sqrt{2}}, -\frac{1}{\sqrt{2}}, -\sqrt{2}\right) = \frac{3}{2}$. Therefore the absolute maximum is $\frac{3}{2}$ at $\left(\frac{1}{\sqrt{2}}, \frac{1}{\sqrt{2}}, \sqrt{2}\right)$ and $\left(-\frac{1}{\sqrt{2}}, -\frac{1}{\sqrt{2}}, -\sqrt{2}\right)$, and the absolute minimum is $\frac{1}{2}$ at $\left(-\frac{1}{\sqrt{2}}, \frac{1}{\sqrt{2}}, -\sqrt{2}\right)$ and $\left(\frac{1}{\sqrt{2}}, -\frac{1}{\sqrt{2}}, \sqrt{2}\right)$.

83. Note that $x = r\cos\theta$ and $y = r\sin\theta$, $r = \sqrt{x^2 + y^2}$ and $\theta = \tan^{-1}\left(\frac{y}{x}\right)$. Thus,

$$\frac{\partial w}{\partial x} = \frac{\partial w}{\partial r}\frac{\partial r}{\partial x} + \frac{\partial w}{\partial \theta}\frac{\partial \theta}{\partial x} = \left(\frac{\partial w}{\partial r}\right)\left(\frac{x}{\sqrt{x^2+y^2}}\right) + \left(\frac{\partial w}{\partial \theta}\right)\left(\frac{-y}{x^2+y^2}\right) = (\cos\theta)\frac{\partial w}{\partial r} - \left(\frac{\sin\theta}{r}\right)\frac{\partial w}{\partial \theta};$$

$$\frac{\partial w}{\partial y} = \frac{\partial w}{\partial r}\frac{\partial r}{\partial y} + \frac{\partial w}{\partial \theta}\frac{\partial \theta}{\partial y} = \left(\frac{\partial w}{\partial r}\right)\left(\frac{y}{\sqrt{x^2+y^2}}\right) + \left(\frac{\partial w}{\partial \theta}\right)\left(\frac{x}{x^2+y^2}\right) = (\sin\theta)\frac{\partial w}{\partial r} + \left(\frac{\cos\theta}{r}\right)\frac{\partial w}{\partial \theta}$$

85. $\frac{\partial u}{\partial y} = b$ and $\frac{\partial u}{\partial x} = a \Rightarrow \frac{\partial w}{\partial x} = \frac{dw}{du}\frac{\partial u}{\partial x} = a\frac{dw}{du}$ and $\frac{\partial w}{\partial y} = \frac{dw}{du}\frac{\partial u}{\partial y} = b\frac{dw}{du} \Rightarrow \frac{1}{a}\frac{\partial w}{\partial x} = \frac{dw}{du}$ and $\frac{1}{b}\frac{\partial w}{\partial y} = \frac{dw}{du}$
$\Rightarrow \frac{1}{a}\frac{\partial w}{\partial x} = \frac{1}{b}\frac{\partial w}{\partial y} \Rightarrow b\frac{\partial w}{\partial x} = a\frac{\partial w}{\partial y}$

87. $e^u \cos v - x = 0 \Rightarrow (e^u \cos v)\frac{\partial u}{\partial x} - (e^u \sin v)\frac{\partial v}{\partial x} = 1$; $e^u \sin v - y = 0 \Rightarrow (e^u \sin v)\frac{\partial u}{\partial x} + (e^u \cos v)\frac{\partial v}{\partial x} = 0$. Solving this system yields $\frac{\partial u}{\partial x} = e^{-u}\cos v$ and $\frac{\partial v}{\partial x} = -e^{-u}\sin v$. Similarly, $e^u \cos v - x = 0$ $\Rightarrow (e^u \cos v)\frac{\partial u}{\partial y} - (e^u \sin v)\frac{\partial v}{\partial y} = 0$ and $e^u \sin v - y = 0 \Rightarrow (e^u \sin v)\frac{\partial u}{\partial y} + (e^u \cos v)\frac{\partial v}{\partial y} = 1$. Solving this second system yields $\frac{\partial u}{\partial y} = e^{-u}\sin v$ and $\frac{\partial v}{\partial y} = e^{-u}\cos v$. Therefore $\left(\frac{\partial u}{\partial x}\mathbf{i} + \frac{\partial u}{\partial y}\mathbf{j}\right) \cdot \left(\frac{\partial v}{\partial x}\mathbf{i} + \frac{\partial v}{\partial y}\mathbf{j}\right)$ $= [(e^{-u}\cos v)\mathbf{i} + (e^{-u}\sin v)\mathbf{j}] \cdot [(-e^{-u}\sin v)\mathbf{i} + (e^{-u}\cos v)\mathbf{j}] = 0 \Rightarrow$ the vectors are orthogonal $\Rightarrow$ the angle between the vectors is the constant $\frac{\pi}{2}$.

89. $(y+z)^2 + (z-x)^2 = 16 \Rightarrow \nabla f = -2(z-x)\mathbf{i} + 2(y+z)\mathbf{j} + 2(y+2z-x)\mathbf{k}$; if the normal line is parallel to the yz-plane, then x is constant $\Rightarrow \frac{\partial f}{\partial x} = 0 \Rightarrow -2(z-x) = 0 \Rightarrow z = x \Rightarrow (y+z)^2 + (z-z)^2 = 16 \Rightarrow y+z = \pm 4$. Let $x = t \Rightarrow z = t \Rightarrow y = -t \pm 4$. Therefore the points are $(t, -t \pm 4, t)$, t a real number.

91. $\nabla f = \lambda(x\mathbf{i} + y\mathbf{j} + z\mathbf{k}) \Rightarrow \frac{\partial f}{\partial x} = \lambda x \Rightarrow f(x, y, z) = \frac{1}{2}\lambda x^2 + g(y, z)$ for some function g $\Rightarrow \lambda y = \frac{\partial f}{\partial y} = \frac{\partial g}{\partial y}$ $\Rightarrow g(y, z) = \frac{1}{2}\lambda y^2 + h(z)$ for some function h $\Rightarrow \lambda z = \frac{\partial f}{\partial z} = \frac{\partial g}{\partial z} = h'(z) \Rightarrow h(z) = \frac{1}{2}\lambda z^2 + C$ for some arbitrary constant C $\Rightarrow g(y, z) = \frac{1}{2}\lambda y^2 + \left(\frac{1}{2}\lambda z^2 + C\right) \Rightarrow f(x, y, z) = \frac{1}{2}\lambda x^2 + \frac{1}{2}\lambda y^2 + \frac{1}{2}\lambda z^2 + C \Rightarrow f(0, 0, a) = \frac{1}{2}\lambda a^2 + C$ and $f(0, 0, -a) = \frac{1}{2}\lambda(-a)^2 + C \Rightarrow f(0, 0, a) = f(0, 0, -a)$ for any constant a, as claimed.

CHAPTER 12 MULTIPLE INTEGRALS

12.1 DOUBLE AND ITERATED INTEGRALS OVER RECTANGLES

1. $\int_1^2 \int_0^4 2xy\, dy\, dx = \int_1^2 \left[x\, y^2\right]_0^4 dx = \int_1^2 16x\, dx = \left[8\, x^2\right]_1^2 = 24$

3. $\int_{-1}^{0} \int_{-1}^{1} (x + y + 1)\, dx\, dy = \int_{-1}^{0} \left[\frac{x^2}{2} + yx + x\right]_{-1}^{1} dy = \int_{-1}^{0} (2y + 2)\, dy = \left[y^2 + 2y\right]_{-1}^{0} = 1$

5. $\int_0^3 \int_0^2 (4 - y^2)\, dy\, dx = \int_0^3 \left[4y - \frac{y^3}{3}\right]_0^2 dx = \int_0^3 \frac{16}{3}\, dx = \left[\frac{16}{3}\, x\right]_0^3 = 16$

7. $\int_0^1 \int_0^1 \frac{y}{1+xy}\, dx\, dy = \int_0^1 \left[\ln|1 + x\, y|\right]_0^1 dy = \int_0^1 \ln|1 + y| dy = \left[y \ln|1 + y| - y + \ln|1 + y|\right]_0^1 = 2\ln 2 - 1$

9. $\int_0^{\ln 2} \int_1^{\ln 5} e^{2x+y}\, dy\, dx = \int_0^{\ln 2} \left[e^{2x+y}\right]_1^{\ln 5} dx = \int_0^{\ln 2} (5e^{2x} - e^{2x+1})\, dx = \left[\frac{5}{2}e^{2x} - \frac{1}{2}e^{2x+1}\right]_0^{\ln 2} = \frac{3}{2}(5 - e)$

11. $\int_{-1}^{2} \int_0^{\pi/2} y \sin x\, dx\, dy = \int_{-1}^{2} \left[-y \cos x\right]_0^{\pi/2} dy = \int_{-1}^{2} y\, dy = \left[\frac{1}{2}y^2\right]_{-1}^{2} = \frac{3}{2}$

13. $\iint_R (6\, y^2 - 2\, x) dA = \int_0^1 \int_0^2 (6\, y^2 - 2\, x)\, dy\, dx = \int_0^1 \left[2\, y^3 - 2\, x\, y\right]_0^2 dx = \int_0^1 (16 - 4\, x)\, dx = \left[16\, x - 2\, x^2\right]_0^1 = 14$

15. $\iint_R x\, y \cos y\, dA = \int_{-1}^{1} \int_0^{\pi} x\, y \cos y\, dy\, dx = \int_{-1}^{1} \left[x\, y \sin y + x \cos y\right]_0^{\pi} dx = \int_{-1}^{1} (-2x)\, dx = \left[-x^2\right]_{-1}^{1} = 0$

17. $\iint_R e^{x-y} dA = \int_0^{\ln 2} \int_0^{\ln 2} e^{x-y}\, dy\, dx = \int_0^{\ln 2} \left[-e^{x-y}\right]_0^{\ln 2} dx = \int_0^{\ln 2} (-e^{x - \ln 2} + e^x)\, dx = \left[-e^{x - \ln 2} + e^x\right]_0^{\ln 2} = \frac{1}{2}$

19. $\iint_R \frac{x\, y^3}{x^2+1} dA = \int_0^1 \int_0^2 \frac{x\, y^3}{x^2+1}\, dy\, dx = \int_0^1 \left[\frac{x\, y^4}{4(x^2+1)}\right]_0^2 dx = \int_0^1 \frac{4\, x}{x^2+1}\, dx = \left[2 \ln|x^2 + 1|\right]_0^1 = 2 \ln 2$

21. $\int_1^2 \int_1^2 \frac{1}{xy}\, dy\, dx = \int_1^2 \frac{1}{x} (\ln 2 - \ln 1)\, dx = (\ln 2) \int_1^2 \frac{1}{x}\, dx = (\ln 2)^2$

23. $V = \iint_R f(x, y)\, dA = \int_{-1}^{1} \int_{-1}^{1} (x^2 + y^2)\, dy\, dx = \int_{-1}^{1} \left[x^2\, y + \frac{1}{3}y^3\right]_{-1}^{1} dx = \int_{-1}^{1} \left(2\, x^2 + \frac{2}{3}\right) dx = \left[\frac{2}{3}x^3 + \frac{2}{3}x\right]_{-1}^{1} = \frac{8}{3}$

25. $V = \iint_R f(x, y)\, dA = \int_0^1 \int_0^1 (2 - x - y)\, dy\, dx = \int_0^1 \left[2\, y - x\, y - \frac{1}{2}y^2\right]_0^1 dx = \int_0^1 \left(\frac{3}{2} - x\right) dx = \left[\frac{3}{2}x - \frac{1}{2}x^2\right]_0^1 = 1$

27. $V = \iint_R f(x, y)\, dA = \int_0^{\pi/2} \int_0^{\pi/4} 2 \sin x \cos y\, dy\, dx = \int_0^{\pi/2} \left[2 \sin x \sin y\right]_0^{\pi/4} dx = \int_0^{\pi/2} \left(\sqrt{2} \sin x\right) dx = \left[-\sqrt{2} \cos x\right]_0^{\pi/2}$
$= \sqrt{2}$

12.2 DOUBLE INTEGRALS OVER GENERAL REGIONS

1. $\int_0^{\pi}\int_0^{x}(x\sin y)\,dy\,dx = \int_0^{\pi}[-x\cos y]_0^x\,dx$
$= \int_0^{\pi}(x - x\cos x)\,dx = \left[\frac{x^2}{2} - (\cos x + x\sin x)\right]_0^{\pi}$
$= \frac{\pi^2}{2} + 2$

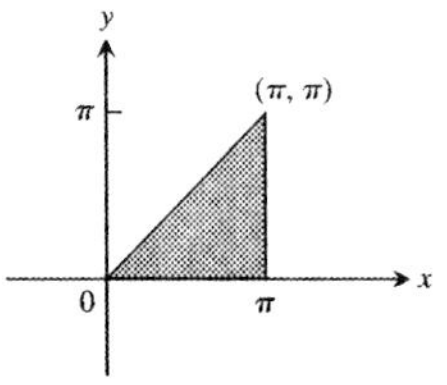

3. $\int_1^{\ln 8}\int_0^{\ln y} e^{x+y}\,dx\,dy = \int_1^{\ln 8}[e^{x+y}]_0^{\ln y}\,dy = \int_1^{\ln 8}(ye^y - e^y)\,dy$
$= [(y-1)e^y - e^y]_1^{\ln 8} = 8(\ln 8 - 1) - 8 + e$
$= 8\ln 8 - 16 + e$

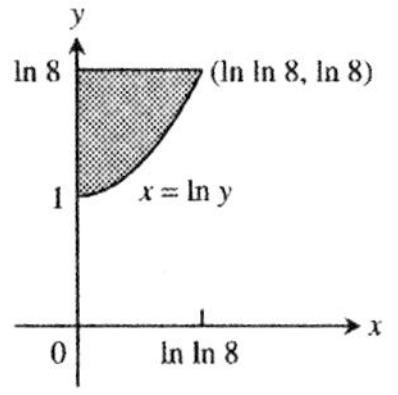

5. $\int_0^1\int_0^{y^2} 3y^3e^{xy}\,dx\,dy = \int_0^1[3y^2e^{xy}]_0^{y^2}\,dy$
$= \int_0^1\left(3y^2e^{y^3} - 3y^2\right)dy = \left[e^{y^3} - y^3\right]_0^1 = e - 2$

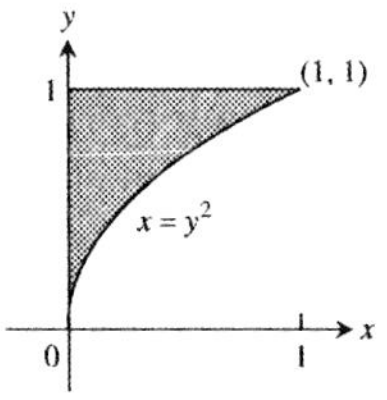

7. $\int_1^2\int_x^{2x}\frac{x}{y}\,dy\,dx = \int_1^2[x\ln y]_x^{2x}\,dx = (\ln 2)\int_1^2 x\,dx = \frac{3}{2}\ln 2$

9. $\int_0^1\int_0^{1-u}\left(v - \sqrt{u}\right)dv\,du = \int_0^1\left[\frac{v^2}{2} - v\sqrt{u}\right]_0^{1-u}du = \int_0^1\left[\frac{1-2u+u^2}{2} - \sqrt{u}(1-u)\right]du$
$= \int_0^1\left(\frac{1}{2} - u + \frac{u^2}{2} - u^{1/2} + u^{3/2}\right)du = \left[\frac{u}{2} - \frac{u^2}{2} + \frac{u^3}{6} - \frac{2}{3}u^{3/2} + \frac{2}{5}u^{5/2}\right]_0^1 = \frac{1}{2} - \frac{1}{2} + \frac{1}{6} - \frac{2}{3} + \frac{2}{5} = -\frac{1}{2} + \frac{2}{5} = -\frac{1}{10}$

11. $\int_{-2}^{0}\int_{v}^{-v}2\,dp\,dv = 2\int_{-2}^{0}[p]_v^{-v}\,dv = 2\int_{-2}^{0}-2v\,dv$
$= -2\left[v^2\right]_{-2}^{0} = 8$

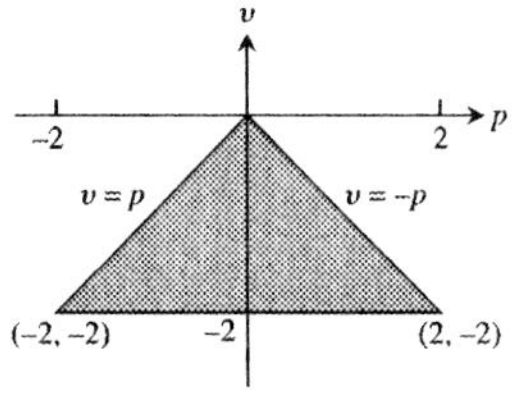

13. $\int_{-\pi/3}^{\pi/3}\int_0^{\sec t}3\cos t\,du\,dt = \int_{-\pi/3}^{\pi/3}[(3\cos t)u]_0^{\sec t}$
$= \int_{-\pi/3}^{\pi/3}3\,dt = 2\pi$

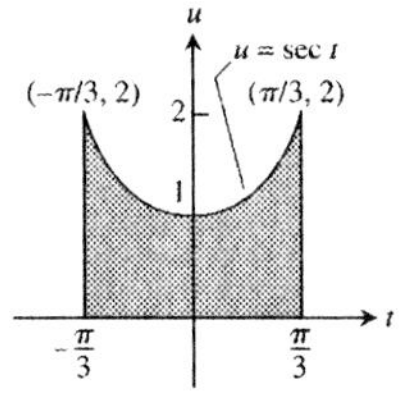

15. $\int_2^4\int_0^{(4-y)/2}dx\,dy$

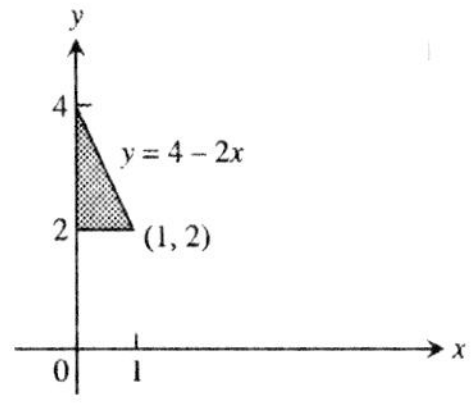

17. $\int_0^1 \int_{x^2}^{x} dy\,dx$

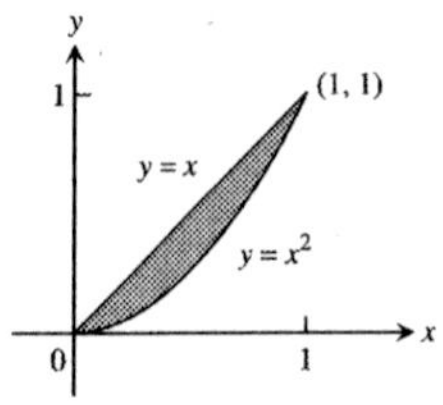

19. $\int_1^e \int_{\ln y}^{1} dx\,dy$

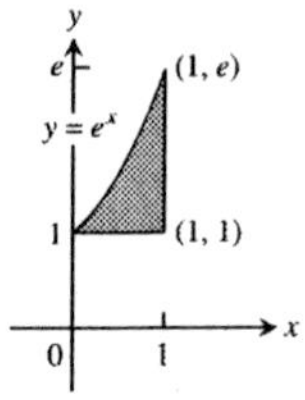

21. $\int_0^9 \int_0^{\frac{1}{2}\sqrt{9-y}} 16x\,dx\,dy$

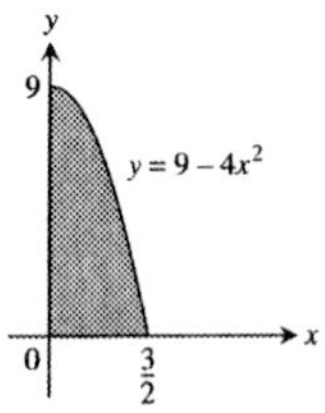

23. $\int_{-1}^1 \int_0^{\sqrt{1-x^2}} 3y\,dy\,dx$

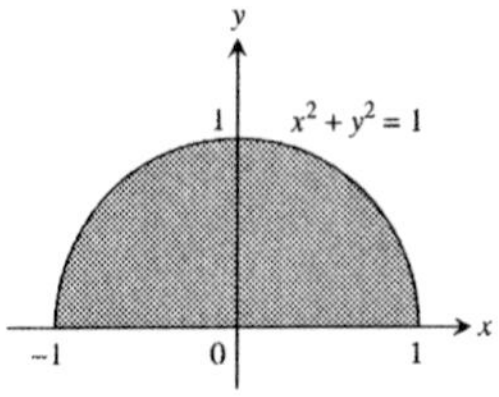

25. $\int_0^\pi \int_x^\pi \frac{\sin y}{y}\,dy\,dx = \int_0^\pi \int_0^y \frac{\sin y}{y}\,dx\,dy = \int_0^\pi \sin y\,dy = 2$

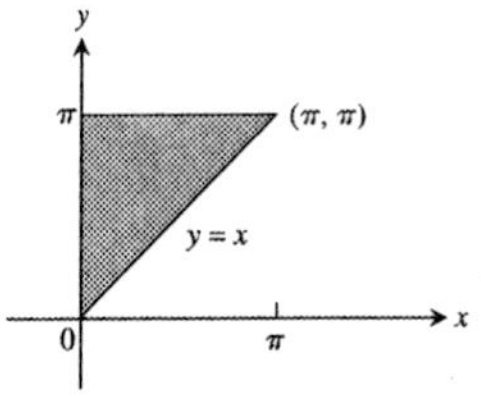

27. $\int_0^1 \int_y^1 x^2 e^{xy}\,dx\,dy = \int_0^1 \int_0^x x^2 e^{xy}\,dy\,dx = \int_0^1 \left[xe^{xy}\right]_0^x dx$
$= \int_0^1 \left(xe^{x^2} - x\right) dx = \left[\frac{1}{2}e^{x^2} - \frac{x^2}{2}\right]_0^1 = \frac{e-2}{2}$

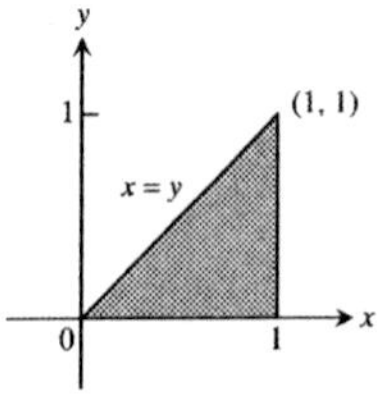

29. $\int_0^{2\sqrt{\ln 3}} \int_{y/2}^{\sqrt{\ln 3}} e^{x^2}\,dx\,dy = \int_0^{\sqrt{\ln 3}} \int_0^{2x} e^{x^2}\,dy\,dx$
$= \int_0^{\sqrt{\ln 3}} 2xe^{x^2}\,dx = \left[e^{x^2}\right]_0^{\sqrt{\ln 3}} = e^{\ln 3} - 1 = 2$

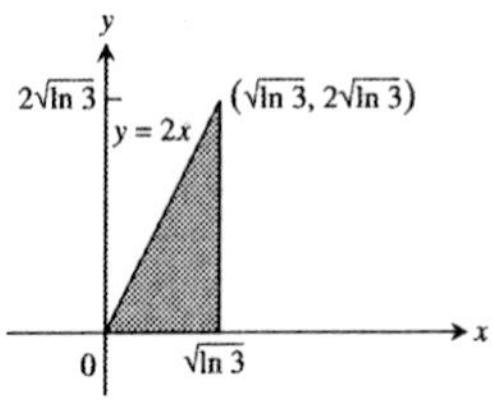

31. $\int_0^{1/16}\int_{y^{1/4}}^{1/2}\cos(16\pi x^5)\,dx\,dy = \int_0^{1/2}\int_0^{x^4}\cos(16\pi x^5)\,dy\,dx$
$= \int_0^{1/2} x^4\cos(16\pi x^5)\,dx = \left[\frac{\sin(16\pi x^5)}{80\pi}\right]_0^{1/2} = \frac{1}{80\pi}$

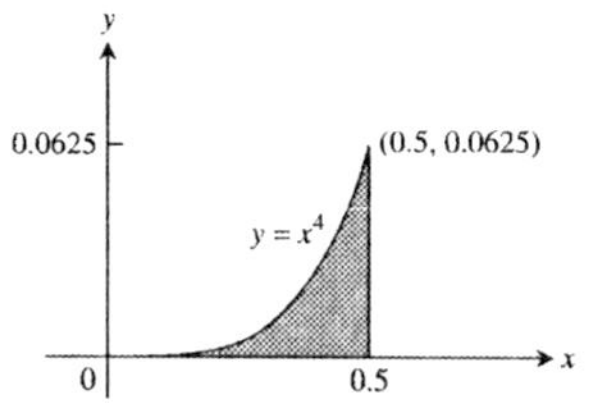

33. $\iint\limits_R (y - 2x^2)\,dA$
$= \int_{-1}^{0}\int_{-x-1}^{x+1}(y - 2x^2)\,dy\,dx + \int_0^1\int_{x-1}^{1-x}(y - 2x^2)\,dy\,dx$
$= \int_{-1}^{0}\left[\frac{1}{2}y^2 - 2x^2y\right]_{-x-1}^{x+1}dx + \int_0^1\left[\frac{1}{2}y^2 - 2x^2y\right]_{x-1}^{1-x}dx$
$= \int_{-1}^{0}\left[\frac{1}{2}(x+1)^2 - 2x^2(x+1) - \frac{1}{2}(-x-1)^2 + 2x^2(-x-1)\right]dx$
$\quad + \int_0^1\left[\frac{1}{2}(1-x)^2 - 2x^2(1-x) - \frac{1}{2}(x-1)^2 + 2x^2(x-1)\right]dx$
$= -4\int_{-1}^{0}(x^3 + x^2)\,dx + 4\int_0^1(x^3 - x^2)\,dx$
$= -4\left[\frac{x^4}{4} + \frac{x^3}{3}\right]_{-1}^{0} + 4\left[\frac{x^4}{4} - \frac{x^3}{3}\right]_0^1 = 4\left[\frac{(-1)^4}{4} + \frac{(-1)^3}{3}\right] + 4\left(\frac{1}{4} - \frac{1}{3}\right) = 8\left(\frac{3}{12} - \frac{4}{12}\right) = -\frac{8}{12} = -\frac{2}{3}$

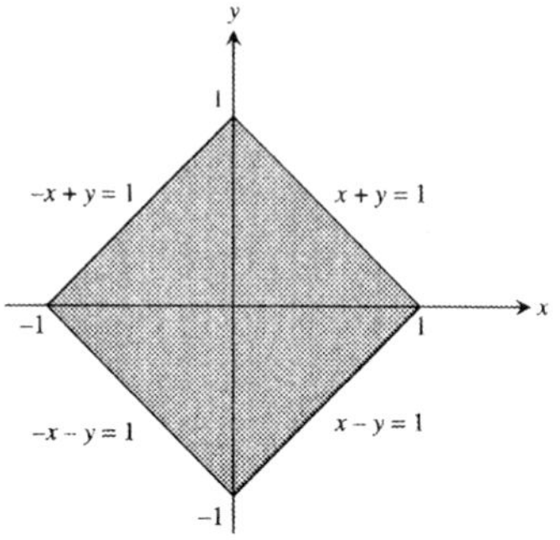

35. $V = \int_0^1\int_x^{2-x}(x^2 + y^2)\,dy\,dx = \int_0^1\left[x^2y + \frac{y^3}{3}\right]_x^{2-x}dx = \int_0^1\left[2x^2 - \frac{7x^3}{3} + \frac{(2-x)^3}{3}\right]dx = \left[\frac{2x^3}{3} - \frac{7x^4}{12} - \frac{(2-x)^4}{12}\right]_0^1$
$= \left(\frac{2}{3} - \frac{7}{12} - \frac{1}{12}\right) - \left(0 - 0 - \frac{16}{12}\right) = \frac{4}{3}$

37. $V = \int_{-4}^{1}\int_{3x}^{4-x^2}(x + 4)\,dy\,dx = \int_{-4}^{1}[xy + 4y]_{3x}^{4-x^2}dx = \int_{-4}^{1}[x(4 - x^2) + 4(4 - x^2) - 3x^2 - 12x]\,dx$
$= \int_{-4}^{1}(-x^3 - 7x^2 - 8x + 16)\,dx = \left[-\frac{1}{4}x^4 - \frac{7}{3}x^3 - 4x^2 + 16x\right]_{-4}^{1} = \left(-\frac{1}{4} - \frac{7}{3} + 12\right) - \left(\frac{64}{3} - 64\right) = \frac{157}{3} - \frac{1}{4} = \frac{625}{12}$

39. $V = \int_0^2\int_0^3(4 - y^2)\,dx\,dy = \int_0^2[4x - y^2x]_0^3\,dy = \int_0^2(12 - 3y^2)\,dy = [12y - y^3]_0^2 = 24 - 8 = 16$

41. $V = \int_0^2\int_0^{2-x}(12 - 3y^2)\,dy\,dx = \int_0^2[12y - y^3]_0^{2-x}dx = \int_0^2[24 - 12x - (2 - x)^3]\,dx = \left[24x - 6x^2 + \frac{(2-x)^4}{4}\right]_0^2 = 20$

43. $V = \int_1^2\int_{-1/x}^{1/x}(x + 1)\,dy\,dx = \int_1^2[xy + y]_{-1/x}^{1/x}dx = \int_1^2\left[1 + \frac{1}{x} - \left(-1 - \frac{1}{x}\right)\right]dx = 2\int_1^2\left(1 + \frac{1}{x}\right)dx = 2[x + \ln x]_1^2$
$= 2(1 + \ln 2)$

45. $\int_1^{\infty}\int_{e^{-x}}^{1}\frac{1}{x^3y}\,dy\,dx = \int_1^{\infty}\left[\frac{\ln y}{x^3}\right]_{e^{-x}}^{1}dx = \int_1^{\infty} -\left(\frac{-x}{x^3}\right)dx = -\lim_{b\to\infty}\left[\frac{1}{x}\right]_1^b = -\lim_{b\to\infty}\left(\frac{1}{b} - 1\right) = 1$

47. $\int_{-\infty}^{\infty}\int_{-\infty}^{\infty}\frac{1}{(x^2+1)(y^2+1)}\,dx\,dy = 2\int_0^{\infty}\left(\frac{2}{y^2+1}\right)\left(\lim_{b\to\infty}\tan^{-1}b - \tan^{-1}0\right)dy = 2\pi\lim_{b\to\infty}\int_0^b\frac{1}{y^2+1}\,dy$
$= 2\pi\left(\lim_{b\to\infty}\tan^{-1}b - \tan^{-1}0\right) = (2\pi)\left(\frac{\pi}{2}\right) = \pi^2$

49. $\iint\limits_R f(x, y)\,dA \approx \frac{1}{4}f\left(-\frac{1}{2}, 0\right) + \frac{1}{8}f(0, 0) + \frac{1}{8}f\left(\frac{1}{4}, 0\right) = \frac{1}{4}\left(-\frac{1}{2}\right) + \frac{1}{8}\left(0 + \frac{1}{4}\right) = -\frac{3}{32}$

51. The ray $\theta = \frac{\pi}{6}$ meets the circle $x^2 + y^2 = 4$ at the point $\left(\sqrt{3}, 1\right)$ $\Rightarrow$ the ray is represented by the line $y = \frac{x}{\sqrt{3}}$. Thus,
$\iint\limits_R f(x, y)\,dA = \int_0^{\sqrt{3}}\int_{x/\sqrt{3}}^{\sqrt{4-x^2}}\sqrt{4-x^2}\,dy\,dx = \int_0^{\sqrt{3}}\left[(4 - x^2) - \frac{x}{\sqrt{3}}\sqrt{4 - x^2}\right]dx = \left[4x - \frac{x^3}{3} + \frac{(4-x^2)^{3/2}}{3\sqrt{3}}\right]_0^{\sqrt{3}} = \frac{20\sqrt{3}}{9}$

53. $V = \int_0^1 \int_x^{2-x} (x^2 + y^2)\,dy\,dx = \int_0^1 \left[x^2y + \frac{y^3}{3}\right]_x^{2-x} dx$
$= \int_0^1 \left[2x^2 - \frac{7x^3}{3} + \frac{(2-x)^3}{3}\right] dx = \left[\frac{2x^3}{3} - \frac{7x^4}{12} - \frac{(2-x)^4}{12}\right]_0^1$
$= \left(\frac{2}{3} - \frac{7}{12} - \frac{1}{12}\right) - \left(0 - 0 - \frac{16}{12}\right) = \frac{4}{3}$

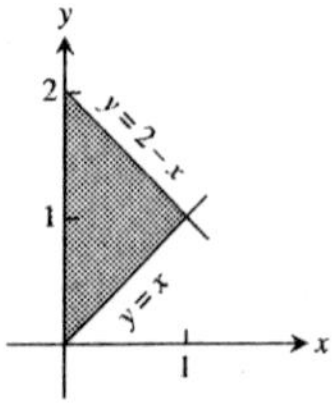

55. To maximize the integral, we want the domain to include all points where the integrand is positive and to exclude all points where the integrand is negative. These criteria are met by the points (x, y) such that $4 - x^2 - 2y^2 \ge 0$ or $x^2 + 2y^2 \le 4$, which is the ellipse $x^2 + 2y^2 = 4$ together with its interior.

57. No, it is not possible. By Fubini's theorem, the two orders of integration must give the same result.

59. $\int_{-b}^{b}\int_{-b}^{b} e^{-x^2-y^2}\,dx\,dy = \int_{-b}^{b}\int_{-b}^{b} e^{-y^2}e^{-x^2}\,dx\,dy = \int_{-b}^{b} e^{-y^2}\left(\int_{-b}^{b} e^{-x^2}\,dx\right) dy = \left(\int_{-b}^{b} e^{-x^2}\,dx\right)\left(\int_{-b}^{b} e^{-y^2}\,dy\right)$
$= \left(\int_{-b}^{b} e^{-x^2}\,dx\right)^2 = \left(2\int_0^b e^{-x^2}\,dx\right)^2 = 4\left(\int_0^b e^{-x^2}\,dx\right)^2$; taking limits as $b \to \infty$ gives the stated result.

12.3 AREA BY DOUBLE INTEGRATION

1. $\int_0^2 \int_0^{2-x} dy\,dx = \int_0^2 (2 - x)\,dx = \left[2x - \frac{x^2}{2}\right]_0^2 = 2$,
or $\int_0^2 \int_0^{2-y} dx\,dy = \int_0^2 (2 - y)\,dy = 2$

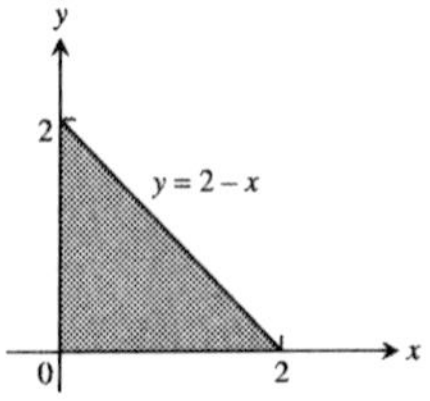

3. $\int_{-2}^{1}\int_{y-2}^{-y^2} dx\,dy = \int_{-2}^{1} (-y^2 - y + 2)\,dy$
$= \left[-\frac{y^3}{3} - \frac{y^2}{2} + 2y\right]_{-2}^{1}$
$= \left(-\frac{1}{3} - \frac{1}{2} + 2\right) - \left(\frac{8}{3} - 2 - 4\right) = \frac{9}{2}$

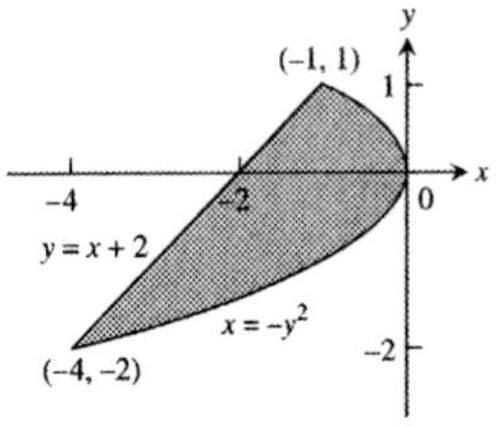

5. $\int_0^{\ln 2}\int_0^{e^x} dy\,dx = \int_0^{\ln 2} e^x\,dx = \left[e^x\right]_0^{\ln 2} = 2 - 1 = 1$

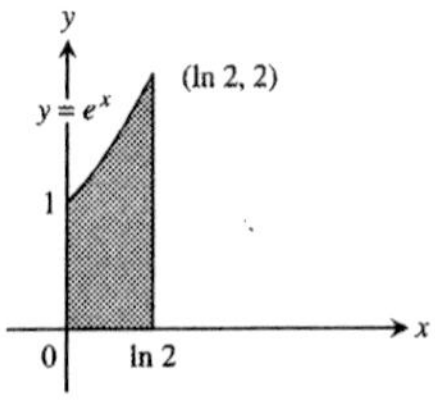

7. $\int_0^1 \int_{y^2}^{2y-y^2} dx\,dy = \int_0^1 (2y - 2y^2)\,dy = \left[y^2 - \frac{2}{3}y^3\right]_0^1$
$= \frac{1}{3}$

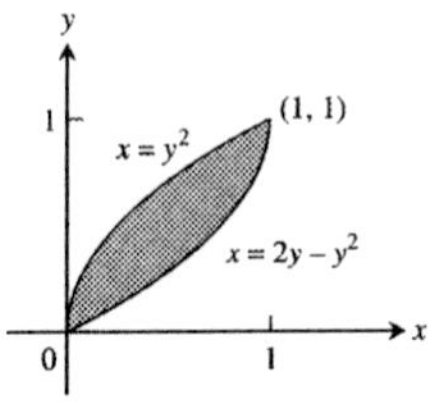

9. $\int_0^6 \int_{y^2/3}^{2y} dx\, dy = \int_0^6 \left(2y - \frac{y^2}{3}\right) dy = \left[y^2 - \frac{y^3}{9}\right]_0^6$
$= 36 - \frac{216}{9} = 12$

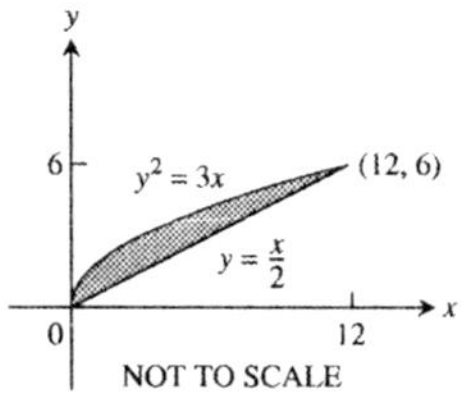

11. $\int_0^{\pi/4} \int_{\sin x}^{\cos x} dy\, dx$
$= \int_0^{\pi/4} (\cos x - \sin x)\, dx = [\sin x + \cos x]_0^{\pi/4}$
$= \left(\frac{\sqrt{2}}{2} + \frac{\sqrt{2}}{2}\right) - (0 + 1) = \sqrt{2} - 1$

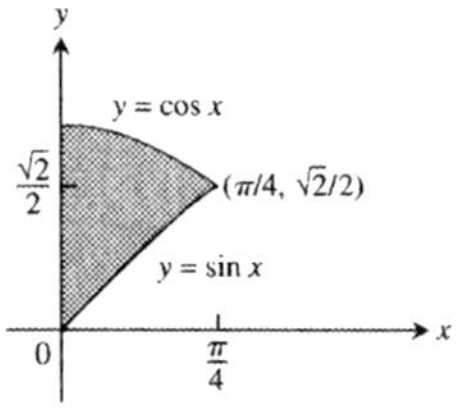

13. $\int_{-1}^0 \int_{-2x}^{1-x} dy\, dx + \int_0^2 \int_{-x/2}^{1-x} dy\, dx$
$= \int_{-1}^0 (1 + x)\, dx + \int_0^2 \left(1 - \frac{x}{2}\right) dx$
$= \left[x + \frac{x^2}{2}\right]_{-1}^0 + \left[x - \frac{x^2}{4}\right]_0^2 = -\left(-1 + \frac{1}{2}\right) + (2 - 1) = \frac{3}{2}$

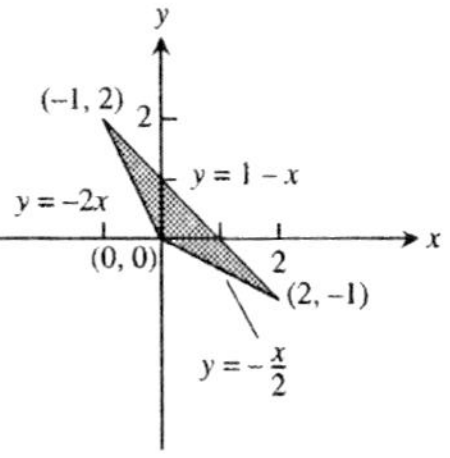

15. (a) average $= \frac{1}{\pi^2} \int_0^\pi \int_0^\pi \sin(x + y)\, dy\, dx = \frac{1}{\pi^2} \int_0^\pi [-\cos(x + y)]_0^\pi\, dx = \frac{1}{\pi^2} \int_0^\pi [-\cos(x + \pi) + \cos x]\, dx$
$= \frac{1}{\pi^2} [-\sin(x + \pi) + \sin x]_0^\pi = \frac{1}{\pi^2} [(-\sin 2\pi + \sin \pi) - (-\sin \pi + \sin 0)] = 0$

(b) average $= \frac{1}{\left(\frac{\pi^2}{2}\right)} \int_0^\pi \int_0^{\pi/2} \sin(x + y)\, dy\, dx = \frac{2}{\pi^2} \int_0^\pi [-\cos(x + y)]_0^{\pi/2}\, dx = \frac{2}{\pi^2} \int_0^\pi \left[-\cos\left(x + \frac{\pi}{2}\right) + \cos x\right] dx$
$= \frac{2}{\pi^2} \left[-\sin\left(x + \frac{\pi}{2}\right) + \sin x\right]_0^\pi = \frac{2}{\pi^2} \left[\left(-\sin \frac{3\pi}{2} + \sin \pi\right) - \left(-\sin \frac{\pi}{2} + \sin 0\right)\right] = \frac{4}{\pi^2}$

17. average height $= \frac{1}{4} \int_0^2 \int_0^2 (x^2 + y^2)\, dy\, dx = \frac{1}{4} \int_0^2 \left[x^2 y + \frac{y^3}{3}\right]_0^2 dx = \frac{1}{4} \int_0^2 \left(2x^2 + \frac{8}{3}\right) dx = \frac{1}{2} \left[\frac{x^3}{3} + \frac{4x}{3}\right]_0^2 = \frac{8}{3}$

19. $\int_{-5}^5 \int_{-2}^0 \frac{10{,}000 e^y}{1 + \frac{|x|}{2}}\, dy\, dx = 10{,}000\left(1 - e^{-2}\right) \int_{-5}^5 \frac{dx}{1 + \frac{|x|}{2}} = 10{,}000\left(1 - e^{-2}\right) \left[\int_{-5}^0 \frac{dx}{1 - \frac{x}{2}} + \int_0^5 \frac{dx}{1 + \frac{x}{2}}\right]$
$= 10{,}000\left(1 - e^{-2}\right) \left[-2 \ln\left(1 - \frac{x}{2}\right)\right]_{-5}^0 + 10{,}000\left(1 - e^{-2}\right) \left[2 \ln\left(1 + \frac{x}{2}\right)\right]_0^5$
$= 10{,}000\left(1 - e^{-2}\right) \left[2 \ln\left(1 + \frac{5}{2}\right)\right] + 10{,}000\left(1 - e^{-2}\right) \left[2 \ln\left(1 + \frac{5}{2}\right)\right] = 40{,}000\left(1 - e^{-2}\right) \ln\left(\frac{7}{2}\right) \approx 43{,}329$

21. Let (x_i, y_i) be the location of the weather station in county i for $i = 1, \ldots, 254$. The average temperature in Texas at time t_0 is approximately $\frac{\sum_{i=1}^{254} T(x_i, y_i)\, \Delta_i A}{A}$, where $T(x_i, y_i)$ is the temperature at time t_0 at the weather station in county i, $\Delta_i A$ is the area of county i, and A is the area of Texas.

12.4 DOUBLE INTEGRALS IN POLAR FORM

1. $\int_{-1}^1 \int_0^{\sqrt{1-x^2}} dy\, dx = \int_0^\pi \int_0^1 r\, dr\, d\theta = \frac{1}{2} \int_0^\pi d\theta = \frac{\pi}{2}$

3. $\int_0^1 \int_0^{\sqrt{1-y^2}} (x^2 + y^2)\, dx\, dy = \int_0^{\pi/2} \int_0^1 r^3\, dr\, d\theta = \frac{1}{4} \int_0^{\pi/2} d\theta = \frac{\pi}{8}$

5. $\int_{-a}^a \int_{-\sqrt{a^2-x^2}}^{\sqrt{a^2-x^2}} dy\, dx = \int_0^{2\pi} \int_0^a r\, dr\, d\theta = \frac{a^2}{2} \int_0^{2\pi} d\theta = \pi a^2$

7. $\int_0^6 \int_0^y x\,dx\,dy = \int_{\pi/4}^{\pi/2} \int_0^{6\csc\theta} r^2 \cos\theta\,dr\,d\theta = 72 \int_{\pi/4}^{\pi/2} \cot\theta \csc^2\theta\,d\theta = -36\left[\cot^2\theta\right]_{\pi/4}^{\pi/2} = 36$

9. $\int_{-1}^0 \int_{-\sqrt{1-x^2}}^0 \frac{2}{1+\sqrt{x^2+y^2}}\,dy\,dx = \int_{\pi}^{3\pi/2} \int_0^1 \frac{2r}{1+r}\,dr\,d\theta = 2\int_{\pi}^{3\pi/2} \int_0^1 \left(1 - \frac{1}{1+r}\right) dr\,d\theta = 2\int_{\pi}^{3\pi/2} (1-\ln 2)\,d\theta$
$= (1-\ln 2)\pi$

11. $\int_0^{\ln 2} \int_0^{\sqrt{(\ln 2)^2 - y^2}} e^{\sqrt{x^2+y^2}}\,dx\,dy = \int_0^{\pi/2} \int_0^{\ln 2} re^r\,dr\,d\theta = \int_0^{\pi/2} (2\ln 2 - 1)\,d\theta = \frac{\pi}{2}(2\ln 2 - 1)$

13. $\int_0^2 \int_0^{\sqrt{1-(x-1)^2}} \frac{x+y}{x^2+y^2}\,dy\,dx = \int_0^{\pi/2} \int_0^{2\cos\theta} \frac{r(\cos\theta+\sin\theta)}{r^2}\,r\,dr\,d\theta = \int_0^{\pi/2} (2\cos^2\theta + 2\sin\theta\cos\theta)\,d\theta$
$= \left[\theta + \frac{\sin 2\theta}{2} + \sin^2\theta\right]_0^{\pi/2} = \frac{\pi+2}{2} = \frac{\pi}{2} + 1$

15. $\int_{-1}^1 \int_{-\sqrt{1-y^2}}^{\sqrt{1-y^2}} \ln(x^2+y^2+1)\,dx\,dy = 4\int_0^{\pi/2} \int_0^1 \ln(r^2+1)\,r\,dr\,d\theta = 2\int_0^{\pi/2} (\ln 4 - 1)\,d\theta = \pi(\ln 4 - 1)$

17. $\int_0^{\pi/2} \int_0^{2\sqrt{2-\sin 2\theta}} r\,dr\,d\theta = 2\int_0^{\pi/2} (2-\sin 2\theta)\,d\theta = 2(\pi - 1)$

19. $A = 2\int_0^{\pi/6} \int_0^{12\cos 3\theta} r\,dr\,d\theta = 144\int_0^{\pi/6} \cos^2 3\theta\,d\theta = 12\pi$

21. $A = \int_0^{\pi/2} \int_0^{1+\sin\theta} r\,dr\,d\theta = \frac{1}{2}\int_0^{\pi/2} \left(\frac{3}{2} + 2\sin\theta - \frac{\cos 2\theta}{2}\right) d\theta = \frac{3\pi}{8} + 1$

23. $\text{average} = \frac{4}{\pi a^2} \int_0^{\pi/2} \int_0^a r\sqrt{a^2 - r^2}\,dr\,d\theta = \frac{4}{3\pi a^2} \int_0^{\pi/2} a^3\,d\theta = \frac{2a}{3}$

25. $\text{average} = \frac{1}{\pi a^2} \int_{-a}^a \int_{-\sqrt{a^2-x^2}}^{\sqrt{a^2-x^2}} \sqrt{x^2+y^2}\,dy\,dx = \frac{1}{\pi a^2} \int_0^{2\pi} \int_0^a r^2\,dr\,d\theta = \frac{a}{3\pi} \int_0^{2\pi} d\theta = \frac{2a}{3}$

27. $\int_0^{2\pi} \int_1^{\sqrt{e}} \left(\frac{\ln r^2}{r}\right) r\,dr\,d\theta = \int_0^{2\pi} \int_1^{\sqrt{e}} 2\ln r\,dr\,d\theta = 2\int_0^{2\pi} \left[r\ln r - r\right]_1^{e^{1/2}} d\theta = 2\int_0^{2\pi} \sqrt{e}\left[\left(\frac{1}{2} - 1\right) + 1\right] d\theta = 2\pi\left(2 - \sqrt{e}\right)$

29. $V = 2\int_0^{\pi/2} \int_1^{1+\cos\theta} r^2\cos\theta\,dr\,d\theta = \frac{2}{3}\int_0^{\pi/2} (3\cos^2\theta + 3\cos^3\theta + \cos^4\theta)\,d\theta$
$= \frac{2}{3}\left[\frac{15\theta}{8} + \sin 2\theta + 3\sin\theta - \sin^3\theta + \frac{\sin 4\theta}{32}\right]_0^{\pi/2} = \frac{4}{3} + \frac{5\pi}{8}$

31. (a) $I^2 = \int_0^\infty \int_0^\infty e^{-(x^2+y^2)}\,dx\,dy = \int_0^{\pi/2} \int_0^\infty \left(e^{-r^2}\right) r\,dr\,d\theta = \int_0^{\pi/2} \left[\lim_{b\to\infty} \int_0^b re^{-r^2}\,dr\right] d\theta$
$= -\frac{1}{2}\int_0^{\pi/2} \lim_{b\to\infty} \left(e^{-b^2} - 1\right) d\theta = \frac{1}{2}\int_0^{\pi/2} d\theta = \frac{\pi}{4} \Rightarrow I = \frac{\sqrt{\pi}}{2}$

(b) $\lim_{x\to\infty} \int_0^x \frac{2e^{-t^2}}{\sqrt{\pi}}\,dt = \frac{2}{\sqrt{\pi}} \int_0^\infty e^{-t^2}\,dt = \left(\frac{2}{\sqrt{\pi}}\right)\left(\frac{\sqrt{\pi}}{2}\right) = 1$, from part (a)

33. Over the disk $x^2 + y^2 \le \frac{3}{4}$: $\iint_R \frac{1}{1-x^2-y^2}\,dA = \int_0^{2\pi} \int_0^{\sqrt{3}/2} \frac{r}{1-r^2}\,dr\,d\theta = \int_0^{2\pi} \left[-\frac{1}{2}\ln(1-r^2)\right]_0^{\sqrt{3}/2} d\theta$
$= \int_0^{2\pi} \left(-\frac{1}{2}\ln\frac{1}{4}\right) d\theta = (\ln 2)\int_0^{2\pi} d\theta = \pi\ln 4$

Over the disk $x^2 + y^2 \le 1$: $\iint_R \frac{1}{1-x^2-y^2}\,dA = \int_0^{2\pi} \int_0^1 \frac{r}{1-r^2}\,dr\,d\theta = \int_0^{2\pi} \left[\lim_{a\to 1^-} \int_0^a \frac{r}{1-r^2}\,dr\right] d\theta$
$= \int_0^{2\pi} \lim_{a\to 1^-} \left[-\frac{1}{2}\ln(1-a^2)\right] d\theta = 2\pi \cdot \lim_{a\to 1^-} \left[-\frac{1}{2}\ln(1-a^2)\right] = 2\pi\cdot\infty$, so the integral does not exist over $x^2 + y^2 \le 1$

35. average $= \frac{1}{\pi a^2} \int_0^{2\pi} \int_0^a [(r \cos \theta - h)^2 + r^2 \sin^2 \theta]\, r\, dr\, d\theta = \frac{1}{\pi a^2} \int_0^{2\pi} \int_0^a (r^3 - 2r^2 h \cos \theta + rh^2)\, dr\, d\theta$

$= \frac{1}{\pi a^2} \int_0^{2\pi} \left(\frac{a^4}{4} - \frac{2a^3 h \cos \theta}{3} + \frac{a^2 h^2}{2} \right) d\theta = \frac{1}{\pi} \int_0^{2\pi} \left(\frac{a^2}{4} - \frac{2ah \cos \theta}{3} + \frac{h^2}{2} \right) d\theta = \frac{1}{\pi} \left[\frac{a^2 \theta}{4} - \frac{2ah \sin \theta}{3} + \frac{h^2 \theta}{2} \right]_0^{2\pi}$

$= \frac{1}{2} (a^2 + 2h^2)$

12.5 TRIPLE INTEGRALS IN RECTANGULAR COORDINATES

1. $\int_0^1 \int_0^{1-x} \int_{x+z}^1 F(x, y, z)\, dy\, dz\, dx = \int_0^1 \int_0^{1-x} \int_{x+z}^1 dy\, dz\, dx = \int_0^1 \int_0^{1-x} (1 - x - z)\, dz\, dx$

$= \int_0^1 \left[(1 - x) - x(1 - x) - \frac{(1-x)^2}{2} \right] dx = \int_0^1 \frac{(1-x)^2}{2} dx = \left[-\frac{(1-x)^3}{6} \right]_0^1 = \frac{1}{6}$

3. $\int_0^1 \int_0^{2-2x} \int_0^{3-3x-3y/2} dz\, dy\, dx$

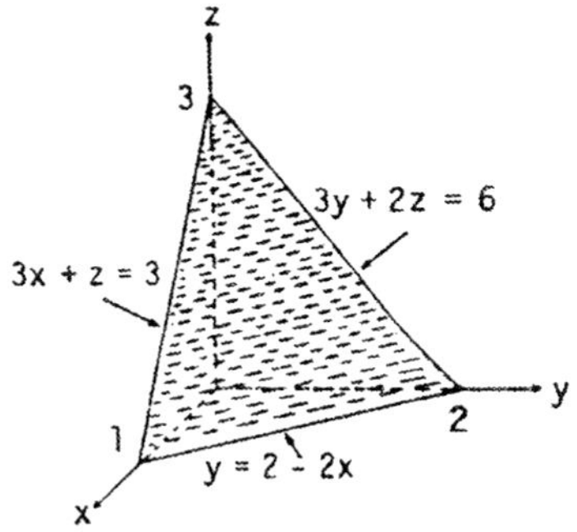

$= \int_0^1 \int_0^{2-2x} \left(3 - 3x - \frac{3}{2} y \right) dy\, dx$

$= \int_0^1 \left[3(1 - x) \cdot 2(1 - x) - \frac{3}{4} \cdot 4(1 - x)^2 \right] dx$

$= 3 \int_0^1 (1 - x)^2\, dx = \left[-(1 - x)^3 \right]_0^1 = 1,$

$\int_0^2 \int_0^{1-y/2} \int_0^{3-3x-3y/2} dz\, dx\, dy,\ \int_0^1 \int_0^{3-3x} \int_0^{2-2x-2z/3} dy\, dz\, dx,$

$\int_0^3 \int_0^{1-z/3} \int_0^{2-2x-2z/3} dy\, dx\, dz,\ \int_0^2 \int_0^{3-3y/2} \int_0^{1-y/2-z/3} dx\, dz\, dy,$

$\int_0^3 \int_0^{2-2z/3} \int_0^{1-y/2-z/3} dx\, dy\, dz$

5. $\int_{-2}^2 \int_{-\sqrt{4-x^2}}^{\sqrt{4-x^2}} \int_{x^2+y^2}^{8-x^2-y^2} dz\, dy\, dx = 4 \int_0^2 \int_0^{\sqrt{4-x^2}} \int_{x^2+y^2}^{8-x^2-y^2} dz\, dy\, dx$

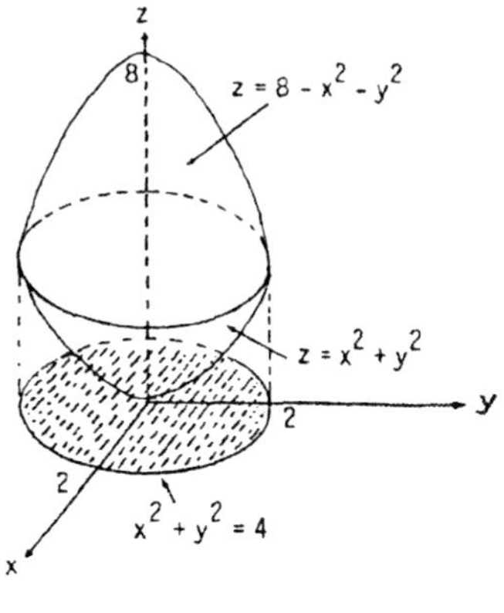

$= 4 \int_0^2 \int_0^{\sqrt{4-x^2}} [8 - 2(x^2 + y^2)]\, dy\, dx$

$= 8 \int_0^2 \int_0^{\sqrt{4-x^2}} (4 - x^2 - y^2)\, dy\, dx$

$= 8 \int_0^{\pi/2} \int_0^2 (4 - r^2)\, r\, dr\, d\theta = 8 \int_0^{\pi/2} \left[2r^2 - \frac{r^4}{4} \right]_0^2 d\theta$

$= 32 \int_0^{\pi/2} d\theta = 32 \left(\frac{\pi}{2} \right) = 16\pi,$

$\int_{-2}^2 \int_{-\sqrt{4-y^2}}^{\sqrt{4-y^2}} \int_{x^2+y^2}^{8-x^2-y^2} dz\, dx\, dy,$

$\int_{-2}^2 \int_{y^2}^4 \int_{-\sqrt{z-y^2}}^{\sqrt{z-y^2}} dx\, dz\, dy + \int_{-2}^2 \int_4^{8-y^2} \int_{-\sqrt{8-z-y^2}}^{\sqrt{8-z-y^2}} dx\, dz\, dy,$

$\int_0^4 \int_{-\sqrt{z}}^{\sqrt{z}} \int_{-\sqrt{z-y^2}}^{\sqrt{z-y^2}} dx\, dy\, dz + \int_4^8 \int_{-\sqrt{8-z}}^{\sqrt{8-z}} \int_{-\sqrt{8-z-y^2}}^{\sqrt{8-z-y^2}} dx\, dy\, dz,\ \int_{-2}^2 \int_{x^2}^4 \int_{-\sqrt{z-x^2}}^{\sqrt{z-x^2}} dy\, dz\, dx + \int_{-2}^2 \int_4^{8-x^2} \int_{-\sqrt{8-z-x^2}}^{\sqrt{8-z-x^2}} dy\, dz\, dx,$

$\int_0^4 \int_{-\sqrt{z}}^{\sqrt{z}} \int_{-\sqrt{z-x^2}}^{\sqrt{z-x^2}} dy\, dx\, dz + \int_4^8 \int_{-\sqrt{8-z}}^{\sqrt{8-z}} \int_{-\sqrt{8-z-x^2}}^{\sqrt{8-z-x^2}} dy\, dx\, dz$

7. $\int_0^1 \int_0^1 \int_0^1 (x^2 + y^2 + z^2)\, dz\, dy\, dx = \int_0^1 \int_0^1 \left(x^2 + y^2 + \frac{1}{3} \right) dy\, dx = \int_0^1 \left(x^2 + \frac{2}{3} \right) dx = 1$

9. $\int_1^e \int_1^e \int_1^e \frac{1}{xyz}\, dx\, dy\, dz = \int_1^e \int_1^e \left[\frac{\ln x}{yz} \right]_1^e dy\, dz = \int_1^e \int_1^e \frac{1}{yz}\, dy\, dz = \int_1^e \left[\frac{\ln y}{z} \right]_1^e dz = \int_1^e \frac{1}{z}\, dz = 1$

11. $\int_0^1 \int_0^\pi \int_0^\pi y \sin z\, dx\, dy\, dz = \int_0^1 \int_0^\pi \pi y \sin z\, dy\, dz = \frac{\pi^3}{2} \int_0^1 \sin z\, dz = \frac{\pi^3}{2} (1 - \cos 1)$

13. $\int_0^3 \int_0^{\sqrt{9-x^2}} \int_0^{\sqrt{9-x^2}} dz\, dy\, dx = \int_0^3 \int_0^{\sqrt{9-x^2}} \sqrt{9 - x^2}\, dy\, dx = \int_0^3 (9 - x^2)\, dx = \left[9x - \frac{x^3}{3} \right]_0^3 = 18$

15. $\int_0^1 \int_0^{2-x} \int_0^{2-x-y} dz\,dy\,dx = \int_0^1 \int_0^{2-x} (2-x-y)\,dy\,dx = \int_0^1 \left[(2-x)^2 - \frac{1}{2}(2-x)^2\right] dx = \frac{1}{2}\int_0^1 (2-x)^2\,dx$
$= \left[-\frac{1}{6}(2-x)^3\right]_0^1 = -\frac{1}{6} + \frac{8}{6} = \frac{7}{6}$

17. $\int_0^\pi \int_0^\pi \int_0^\pi \cos(u+v+w)\,du\,dv\,dw = \int_0^\pi \int_0^\pi [\sin(w+v+\pi) - \sin(w+v)]\,dv\,dw$
$= \int_0^\pi [(-\cos(w+2\pi) + \cos(w+\pi)) + (\cos(w+\pi) - \cos w)]\,dw$
$= [-\sin(w+2\pi) + \sin(w+\pi) - \sin w + \sin(w+\pi)]_0^\pi = 0$

19. $\int_0^{\pi/4} \int_0^{\ln \sec v} \int_{-\infty}^{2t} e^x\,dx\,dt\,dv = \int_0^{\pi/4} \int_0^{\ln \sec v} \lim_{b \to -\infty} (e^{2t} - e^b)\,dt\,dv = \int_0^{\pi/4} \int_0^{\ln \sec v} e^{2t}\,dt\,dv = \int_0^{\pi/4} \left(\frac{1}{2}e^{2\ln \sec v} - \frac{1}{2}\right) dv$
$= \int_0^{\pi/4} \left(\frac{\sec^2 v}{2} - \frac{1}{2}\right) dv = \left[\frac{\tan v}{2} - \frac{v}{2}\right]_0^{\pi/4} = \frac{1}{2} - \frac{\pi}{8}$

21. (a) $\int_{-1}^1 \int_0^{1-x^2} \int_{x^2}^{1-z} dy\,dz\,dx$ (b) $\int_0^1 \int_{-\sqrt{1-z}}^{\sqrt{1-z}} \int_{x^2}^{1-z} dy\,dx\,dz$ (c) $\int_0^1 \int_0^{1-z} \int_{-\sqrt{y}}^{\sqrt{y}} dx\,dy\,dz$

(d) $\int_0^1 \int_0^{1-y} \int_{-\sqrt{y}}^{\sqrt{y}} dx\,dz\,dy$ (e) $\int_0^1 \int_{-\sqrt{y}}^{\sqrt{y}} \int_0^{1-y} dz\,dx\,dy$

23. $V = \int_0^1 \int_{-1}^1 \int_0^{y^2} dz\,dy\,dx = \int_0^1 \int_{-1}^1 y^2\,dy\,dx = \frac{2}{3}\int_0^1 dx = \frac{2}{3}$

25. $V = \int_0^4 \int_0^{\sqrt{4-x}} \int_0^{2-y} dz\,dy\,dx = \int_0^4 \int_0^{\sqrt{4-x}} (2-y)\,dy\,dx = \int_0^4 \left[2\sqrt{4-x} - \left(\frac{4-x}{2}\right)\right] dx$
$= \left[-\frac{4}{3}(4-x)^{3/2} + \frac{1}{4}(4-x)^2\right]_0^4 = \frac{4}{3}(4)^{3/2} - \frac{1}{4}(16) = \frac{32}{3} - 4 = \frac{20}{3}$

27. $V = \int_0^1 \int_0^{2-2x} \int_0^{3-3x-3y/2} dz\,dy\,dx = \int_0^1 \int_0^{2-2x} \left(3 - 3x - \frac{3}{2}y\right) dy\,dx = \int_0^1 \left[6(1-x)^2 - \frac{3}{4}\cdot 4(1-x)^2\right] dx$
$= \int_0^1 3(1-x)^2\,dx = [-(1-x)^3]_0^1 = 1$

29. $V = 8\int_0^1 \int_0^{\sqrt{1-x^2}} \int_0^{\sqrt{1-x^2}} dz\,dy\,dx = 8\int_0^1 \int_0^{\sqrt{1-x^2}} \sqrt{1-x^2}\,dy\,dx = 8\int_0^1 (1-x^2)\,dx = \frac{16}{3}$

31. $V = \int_0^4 \int_0^{(\sqrt{16-y^2})/2} \int_0^{4-y} dx\,dz\,dy = \int_0^4 \int_0^{(\sqrt{16-y^2})/2} (4-y)\,dz\,dy = \int_0^4 \frac{\sqrt{16-y^2}}{2}(4-y)\,dy$
$= \int_0^4 2\sqrt{16-y^2}\,dy - \frac{1}{2}\int_0^4 y\sqrt{16-y^2}\,dy = \left[y\sqrt{16-y^2} + 16\sin^{-1}\frac{y}{4}\right]_0^4 + \left[\frac{1}{6}(16-y^2)^{3/2}\right]_0^4$
$= 16\left(\frac{\pi}{2}\right) - \frac{1}{6}(16)^{3/2} = 8\pi - \frac{32}{3}$

33. $\int_0^2 \int_0^{2-x} \int_{(2-x-y)/2}^{4-2x-2y} dz\,dy\,dx = \int_0^2 \int_0^{2-x} \left(3 - \frac{3x}{2} - \frac{3y}{2}\right) dy\,dx$
$= \int_0^2 \left[3\left(1 - \frac{x}{2}\right)(2-x) - \frac{3}{4}(2-x)^2\right] dx$
$= \int_0^2 \left[6 - 6x + \frac{3x^2}{2} - \frac{3(2-x)^2}{4}\right] dx$
$= \left[6x - 3x^2 + \frac{x^3}{2} + \frac{(2-x)^3}{4}\right]_0^2 = (12 - 12 + 4 + 0) - \frac{2^3}{4} = 2$

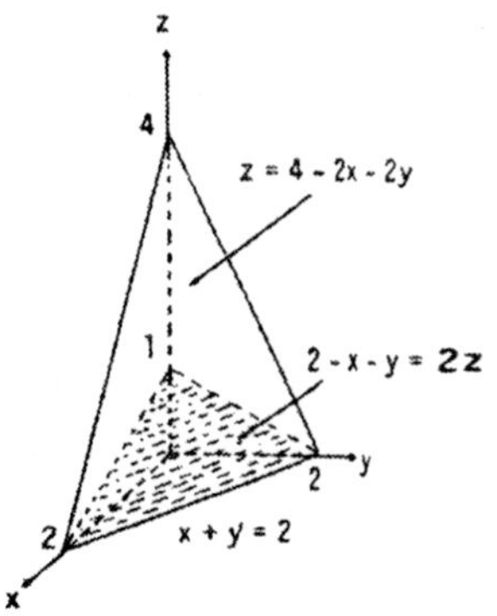

35. $V = 2\int_{-2}^{2}\int_{0}^{\sqrt{4-x^2}/2}\int_{0}^{x+2} dz\,dy\,dx = 2\int_{-2}^{2}\int_{0}^{\sqrt{4-x^2}/2}(x+2)\,dy\,dx = \int_{-2}^{2}(x+2)\sqrt{4-x^2}\,dx$
$= \int_{-2}^{2} 2\sqrt{4-x^2}\,dx + \int_{-2}^{2} x\sqrt{4-x^2}\,dx = \left[x\sqrt{4-x^2} + 4\sin^{-1}\frac{x}{2}\right]_{-2}^{2} + \left[-\frac{1}{3}(4-x^2)^{3/2}\right]_{-2}^{2}$
$= 4\left(\frac{\pi}{2}\right) - 4\left(-\frac{\pi}{2}\right) = 4\pi$

37. $\text{average} = \frac{1}{8}\int_{0}^{2}\int_{0}^{2}\int_{0}^{2}(x^2+9)\,dz\,dy\,dx = \frac{1}{8}\int_{0}^{2}\int_{0}^{2}(2x^2+18)\,dy\,dx = \frac{1}{8}\int_{0}^{2}(4x^2+36)\,dx = \frac{31}{3}$

39. $\text{average} = \int_{0}^{1}\int_{0}^{1}\int_{0}^{1}(x^2+y^2+z^2)\,dz\,dy\,dx = \int_{0}^{1}\int_{0}^{1}\left(x^2+y^2+\frac{1}{3}\right)dy\,dx = \int_{0}^{1}\left(x^2+\frac{2}{3}\right)dx = 1$

41. $\int_{0}^{4}\int_{0}^{1}\int_{2y}^{2}\frac{4\cos(x^2)}{2\sqrt{z}}\,dx\,dy\,dz = \int_{0}^{4}\int_{0}^{2}\int_{0}^{x/2}\frac{4\cos(x^2)}{2\sqrt{z}}\,dy\,dx\,dz = \int_{0}^{4}\int_{0}^{2}\frac{x\cos(x^2)}{\sqrt{z}}\,dx\,dz = \int_{0}^{4}\left(\frac{\sin 4}{2}\right)z^{-1/2}\,dz$
$= \left[(\sin 4)z^{1/2}\right]_{0}^{4} = 2\sin 4$

43. $\int_{0}^{1}\int_{\sqrt[3]{z}}^{1}\int_{0}^{\ln 3}\frac{\pi e^{2x}\sin(\pi y^2)}{y^2}\,dx\,dy\,dz = \int_{0}^{1}\int_{\sqrt[3]{z}}^{1}\frac{4\pi\sin(\pi y^2)}{y^2}\,dy\,dz = \int_{0}^{1}\int_{0}^{y^3}\frac{4\pi\sin(\pi y^2)}{y^2}\,dz\,dy$
$= \int_{0}^{1}4\pi y\sin(\pi y^2)\,dy = \left[-2\cos(\pi y^2)\right]_{0}^{1} = -2(-1)+2(1) = 4$

45. $\int_{0}^{1}\int_{0}^{4-a-x^2}\int_{a}^{4-x^2-y}dz\,dy\,dx = \frac{4}{15} \Rightarrow \int_{0}^{1}\int_{0}^{4-a-x^2}(4-x^2-y-a)\,dy\,dx = \frac{4}{15}$
$\Rightarrow \int_{0}^{1}\left[(4-a-x^2)^2 - \frac{1}{2}(4-a-x^2)^2\right]dx = \frac{4}{15} \Rightarrow \frac{1}{2}\int_{0}^{1}(4-a-x^2)^2\,dx = \frac{4}{15} \Rightarrow \int_{0}^{1}\left[(4-a)^2 - 2x^2(4-a) + x^4\right]dx$
$= \frac{8}{15} \Rightarrow \left[(4-a)^2x - \frac{2}{3}x^3(4-a) + \frac{x^5}{5}\right]_{0}^{1} = \frac{8}{15} \Rightarrow (4-a)^2 - \frac{2}{3}(4-a) + \frac{1}{5} = \frac{8}{15} \Rightarrow 15(4-a)^2 - 10(4-a) - 5 = 0$
$\Rightarrow 3(4-a)^2 - 2(4-a) - 1 = 0 \Rightarrow [3(4-a)+1][(4-a)-1] = 0 \Rightarrow 4-a = -\frac{1}{3} \text{ or } 4-a = 1 \Rightarrow a = \frac{13}{3} \text{ or } a = 3$

47. To minimize the integral, we want the domain to include all points where the integrand is negative and to exclude all points where it is positive. These criteria are met by the points (x, y, z) such that $4x^2 + 4y^2 + z^2 - 4 \le 0$ or $4x^2 + 4y^2 + z^2 \le 4$, which is a solid ellipsoid centered at the origin.

12.6 MOMENTS AND CENTERS OF MASS

1. $M = \int_{0}^{1}\int_{x}^{2-x^2}3\,dy\,dx = 3\int_{0}^{1}(2-x^2-x)\,dx = \frac{7}{2}$; $M_y = \int_{0}^{1}\int_{x}^{2-x^2}3x\,dy\,dx = 3\int_{0}^{1}[xy]_{x}^{2-x^2}\,dx$
$= 3\int_{0}^{1}(2x-x^3-x^2)\,dx = \frac{5}{4}$; $M_x = \int_{0}^{1}\int_{x}^{2-x^2}3y\,dy\,dx = \frac{3}{2}\int_{0}^{1}[y^2]_{x}^{2-x^2}\,dx = \frac{3}{2}\int_{0}^{1}(4-5x^2+x^4)\,dx = \frac{19}{5}$
$\Rightarrow \overline{x} = \frac{5}{14}$ and $\overline{y} = \frac{38}{35}$

3. $M = \int_{0}^{2}\int_{y^2/2}^{4-y}dx\,dy = \int_{0}^{2}\left(4-y-\frac{y^2}{2}\right)dy = \frac{14}{3}$; $M_y = \int_{0}^{2}\int_{y^2/2}^{4-y}x\,dx\,dy = \frac{1}{2}\int_{0}^{2}[x^2]_{y^2/2}^{4-y}\,dy$
$= \frac{1}{2}\int_{0}^{2}\left(16-8y+y^2-\frac{y^4}{4}\right)dy = \frac{128}{15}$; $M_x = \int_{0}^{2}\int_{y^2/2}^{4-y}y\,dx\,dy = \int_{0}^{2}\left(4y-y^2-\frac{y^3}{2}\right)dy = \frac{10}{3}$
$\Rightarrow \overline{x} = \frac{64}{35}$ and $\overline{y} = \frac{5}{7}$

5. $M = \int_{0}^{a}\int_{0}^{\sqrt{a^2-x^2}}dy\,dx = \frac{\pi a^2}{4}$; $M_y = \int_{0}^{a}\int_{0}^{\sqrt{a^2-x^2}}x\,dy\,dx = \int_{0}^{a}[xy]_{0}^{\sqrt{a^2-x^2}}\,dx = \int_{0}^{a}x\sqrt{a^2-x^2}\,dx = \frac{a^3}{3}$
$\Rightarrow \overline{x} = \overline{y} = \frac{4a}{3\pi}$, by symmetry

7. $I_x = \int_{-2}^{2}\int_{-\sqrt{4-x^2}}^{\sqrt{4-x^2}}y^2\,dy\,dx = \int_{-2}^{2}\left[\frac{y^3}{3}\right]_{-\sqrt{4-x^2}}^{\sqrt{4-x^2}}dx = \frac{2}{3}\int_{-2}^{2}(4-x^2)^{3/2}\,dx = 4\pi$; $I_y = 4\pi$, by symmetry;
$I_o = I_x + I_y = 8\pi$

9. $M = \int_{-\infty}^{0}\int_{0}^{e^x} dy\,dx = \int_{-\infty}^{0} e^x\,dx = \lim_{b \to -\infty} \int_{b}^{0} e^x\,dx = 1 - \lim_{b \to -\infty} e^b = 1;\ M_y = \int_{-\infty}^{0}\int_{0}^{e^x} x\,dy\,dx = \int_{-\infty}^{0} xe^x\,dx$
$= \lim_{b \to -\infty} \int_{b}^{0} xe^x\,dx = \lim_{b \to -\infty} \left[xe^x - e^x\right]_b^0 = -1 - \lim_{b \to -\infty} (be^b - e^b) = -1;\ M_x = \int_{-\infty}^{0}\int_{0}^{e^x} y\,dy\,dx$
$= \frac{1}{2}\int_{-\infty}^{0} e^{2x}\,dx = \frac{1}{2}\lim_{b \to -\infty} \int_{b}^{0} e^{2x}\,dx = \frac{1}{4} \Rightarrow \overline{x} = -1 \text{ and } \overline{y} = \frac{1}{4}$

11. $M = \int_{0}^{2}\int_{-y}^{y-y^2} (x+y)\,dx\,dy = \int_{0}^{2}\left[\frac{x^2}{2} + xy\right]_{-y}^{y-y^2} dy = \int_{0}^{2}\left(\frac{y^4}{2} - 2y^3 + 2y^2\right) dy = \left[\frac{y^5}{10} - \frac{y^4}{2} + \frac{2y^3}{3}\right]_0^2 = \frac{8}{15};$
$I_x = \int_{0}^{2}\int_{-y}^{y-y^2} y^2(x+y)\,dx\,dy = \int_{0}^{2}\left[\frac{x^2y^2}{2} + xy^3\right]_{-y}^{y-y^2} dy = \int_{0}^{2}\left(\frac{y^6}{2} - 2y^5 + 2y^4\right) dy = \frac{64}{105};$

13. $M = \int_{0}^{1}\int_{x}^{2-x} (6x+3y+3)\,dy\,dx = \int_{0}^{1}\left[6xy + \frac{3}{2}y^2 + 3y\right]_x^{2-x} dx = \int_{0}^{1} (12 - 12x^2)\,dx = 8;$
$M_y = \int_{0}^{1}\int_{x}^{2-x} x(6x+3y+3)\,dy\,dx = \int_{0}^{1} (12x - 12x^3)\,dx = 3;\ M_x = \int_{0}^{1}\int_{x}^{2-x} y(6x+3y+3)\,dy\,dx$
$= \int_{0}^{1} (14 - 6x - 6x^2 - 2x^3)\,dx = \frac{17}{2} \Rightarrow \overline{x} = \frac{3}{8} \text{ and } \overline{y} = \frac{17}{16}$

15. $M = \int_{0}^{1}\int_{0}^{6} (x+y+1)\,dx\,dy = \int_{0}^{1} (6y+24)\,dy = 27;\ M_x = \int_{0}^{1}\int_{0}^{6} y(x+y+1)\,dx\,dy = \int_{0}^{1} y(6y+24)\,dy = 14;$
$M_y = \int_{0}^{1}\int_{0}^{6} x(x+y+1)\,dx\,dy = \int_{0}^{1} (18y+90)\,dy = 99 \Rightarrow \overline{x} = \frac{11}{3} \text{ and } \overline{y} = \frac{14}{27};\ I_y = \int_{0}^{1}\int_{0}^{6} x^2(x+y+1)\,dx\,dy$
$= 216\int_{0}^{1}\left(\frac{y}{3} + \frac{11}{6}\right) dy = 432$

17. $M = \int_{-1}^{1}\int_{0}^{x^2} (7y+1)\,dy\,dx = \int_{-1}^{1}\left(\frac{7x^4}{2} + x^2\right) dx = \frac{31}{15};\ M_x = \int_{-1}^{1}\int_{0}^{x^2} y(7y+1)\,dy\,dx = \int_{-1}^{1}\left(\frac{7x^6}{3} + \frac{x^4}{2}\right) dx = \frac{13}{15};$
$M_y = \int_{-1}^{1}\int_{0}^{x^2} x(7y+1)\,dy\,dx = \int_{-1}^{1}\left(\frac{7x^5}{2} + x^3\right) dx = 0 \Rightarrow \overline{x} = 0 \text{ and } \overline{y} = \frac{13}{31};\ I_y = \int_{-1}^{1}\int_{0}^{x^2} x^2(7y+1)\,dy\,dx$
$= \int_{-1}^{1}\left(\frac{7x^6}{2} + x^4\right) dx = \frac{7}{5}$

19. $M = \int_{0}^{1}\int_{-y}^{y} (y+1)\,dx\,dy = \int_{0}^{1} (2y^2+2y)\,dy = \frac{5}{3};\ M_x = \int_{0}^{1}\int_{-y}^{y} y(y+1)\,dx\,dy = 2\int_{0}^{1} (y^3+y^2)\,dy = \frac{7}{6};$
$M_y = \int_{0}^{1}\int_{-y}^{y} x(y+1)\,dx\,dy = \int_{0}^{1} 0\,dy = 0 \Rightarrow \overline{x} = 0 \text{ and } \overline{y} = \frac{7}{10};\ I_x = \int_{0}^{1}\int_{-y}^{y} y^2(y+1)\,dx\,dy = \int_{0}^{1} (2y^4+2y^3)\,dy$
$= \frac{9}{10};\ I_y = \int_{0}^{1}\int_{-y}^{y} x^2(y+1)\,dx\,dy = \frac{1}{3}\int_{0}^{1} (2y^4+2y^3)\,dy = \frac{3}{10} \Rightarrow I_o = I_x + I_y = \frac{6}{5}$

21. $I_x = \int_{0}^{a}\int_{0}^{b}\int_{0}^{c} (y^2+z^2)\,dz\,dy\,dx = \int_{0}^{a}\int_{0}^{b}\left(cy^2 + \frac{c^3}{3}\right) dy\,dx = \int_{0}^{a}\left(\frac{cb^3}{3} + \frac{c^3b}{3}\right) dx = \frac{abc\,(b^2+c^2)}{3}$
$= \frac{M}{3}(b^2+c^2)$ where $M = abc$; $I_y = \frac{M}{3}(a^2+c^2)$ and $I_z = \frac{M}{3}(a^2+b^2)$, by symmetry

23. $M = 4\int_{0}^{1}\int_{0}^{1}\int_{4y^2}^{4} dz\,dy\,dx = 4\int_{0}^{1}\int_{0}^{1} (4-4y^2)\,dy\,dx = 16\int_{0}^{1} \frac{2}{3}\,dx = \frac{32}{3};\ M_{xy} = 4\int_{0}^{1}\int_{0}^{1}\int_{4y^2}^{4} z\,dz\,dy\,dx$
$= 2\int_{0}^{1}\int_{0}^{1} (16 - 16y^4)\,dy\,dx = \frac{128}{5}\int_{0}^{1} dx = \frac{128}{5} \Rightarrow \overline{z} = \frac{12}{5}$, and $\overline{x} = \overline{y} = 0$, by symmetry;
$I_x = 4\int_{0}^{1}\int_{0}^{1}\int_{4y^2}^{4} (y^2+z^2)\,dz\,dy\,dx = 4\int_{0}^{1}\int_{0}^{1}\left[\left(4y^2 + \frac{64}{3}\right) - \left(4y^4 + \frac{64y^6}{3}\right)\right] dy\,dx = 4\int_{0}^{1} \frac{1976}{105}\,dx = \frac{7904}{105};$
$I_y = 4\int_{0}^{1}\int_{0}^{1}\int_{4y^2}^{4} (x^2+z^2)\,dz\,dy\,dx = 4\int_{0}^{1}\int_{0}^{1}\left[\left(4x^2 + \frac{64}{3}\right) - \left(4x^2y^2 + \frac{64y^6}{3}\right)\right] dy\,dx = 4\int_{0}^{1}\left(\frac{8}{3}x^2 + \frac{128}{7}\right) dx$
$= \frac{4832}{63};\ I_z = 4\int_{0}^{1}\int_{0}^{1}\int_{4y^2}^{4} (x^2+y^2)\,dz\,dy\,dx = 16\int_{0}^{1}\int_{0}^{1} (x^2 - x^2y^2 + y^2 - y^4)\,dy\,dx$
$= 16\int_{0}^{1}\left(\frac{2x^2}{3} + \frac{2}{15}\right) dx = \frac{256}{45}$

25. (a) $M = 4\int_0^2\int_0^{\sqrt{4-x^2}}\int_{x^2+y^2}^4 dz\,dy\,dx = 4\int_0^{\pi/2}\int_0^2\int_{r^2}^4 r\,dz\,dr\,d\theta = 4\int_0^{\pi/2}\int_0^2 (4r - r^3)\,dr\,d\theta = 4\int_0^{\pi/2} 4\,d\theta = 8\pi$;
$M_{xy} = \int_0^{2\pi}\int_0^2\int_{r^2}^4 zr\,dz\,dr\,d\theta = \int_0^{2\pi}\int_0^2 \frac{r}{2}(16 - r^4)\,dr\,d\theta = \frac{32}{3}\int_0^{2\pi} d\theta = \frac{64\pi}{3} \Rightarrow \bar{z} = \frac{8}{3}$, and $\bar{x} = \bar{y} = 0$, by symmetry

(b) $M = 8\pi \Rightarrow 4\pi = \int_0^{2\pi}\int_0^{\sqrt{c}}\int_{r^2}^c r\,dz\,dr\,d\theta = \int_0^{2\pi}\int_0^{\sqrt{c}} (cr - r^3)\,dr\,d\theta = \int_0^{2\pi} \frac{c^2}{4}\,d\theta = \frac{c^2\pi}{2} \Rightarrow c^2 = 8 \Rightarrow c = 2\sqrt{2}$, since $c > 0$

27. The plane $y + 2z = 2$ is the top of the wedge $\Rightarrow I_L = \int_{-2}^2\int_{-2}^4\int_{-1}^{(2-y)/2} [(y-6)^2 + z^2]\,dz\,dy\,dx$
$= \int_{-2}^2\int_{-2}^4 \left[\frac{(y-6)^2(4-y)}{2} + \frac{(2-y)^3}{24} + \frac{1}{3}\right] dy\,dx$; let $t = 2 - y \Rightarrow I_L = 4\int_{-2}^4 \left(\frac{13t^3}{24} + 5t^2 + 16t + \frac{49}{3}\right) dt = 1386$;
$M = \frac{1}{2}(3)(6)(4) = 36$

29. (a) $M = \int_0^2\int_0^{2-x}\int_0^{2-x-y} 2x\,dz\,dy\,dx = \int_0^2\int_0^{2-x} (4x - 2x^2 - 2xy)\,dy\,dx = \int_0^2 (x^3 - 4x^2 + 4x)\,dx = \frac{4}{3}$

(b) $M_{xy} = \int_0^2\int_0^{2-x}\int_0^{2-x-y} 2xz\,dz\,dy\,dx = \int_0^2\int_0^{2-x} x(2-x-y)^2\,dy\,dx = \int_0^2 \frac{x(2-x)^3}{3}\,dx = \frac{8}{15}$; $M_{xz} = \frac{8}{15}$ by symmetry; $M_{yz} = \int_0^2\int_0^{2-x}\int_0^{2-x-y} 2x^2\,dz\,dy\,dx = \int_0^2\int_0^{2-x} 2x^2(2-x-y)\,dy\,dx = \int_0^2 (2x - x^2)^2\,dx = \frac{16}{15}$
$\Rightarrow \bar{x} = \frac{4}{5}$, and $\bar{y} = \bar{z} = \frac{2}{5}$

31. (a) $M = \int_0^1\int_0^1\int_0^1 (x + y + z + 1)\,dz\,dy\,dx = \int_0^1\int_0^1 \left(x + y + \frac{3}{2}\right) dy\,dx = \int_0^1 (x + 2)\,dx = \frac{5}{2}$

(b) $M_{xy} = \int_0^1\int_0^1\int_0^1 z(x + y + z + 1)\,dz\,dy\,dx = \frac{1}{2}\int_0^1\int_0^1 \left(x + y + \frac{5}{3}\right) dy\,dx = \frac{1}{2}\int_0^1 \left(x + \frac{13}{6}\right) dx = \frac{4}{3}$
$\Rightarrow M_{xy} = M_{yz} = M_{xz} = \frac{4}{3}$, by symmetry $\Rightarrow \bar{x} = \bar{y} = \bar{z} = \frac{8}{15}$

(c) $I_z = \int_0^1\int_0^1\int_0^1 (x^2 + y^2)(x + y + z + 1)\,dz\,dy\,dx = \int_0^1\int_0^1 (x^2 + y^2)\left(x + y + \frac{3}{2}\right) dy\,dx$
$= \int_0^1 \left(x^3 + 2x^2 + \frac{1}{3}x + \frac{3}{4}\right) dx = \frac{11}{6} \Rightarrow I_x = I_y = I_z = \frac{11}{6}$, by symmetry

33. $M = \int_0^1\int_{z-1}^{1-z}\int_0^{\sqrt{z}} (2y + 5)\,dy\,dx\,dz = \int_0^1\int_{z-1}^{1-z} \left(z + 5\sqrt{z}\right) dx\,dz = \int_0^1 2\left(z + 5\sqrt{z}\right)(1 - z)\,dz$
$= 2\int_0^1 \left(5z^{1/2} + z - 5z^{3/2} - z^2\right) dz = 2\left[\frac{10}{3}z^{3/2} + \frac{1}{2}z^2 - 2z^{5/2} - \frac{1}{3}z^3\right]_0^1 = 2\left(\frac{9}{3} - \frac{3}{2}\right) = 3$

12.7 TRIPLE INTEGRALS IN CYLINDRICAL AND SPHERICAL COORDINATES

1. $\int_0^{2\pi}\int_0^1\int_r^{\sqrt{2-r^2}} dz\,r\,dr\,d\theta = \int_0^{2\pi}\int_0^1 \left[r(2 - r^2)^{1/2} - r^2\right] dr\,d\theta = \int_0^{2\pi} \left[-\frac{1}{3}(2 - r^2)^{3/2} - \frac{r^3}{3}\right]_0^1 d\theta$
$= \int_0^{2\pi} \left(\frac{2^{3/2}}{3} - \frac{2}{3}\right) d\theta = \frac{4\pi\left(\sqrt{2} - 1\right)}{3}$

3. $\int_0^{2\pi}\int_0^{\theta/2\pi}\int_0^{3+24r^2} dz\,r\,dr\,d\theta = \int_0^{2\pi}\int_0^{\theta/2\pi} (3r + 24r^3)\,dr\,d\theta = \int_0^{2\pi} \left[\frac{3}{2}r^2 + 6r^4\right]_0^{\theta/2\pi} d\theta = \frac{3}{2}\int_0^{2\pi} \left(\frac{\theta^2}{4\pi^2} + \frac{4\theta^4}{16\pi^4}\right) d\theta$
$= \frac{3}{2}\left[\frac{\theta^3}{12\pi^2} + \frac{\theta^5}{20\pi^4}\right]_0^{2\pi} = \frac{17\pi}{5}$

5. $\int_0^{2\pi}\int_0^1\int_r^{(2-r^2)^{-1/2}} 3\,dz\,r\,dr\,d\theta = 3\int_0^{2\pi}\int_0^1 \left[r(2 - r^2)^{-1/2} - r^2\right] dr\,d\theta = 3\int_0^{2\pi} \left[-(2 - r^2)^{1/2} - \frac{r^3}{3}\right]_0^1 d\theta$
$= 3\int_0^{2\pi} \left(\sqrt{2} - \frac{4}{3}\right) d\theta = \pi\left(6\sqrt{2} - 8\right)$

7. $\int_0^{2\pi}\int_0^3\int_0^{z/3} r^3\,dr\,dz\,d\theta = \int_0^{2\pi}\int_0^3 \frac{z^4}{324}\,dz\,d\theta = \int_0^{2\pi} \frac{3}{20}\,d\theta = \frac{3\pi}{10}$

9. $\int_0^1\int_0^{\sqrt{z}}\int_0^{2\pi}(r^2\cos^2\theta+z^2)\,r\,d\theta\,dr\,dz = \int_0^1\int_0^{\sqrt{z}}\left[\frac{r^2\theta}{2}+\frac{r^2\sin 2\theta}{4}+z^2\theta\right]_0^{2\pi} r\,dr\,dz = \int_0^1\int_0^{\sqrt{z}}(\pi r^3+2\pi r z^2)\,dr\,dz$
$= \int_0^1\left[\frac{\pi r^4}{4}+\pi r^2 z^2\right]_0^{\sqrt{z}} dz = \int_0^1\left(\frac{\pi z^2}{4}+\pi z^3\right)dz = \left[\frac{\pi z^3}{12}+\frac{\pi z^4}{4}\right]_0^1 = \frac{\pi}{3}$

11. (a) $\int_0^{2\pi}\int_0^1\int_0^{\sqrt{4-r^2}} dz\,r\,dr\,d\theta$
(b) $\int_0^{2\pi}\int_0^{\sqrt{3}}\int_0^1 r\,dr\,dz\,d\theta + \int_0^{2\pi}\int_{\sqrt{3}}^2\int_0^{\sqrt{4-z^2}} r\,dr\,dz\,d\theta$
(c) $\int_0^1\int_0^{\sqrt{4-r^2}}\int_0^{2\pi} r\,d\theta\,dz\,dr$

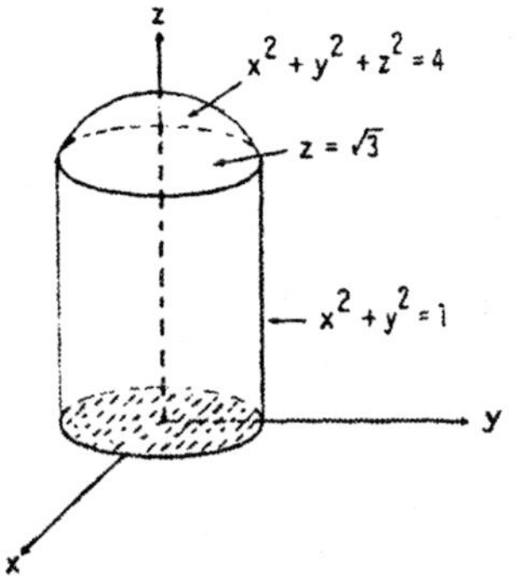

13. $\int_{-\pi/2}^{\pi/2}\int_0^{\cos\theta}\int_0^{3r^2} f(r,\theta,z)\,dz\,r\,dr\,d\theta$

15. $\int_0^{\pi}\int_0^{2\sin\theta}\int_0^{4-r\sin\theta} f(r,\theta,z)\,dz\,r\,dr\,d\theta$

17. $\int_{-\pi/2}^{\pi/2}\int_1^{1+\cos\theta}\int_0^{4} f(r,\theta,z)\,dz\,r\,dr\,d\theta$

19. $\int_0^{\pi/4}\int_0^{\sec\theta}\int_0^{2-r\sin\theta} f(r,\theta,z)\,dz\,r\,dr\,d\theta$

21. $\int_0^{\pi}\int_0^{\pi}\int_0^{2\sin\phi}\rho^2\sin\phi\,d\rho\,d\phi\,d\theta = \frac{8}{3}\int_0^{\pi}\int_0^{\pi}\sin^4\phi\,d\phi\,d\theta = \frac{8}{3}\int_0^{\pi}\left(\left[-\frac{\sin^3\phi\cos\phi}{4}\right]_0^{\pi} + \frac{3}{4}\int_0^{\pi}\sin^2\phi\,d\phi\right)d\theta$
$= 2\int_0^{\pi}\int_0^{\pi}\sin^2\phi\,d\phi\,d\theta = \int_0^{\pi}\left[\theta-\frac{\sin 2\theta}{2}\right]_0^{\pi} d\theta = \int_0^{\pi}\pi\,d\theta = \pi^2$

23. $\int_0^{2\pi}\int_0^{\pi}\int_0^{(1-\cos\phi)/2}\rho^2\sin\phi\,d\rho\,d\phi\,d\theta = \frac{1}{24}\int_0^{2\pi}\int_0^{\pi}(1-\cos\phi)^3\sin\phi\,d\phi\,d\theta = \frac{1}{96}\int_0^{2\pi}\left[(1-\cos\phi)^4\right]_0^{\pi} d\theta$
$= \frac{1}{96}\int_0^{2\pi}(2^4-0)\,d\theta = \frac{16}{96}\int_0^{2\pi} d\theta = \frac{1}{6}(2\pi) = \frac{\pi}{3}$

25. $\int_0^{2\pi}\int_0^{\pi/3}\int_{\sec\phi}^{2} 3\rho^2\sin\phi\,d\rho\,d\phi\,d\theta = \int_0^{2\pi}\int_0^{\pi/3}(8-\sec^3\phi)\sin\phi\,d\phi\,d\theta = \int_0^{2\pi}\left[-8\cos\phi-\frac{1}{2}\sec^2\phi\right]_0^{\pi/3} d\theta$
$= \int_0^{2\pi}\left[(-4-2)-\left(-8-\frac{1}{2}\right)\right]d\theta = \frac{5}{2}\int_0^{2\pi} d\theta = 5\pi$

27. $\int_0^2\int_{-\pi}^0\int_{\pi/4}^{\pi/2}\rho^3\sin 2\phi\,d\phi\,d\theta\,d\rho = \int_0^2\int_{-\pi}^0\rho^3\left[-\frac{\cos 2\phi}{2}\right]_{\pi/4}^{\pi/2} d\theta\,d\rho = \int_0^2\int_{-\pi}^0\frac{\rho^3}{2}\,d\theta\,d\rho = \int_0^2\frac{\rho^3\pi}{2}\,d\rho = \left[\frac{\pi\rho^4}{8}\right]_0^2 = 2\pi$

29. $\int_0^1\int_0^{\pi}\int_0^{\pi/4} 12\rho\sin^3\phi\,d\phi\,d\theta\,d\rho = \int_0^1\int_0^{\pi}\left(12\rho\left[\frac{-\sin^2\phi\cos\phi}{3}\right]_0^{\pi/4} + 8\rho\int_0^{\pi/4}\sin\phi\,d\phi\right)d\theta\,d\rho$
$= \int_0^1\int_0^{\pi}\left(-\frac{2\rho}{\sqrt{2}} - 8\rho\left[\cos\phi\right]_0^{\pi/4}\right)d\theta\,d\rho = \int_0^1\int_0^{\pi}\left(8\rho-\frac{10\rho}{\sqrt{2}}\right)d\theta\,d\rho = \pi\int_0^1\left(8\rho-\frac{10\rho}{\sqrt{2}}\right)d\rho = \pi\left[4\rho^2-\frac{5\rho^2}{\sqrt{2}}\right]_0^1$
$= \frac{\left(4\sqrt{2}-5\right)\pi}{\sqrt{2}}$

31. (a) $x^2+y^2=1 \Rightarrow \rho^2\sin^2\phi = 1$, and $\rho\sin\phi = 1 \Rightarrow \rho = \csc\phi$; thus
$\int_0^{2\pi}\int_0^{\pi/6}\int_0^{2}\rho^2\sin\phi\,d\rho\,d\phi\,d\theta + \int_0^{2\pi}\int_{\pi/6}^{\pi/2}\int_0^{\csc\phi}\rho^2\sin\phi\,d\rho\,d\phi\,d\theta$
(b) $\int_0^{2\pi}\int_1^{2}\int_{\pi/6}^{\sin^{-1}(1/\rho)}\rho^2\sin\phi\,d\phi\,d\rho\,d\theta + \int_0^{2\pi}\int_0^{2}\int_0^{\pi/6}\rho^2\sin\phi\,d\phi\,d\rho\,d\theta$

33. $V = \int_0^{2\pi} \int_0^{\pi/2} \int_{\cos\phi}^{2} \rho^2 \sin\phi \, d\rho \, d\phi \, d\theta = \frac{1}{3} \int_0^{2\pi} \int_0^{\pi/2} (8 - \cos^3\phi) \sin\phi \, d\phi \, d\theta$
$= \frac{1}{3} \int_0^{2\pi} \left[-8\cos\phi + \frac{\cos^4\phi}{4}\right]_0^{\pi/2} d\theta = \frac{1}{3} \int_0^{2\pi} \left(8 - \frac{1}{4}\right) d\theta = \left(\frac{31}{12}\right)(2\pi) = \frac{31\pi}{6}$

35. $V = \int_0^{2\pi} \int_0^{\pi} \int_0^{1-\cos\phi} \rho^2 \sin\phi \, d\rho \, d\phi \, d\theta = \frac{1}{3} \int_0^{2\pi} \int_0^{\pi} (1 - \cos\phi)^3 \sin\phi \, d\phi \, d\theta = \frac{1}{3} \int_0^{2\pi} \left[\frac{(1-\cos\phi)^4}{4}\right]_0^{\pi} d\theta$
$= \frac{1}{12}(2)^4 \int_0^{2\pi} d\theta = \frac{4}{3}(2\pi) = \frac{8\pi}{3}$

37. $V = \int_0^{2\pi} \int_{\pi/4}^{\pi/2} \int_0^{2\cos\phi} \rho^2 \sin\phi \, d\rho \, d\phi \, d\theta = \frac{8}{3} \int_0^{2\pi} \int_{\pi/4}^{\pi/2} \cos^3\phi \sin\phi \, d\phi \, d\theta = \frac{8}{3} \int_0^{2\pi} \left[-\frac{\cos^4\phi}{4}\right]_{\pi/4}^{\pi/2} d\theta$
$= \left(\frac{8}{3}\right)\left(\frac{1}{16}\right) \int_0^{2\pi} d\theta = \frac{1}{6}(2\pi) = \frac{\pi}{3}$

39. (a) $8 \int_0^{\pi/2} \int_0^{\pi/2} \int_0^{2} \rho^2 \sin\phi \, d\rho \, d\phi \, d\theta$ (b) $8 \int_0^{\pi/2} \int_0^{2} \int_0^{\sqrt{4-r^2}} dz \, r \, dr \, d\theta$
(c) $8 \int_0^{2} \int_0^{\sqrt{4-x^2}} \int_0^{\sqrt{4-x^2-y^2}} dz \, dy \, dx$

41. (a) $V = \int_0^{2\pi} \int_0^{\pi/3} \int_{\sec\phi}^{2} \rho^2 \sin\phi \, d\rho \, d\phi \, d\theta$ (b) $V = \int_0^{2\pi} \int_0^{\sqrt{3}} \int_1^{\sqrt{4-r^2}} dz \, r \, dr \, d\theta$
(c) $V = \int_{-\sqrt{3}}^{\sqrt{3}} \int_{-\sqrt{3-x^2}}^{\sqrt{3-x^2}} \int_1^{\sqrt{4-x^2-y^2}} dz \, dy \, dx$
(d) $V = \int_0^{2\pi} \int_0^{\sqrt{3}} \left[r(4-r^2)^{1/2} - r\right] dr \, d\theta = \int_0^{2\pi} \left[-\frac{(4-r^2)^{3/2}}{3} - \frac{r^2}{2}\right]_0^{\sqrt{3}} d\theta = \int_0^{2\pi} \left(-\frac{1}{3} - \frac{3}{2} + \frac{4^{3/2}}{3}\right) d\theta$
$= \frac{5}{6} \int_0^{2\pi} d\theta = \frac{5\pi}{3}$

43. $V = 4 \int_0^{\pi/2} \int_0^{1} \int_{r^4-1}^{4-4r^2} dz \, r \, dr \, d\theta = 4 \int_0^{\pi/2} \int_0^{1} (5r - 4r^3 - r^5) \, dr \, d\theta = 4 \int_0^{\pi/2} \left(\frac{5}{2} - 1 - \frac{1}{6}\right) d\theta = 4 \int_0^{\pi/2} d\theta = \frac{8\pi}{3}$

45. $V = \int_{3\pi/2}^{2\pi} \int_0^{3\cos\theta} \int_0^{-r\sin\theta} dz \, r \, dr \, d\theta = \int_{3\pi/2}^{2\pi} \int_0^{3\cos\theta} -r^2 \sin\theta \, dr \, d\theta = \int_{3\pi/2}^{2\pi} (-9\cos^3\theta)(\sin\theta) \, d\theta = \left[\frac{9}{4}\cos^4\theta\right]_{3\pi/2}^{2\pi}$
$= \frac{9}{4} - 0 = \frac{9}{4}$

47. $V = \int_0^{\pi/2} \int_0^{\sin\theta} \int_0^{\sqrt{1-r^2}} dz \, r \, dr \, d\theta = \int_0^{\pi/2} \int_0^{\sin\theta} r\sqrt{1-r^2} \, dr \, d\theta = \int_0^{\pi/2} \left[-\frac{1}{3}(1-r^2)^{3/2}\right]_0^{\sin\theta} d\theta$
$= -\frac{1}{3} \int_0^{\pi/2} \left[(1 - \sin^2\theta)^{3/2} - 1\right] d\theta = -\frac{1}{3} \int_0^{\pi/2} (\cos^3\theta - 1) \, d\theta = -\frac{1}{3}\left(\left[\frac{\cos^2\theta \sin\theta}{3}\right]_0^{\pi/2} + \frac{2}{3}\int_0^{\pi/2} \cos\theta \, d\theta\right) + \left[\frac{\theta}{3}\right]_0^{\pi/2}$
$= -\frac{2}{9}\left[\sin\theta\right]_0^{\pi/2} + \frac{\pi}{6} = \frac{-4+3\pi}{18}$

49. $V = \int_0^{2\pi} \int_{\pi/3}^{2\pi/3} \int_0^{a} \rho^2 \sin\phi \, d\rho \, d\phi \, d\theta = \int_0^{2\pi} \int_{\pi/3}^{2\pi/3} \frac{a^3}{3} \sin\phi \, d\phi \, d\theta = \frac{a^3}{3} \int_0^{2\pi} \left[-\cos\phi\right]_{\pi/3}^{2\pi/3} d\theta = \frac{a^3}{3} \int_0^{2\pi} \left(\frac{1}{2} + \frac{1}{2}\right) d\theta = \frac{2\pi a^3}{3}$

51. $V = \int_0^{2\pi} \int_0^{\pi/3} \int_{\sec\phi}^{2} \rho^2 \sin\phi \, d\rho \, d\phi \, d\theta$
$= \frac{1}{3} \int_0^{2\pi} \int_0^{\pi/3} (8\sin\phi - \tan\phi \sec^2\phi) \, d\phi \, d\theta$
$= \frac{1}{3} \int_0^{2\pi} \left[-8\cos\phi - \frac{1}{2}\tan^2\phi\right]_0^{\pi/3} d\theta$
$= \frac{1}{3} \int_0^{2\pi} \left[-4 - \frac{1}{2}(3) + 8\right] d\theta = \frac{1}{3} \int_0^{2\pi} \frac{5}{2} \, d\theta = \frac{5}{6}(2\pi) = \frac{5\pi}{3}$

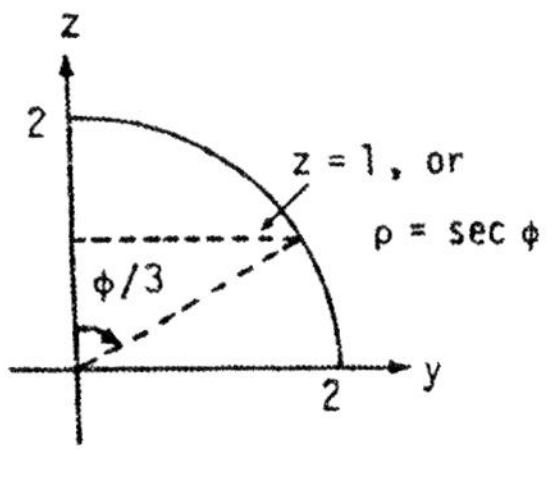

53. $V = 4 \int_0^{\pi/2} \int_0^{1} \int_0^{r^2} dz \, r \, dr \, d\theta = 4 \int_0^{\pi/2} \int_0^{1} r^3 \, dr \, d\theta = \int_0^{\pi/2} d\theta = \frac{\pi}{2}$

55. $V = 8\int_0^{\pi/2}\int_1^{\sqrt{2}}\int_0^r dz\, r\, dr\, d\theta = 8\int_0^{\pi/2}\int_1^{\sqrt{2}} r^2\, dr\, d\theta = 8\left(\frac{2\sqrt{2}-1}{3}\right)\int_0^{\pi/2} d\theta = \frac{4\pi\left(2\sqrt{2}-1\right)}{3}$

57. $V = \int_0^{2\pi}\int_0^2\int_0^{4-r\sin\theta} dz\, r\, dr\, d\theta = \int_0^{2\pi}\int_0^2 (4r - r^2\sin\theta)\, dr\, d\theta = 8\int_0^{2\pi}\left(1 - \frac{\sin\theta}{3}\right) d\theta = 16\pi$

59. The paraboloids intersect when $4x^2 + 4y^2 = 5 - x^2 - y^2 \Rightarrow x^2 + y^2 = 1$ and $z = 4$
$\Rightarrow V = 4\int_0^{\pi/2}\int_0^1\int_{4r^2}^{5-r^2} dz\, r\, dr\, d\theta = 4\int_0^{\pi/2}\int_0^1 (5r - 5r^3)\, dr\, d\theta = 20\int_0^{\pi/2}\left[\frac{r^2}{2} - \frac{r^4}{4}\right]_0^1 d\theta = 5\int_0^{\pi/2} d\theta = \frac{5\pi}{2}$

61. $V = 8\int_0^{2\pi}\int_0^1\int_0^{\sqrt{4-r^2}} dz\, r\, dr\, d\theta = 8\int_0^{2\pi}\int_0^1 r\left(4 - r^2\right)^{1/2} dr\, d\theta = 8\int_0^{2\pi}\left[-\frac{1}{3}\left(4 - r^2\right)^{3/2}\right]_0^1 d\theta$
$= -\frac{8}{3}\int_0^{2\pi}\left(3^{3/2} - 8\right) d\theta = \frac{4\pi\left(8 - 3\sqrt{3}\right)}{3}$

63. $\text{average} = \frac{1}{2\pi}\int_0^{2\pi}\int_0^1\int_{-1}^1 r^2\, dz\, dr\, d\theta = \frac{1}{2\pi}\int_0^{2\pi}\int_0^1 2r^2\, dr\, d\theta = \frac{1}{3\pi}\int_0^{2\pi} d\theta = \frac{2}{3}$

65. $\text{average} = \frac{1}{\left(\frac{4\pi}{3}\right)}\int_0^{2\pi}\int_0^{\pi}\int_0^1 \rho^3 \sin\phi\, d\rho\, d\phi\, d\theta = \frac{3}{16\pi}\int_0^{2\pi}\int_0^{\pi}\sin\phi\, d\phi\, d\theta = \frac{3}{8\pi}\int_0^{2\pi} d\theta = \frac{3}{4}$

67. $M = 4\int_0^{\pi/2}\int_0^1\int_0^r dz\, r\, dr\, d\theta = 4\int_0^{\pi/2}\int_0^1 r^2\, dr\, d\theta = \frac{4}{3}\int_0^{\pi/2} d\theta = \frac{2\pi}{3}$; $M_{xy} = \int_0^{2\pi}\int_0^1\int_0^r z\, dz\, r\, dr\, d\theta$
$= \frac{1}{2}\int_0^{2\pi}\int_0^1 r^3\, dr\, d\theta = \frac{1}{8}\int_0^{2\pi} d\theta = \frac{\pi}{4} \Rightarrow \bar{z} = \frac{M_{xy}}{M} = \left(\frac{\pi}{4}\right)\left(\frac{3}{2\pi}\right) = \frac{3}{8}$, and $\bar{x} = \bar{y} = 0$, by symmetry

69. $M = \frac{8\pi}{3}$; $M_{xy} = \int_0^{2\pi}\int_{\pi/3}^{\pi/2}\int_0^2 z\rho^2 \sin\phi\, d\rho\, d\phi\, d\theta = \int_0^{2\pi}\int_{\pi/3}^{\pi/2}\int_0^2 \rho^3 \cos\phi \sin\phi\, d\rho\, d\phi\, d\theta = 4\int_0^{2\pi}\int_{\pi/3}^{\pi/2} \cos\phi \sin\phi\, d\phi\, d\theta$
$= 4\int_0^{2\pi}\left[\frac{\sin^2\phi}{2}\right]_{\pi/3}^{\pi/2} d\theta = 4\int_0^{2\pi}\left(\frac{1}{2} - \frac{3}{8}\right) d\theta = \frac{1}{2}\int_0^{2\pi} d\theta = \pi \Rightarrow \bar{z} = \frac{M_{xy}}{M} = (\pi)\left(\frac{3}{8\pi}\right) = \frac{3}{8}$, and $\bar{x} = \bar{y} = 0$, by symmetry

71. $M = \int_0^{2\pi}\int_0^4\int_0^{\sqrt{r}} dz\, r\, dr\, d\theta = \int_0^{2\pi}\int_0^4 r^{3/2}\, dr\, d\theta = \frac{64}{5}\int_0^{2\pi} d\theta = \frac{128\pi}{5}$; $M_{xy} = \int_0^{2\pi}\int_0^4\int_0^{\sqrt{r}} z\, dz\, r\, dr\, d\theta$
$= \frac{1}{2}\int_0^{2\pi}\int_0^4 r^2\, dr\, d\theta = \frac{32}{3}\int_0^{2\pi} d\theta = \frac{64\pi}{3} \Rightarrow \bar{z} = \frac{M_{xy}}{M} = \frac{5}{6}$, and $\bar{x} = \bar{y} = 0$, by symmetry

73. We orient the cone with its vertex at the origin and axis along the z-axis $\Rightarrow \phi = \frac{\pi}{4}$. We use the the x-axis which is through the vertex and parallel to the base of the cone $\Rightarrow I_x = \int_0^{2\pi}\int_0^1\int_r^1 (r^2\sin^2\theta + z^2)\, dz\, r\, dr\, d\theta$
$= \int_0^{2\pi}\int_0^1\left(r^3\sin^2\theta - r^4\sin^2\theta + \frac{r}{3} - \frac{r^4}{3}\right) dr\, d\theta = \int_0^{2\pi}\left(\frac{\sin^2\theta}{20} + \frac{1}{10}\right) d\theta = \left[\frac{\theta}{40} - \frac{\sin 2\theta}{80} + \frac{\theta}{10}\right]_0^{2\pi} = \frac{\pi}{20} + \frac{\pi}{5} = \frac{\pi}{4}$

75. (a) $M = \int_0^{2\pi}\int_0^1\int_r^1 z\, dz\, r\, dr\, d\theta = \frac{1}{2}\int_0^{2\pi}\int_0^1 (r - r^3)\, dr\, d\theta = \frac{1}{8}\int_0^{2\pi} d\theta = \frac{\pi}{4}$; $M_{xy} = \int_0^{2\pi}\int_0^1\int_r^1 z^2\, dz\, r\, dr\, d\theta$
$= \frac{1}{3}\int_0^{2\pi}\int_0^1 (r - r^4)\, dr\, d\theta = \frac{1}{10}\int_0^{2\pi} d\theta = \frac{\pi}{5} \Rightarrow \bar{z} = \frac{4}{5}$, and $\bar{x} = \bar{y} = 0$, by symmetry; $I_z = \int_0^{2\pi}\int_0^1\int_r^1 zr^3\, dz\, dr\, d\theta$
$= \frac{1}{2}\int_0^{2\pi}\int_0^1 (r^3 - r^5)\, dr\, d\theta = \frac{1}{24}\int_0^{2\pi} d\theta = \frac{\pi}{12}$

(b) $M = \int_0^{2\pi}\int_0^1\int_r^1 z^2\, dz\, r\, dr\, d\theta = \frac{\pi}{5}$ from part (a); $M_{xy} = \int_0^{2\pi}\int_0^1\int_r^1 z^3\, dz\, r\, dr\, d\theta = \frac{1}{4}\int_0^{2\pi}\int_0^1 (r - r^5)\, dr\, d\theta$
$= \frac{1}{12}\int_0^{2\pi} d\theta = \frac{\pi}{6} \Rightarrow \bar{z} = \frac{5}{6}$, and $\bar{x} = \bar{y} = 0$, by symmetry; $I_z = \int_0^{2\pi}\int_0^1\int_r^1 z^2 r^3\, dz\, dr\, d\theta = \frac{1}{3}\int_0^{2\pi}\int_0^1 (r^3 - r^6)\, dr\, d\theta$
$= \frac{1}{28}\int_0^{2\pi} d\theta = \frac{\pi}{14}$

12.8 SUBSTITUTIONS IN MULTIPLE INTEGRALS

1. (a) $x - y = u$ and $2x + y = v \Rightarrow 3x = u + v$ and $y = x - u \Rightarrow x = \frac{1}{3}(u+v)$ and $y = \frac{1}{3}(-2u+v)$;

$$\frac{\partial(x,y)}{\partial(u,v)} = \begin{vmatrix} \frac{1}{3} & \frac{1}{3} \\ -\frac{2}{3} & \frac{1}{3} \end{vmatrix} = \frac{1}{9} + \frac{2}{9} = \frac{1}{3}$$

(b) The line segment $y = x$ from $(0,0)$ to $(1,1)$ is $x - y = 0$ $\Rightarrow u = 0$; the line segment $y = -2x$ from $(0,0)$ to $(1,-2)$ is $2x + y = 0 \Rightarrow v = 0$; the line segment $x = 1$ from $(1,1)$ to $(1,-2)$ is $(x - y) + (2x + y) = 3$ $\Rightarrow u + v = 3$. The transformed region is sketched at the right.

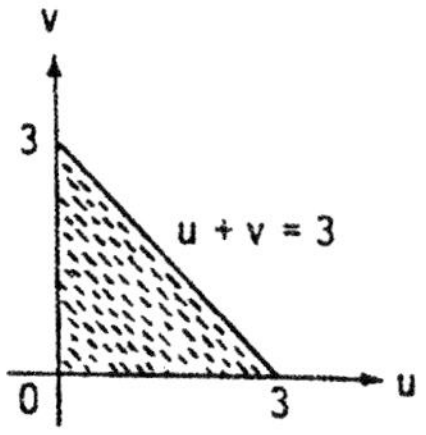

3. (a) $3x + 2y = u$ and $x + 4y = v \Rightarrow -5x = -2u + v$ and $y = \frac{1}{2}(u - 3x) \Rightarrow x = \frac{1}{5}(2u - v)$ and $y = \frac{1}{10}(3v - u)$;

$$\frac{\partial(x,y)}{\partial(u,v)} = \begin{vmatrix} \frac{2}{5} & -\frac{1}{5} \\ -\frac{1}{10} & \frac{3}{10} \end{vmatrix} = \frac{6}{50} - \frac{1}{50} = \frac{1}{10}$$

(b) The x-axis $y = 0 \Rightarrow u = 3v$; the y-axis $x = 0$ $\Rightarrow v = 2u$; the line $x + y = 1$ $\Rightarrow \frac{1}{5}(2u - v) + \frac{1}{10}(3v - u) = 1$ $\Rightarrow 2(2u - v) + (3v - u) = 10 \Rightarrow 3u + v = 10$. The transformed region is sketched at the right.

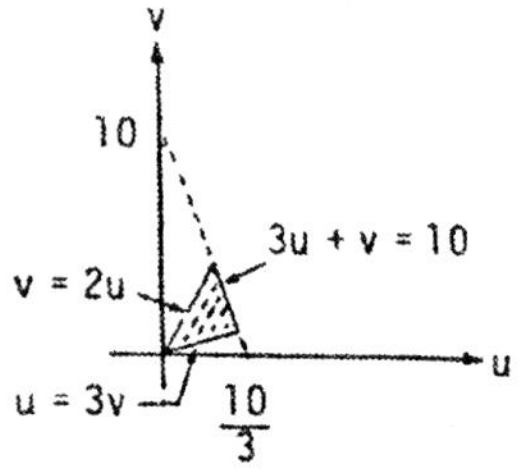

5. $$\int_0^4 \int_{y/2}^{(y/2)+1} \left(x - \frac{y}{2}\right) dx\,dy = \int_0^4 \left[\frac{x^2}{2} - \frac{xy}{2}\right]_{\frac{y}{2}}^{\frac{y}{2}+1} dy = \frac{1}{2}\int_0^4 \left[\left(\frac{y}{2}+1\right)^2 - \left(\frac{y}{2}\right)^2 - \left(\frac{y}{2}+1\right)y + \left(\frac{y}{2}\right)y\right] dy$$
$$= \frac{1}{2}\int_0^4 (y + 1 - y)\,dy = \frac{1}{2}\int_0^4 dy = \frac{1}{2}(4) = 2$$

7. $$\iint_R (3x^2 + 14xy + 8y^2)\,dx\,dy$$
$$= \iint_R (3x + 2y)(x + 4y)\,dx\,dy$$
$$= \iint_G uv \left|\frac{\partial(x,y)}{\partial(u,v)}\right| du\,dv = \frac{1}{10}\iint_G uv\,du\,dv;$$

We find the boundaries of G from the boundaries of R, shown in the accompanying figure:

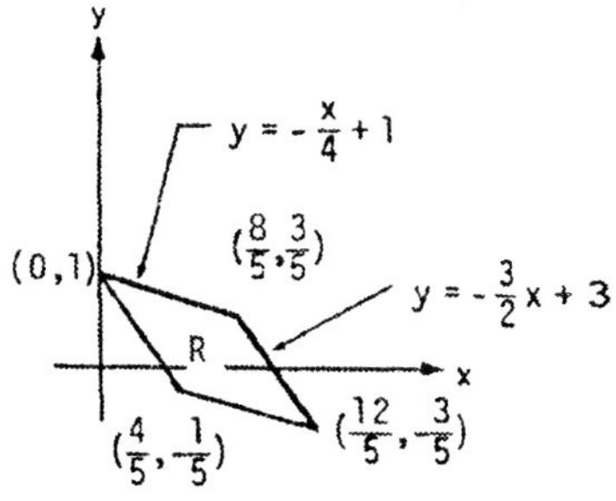

xy-equations for the boundary of R	Corresponding uv-equations for the boundary of G	Simplified uv-equations
$y = -\frac{3}{2}x + 1$	$\frac{1}{10}(3v - u) = -\frac{3}{10}(2u - v) + 1$	$u = 2$
$y = -\frac{3}{2}x + 3$	$\frac{1}{10}(3v - u) = -\frac{3}{10}(2u - v) + 3$	$u = 6$
$y = -\frac{1}{4}x$	$\frac{1}{10}(3v - u) = -\frac{1}{20}(2u - v)$	$v = 0$
$y = -\frac{1}{4}x + 1$	$\frac{1}{10}(3v - u) = -\frac{1}{20}(2u - v) + 1$	$v = 4$

$$\Rightarrow \frac{1}{10}\iint_G uv\,du\,dv = \frac{1}{10}\int_2^6\int_0^4 uv\,dv\,du = \frac{1}{10}\int_2^6 u\left[\frac{v^2}{2}\right]_0^4 du = \frac{4}{5}\int_2^6 u\,du = \left(\frac{4}{5}\right)\left[\frac{u^2}{2}\right]_2^6 = \left(\frac{4}{5}\right)(18 - 2) = \frac{64}{5}$$

9. $x = \frac{u}{v}$ and $y = uv \Rightarrow \frac{y}{x} = v^2$ and $xy = u^2$; $\frac{\partial(x,y)}{\partial(u,v)} = J(u,v) = \begin{vmatrix} v^{-1} & -uv^{-2} \\ v & u \end{vmatrix} = v^{-1}u + v^{-1}u = \frac{2u}{v}$;

$y = x \Rightarrow uv = \frac{u}{v} \Rightarrow v = 1$, and $y = 4x \Rightarrow v = 2$; $xy = 1 \Rightarrow u = 1$, and $xy = 9 \Rightarrow u = 3$; thus

$\iint\limits_R \left(\sqrt{\frac{y}{x}} + \sqrt{xy}\right) dx\,dy = \int_1^3 \int_1^2 (v+u)\left(\frac{2u}{v}\right) dv\,du = \int_1^3 \int_1^2 \left(2u + \frac{2u^2}{v}\right) dv\,du = \int_1^3 \left[2uv + 2u^2 \ln v\right]_1^2 du$

$= \int_1^3 (2u + 2u^2 \ln 2)\,du = \left[u^2 + \frac{2}{3}u^2 \ln 2\right]_1^3 = 8 + \frac{2}{3}(26)(\ln 2) = 8 + \frac{52}{3}(\ln 2)$

11. $x = ar\cos\theta$ and $y = ar\sin\theta \Rightarrow \frac{\partial(x,y)}{\partial(r,\theta)} = J(r,\theta) = \begin{vmatrix} a\cos\theta & -ar\sin\theta \\ b\sin\theta & br\cos\theta \end{vmatrix} = abr\cos^2\theta + abr\sin^2\theta = abr$;

$I_0 = \iint\limits_R (x^2+y^2)\,dA = \int_0^{2\pi}\int_0^1 r^2(a^2\cos^2\theta + b^2\sin^2\theta)\,|J(r,\theta)|\,dr\,d\theta = \int_0^{2\pi}\int_0^1 abr^3(a^2\cos^2\theta + b^2\sin^2\theta)\,dr\,d\theta$

$= \frac{ab}{4}\int_0^{2\pi}(a^2\cos^2\theta + b^2\sin^2\theta)\,d\theta = \frac{ab}{4}\left[\frac{a^2\theta}{2} + \frac{a^2\sin 2\theta}{4} + \frac{b^2\theta}{2} - \frac{b^2\sin 2\theta}{4}\right]_0^{2\pi} = \frac{ab\pi(a^2+b^2)}{4}$

13. The region of integration R in the xy-plane is sketched in the figure at the right. The boundaries of the image G are obtained as follows, with G sketched at the right:

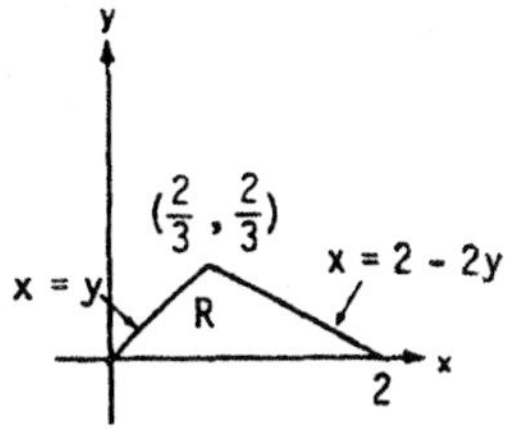

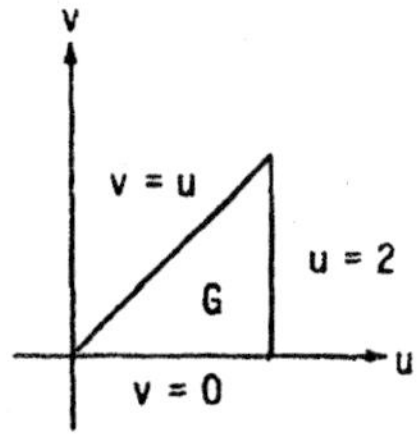

xy-equations for the boundary of R	Corresponding uv-equations for the boundary of G	Simplified uv-equations
$x = y$	$\frac{1}{3}(u+2v) = \frac{1}{3}(u-v)$	$v = 0$
$x = 2 - 2y$	$\frac{1}{3}(u+2v) = 2 - \frac{2}{3}(u-v)$	$u = 2$
$y = 0$	$0 = \frac{1}{3}(u-v)$	$v = u$

Also, from Exercise 2, $\frac{\partial(x,y)}{\partial(u,v)} = J(u,v) = -\frac{1}{3} \Rightarrow \int_0^{2/3}\int_y^{2-2y} (x+2y)e^{(y-x)}\,dx\,dy = \int_0^2\int_0^u ue^{-v}\left|-\frac{1}{3}\right| dv\,du$

$= \frac{1}{3}\int_0^2 u\left[-e^{-v}\right]_0^u du = \frac{1}{3}\int_0^2 u(1-e^{-u})\,du = \frac{1}{3}\left[u(u+e^{-u}) - \frac{u^2}{2} + e^{-u}\right]_0^2 = \frac{1}{3}\left[2(2+e^{-2}) - 2 + e^{-2} - 1\right]$

$= \frac{1}{3}(3e^{-2}+1) \approx 0.4687$

15. (a) $x = u\cos v$ and $y = u\sin v \Rightarrow \frac{\partial(x,y)}{\partial(u,v)} = \begin{vmatrix} \cos v & -u\sin v \\ \sin v & u\cos v \end{vmatrix} = u\cos^2 v + u\sin^2 v = u$

(b) $x = u\sin v$ and $y = u\cos v \Rightarrow \frac{\partial(x,y)}{\partial(u,v)} = \begin{vmatrix} \sin v & u\cos v \\ \cos v & -u\sin v \end{vmatrix} = -u\sin^2 v - u\cos^2 v = -u$

17. $\begin{vmatrix} \sin\phi\cos\theta & \rho\cos\phi\cos\theta & -\rho\sin\phi\sin\theta \\ \sin\phi\sin\theta & \rho\cos\phi\sin\theta & \rho\sin\phi\cos\theta \\ \cos\phi & -\rho\sin\phi & 0 \end{vmatrix}$

$= (\cos\phi)\begin{vmatrix} \rho\cos\phi\cos\theta & -\rho\sin\phi\sin\theta \\ \rho\cos\phi\sin\theta & \rho\sin\phi\cos\theta \end{vmatrix} + (\rho\sin\phi)\begin{vmatrix} \sin\phi\cos\theta & -\rho\sin\phi\sin\theta \\ \sin\phi\sin\theta & \rho\sin\phi\cos\theta \end{vmatrix}$

$= (\rho^2\cos\phi)(\sin\phi\cos\phi\cos^2\theta + \sin\phi\cos\phi\sin^2\theta) + (\rho^2\sin\phi)(\sin^2\phi\cos^2\theta + \sin^2\phi\sin^2\theta)$

$= \rho^2\sin\phi\cos^2\phi + \rho^2\sin^3\phi = (\rho^2\sin\phi)(\cos^2\phi + \sin^2\phi) = \rho^2\sin\phi$

19. $\int_0^3\int_0^4\int_{y/2}^{1+(y/2)}\left(\frac{2x-y}{2}+\frac{z}{3}\right)dx\,dy\,dz = \int_0^3\int_0^4\left[\frac{x^2}{2}-\frac{xy}{2}+\frac{xz}{3}\right]_{y/2}^{1+(y/2)}dy\,dz = \int_0^3\int_0^4\left[\frac{1}{2}(y+1)-\frac{y}{2}+\frac{z}{3}\right]dy\,dz$

$= \int_0^3\left[\frac{(y+1)^2}{4}-\frac{y^2}{4}+\frac{yz}{3}\right]_0^4 dz = \int_0^3\left(\frac{9}{4}+\frac{4z}{3}-\frac{1}{4}\right)dz = \int_0^3\left(2+\frac{4z}{3}\right)dz = \left[2z+\frac{2z^2}{3}\right]_0^3 = 12$

21. $J(u,v,w) = \begin{vmatrix} a & 0 & 0 \\ 0 & b & 0 \\ 0 & 0 & c \end{vmatrix} = abc$; for R and G as in Exercise 19, $\iiint_R |xyz|\,dx\,dy\,dz$

$= \iiint_G a^2b^2c^2uvw\,dw\,dv\,du = 8a^2b^2c^2\int_0^{\pi/2}\int_0^{\pi/2}\int_0^1(\rho\sin\phi\cos\theta)(\rho\sin\phi\sin\theta)(\rho\cos\phi)(\rho^2\sin\phi)\,d\rho\,d\phi\,d\theta$

$= \frac{4a^2b^2c^2}{3}\int_0^{\pi/2}\int_0^{\pi/2}\sin\theta\cos\theta\sin^3\phi\cos\phi\,d\phi\,d\theta = \frac{a^2b^2c^2}{3}\int_0^{\pi/2}\sin\theta\cos\theta\,d\theta = \frac{a^2b^2c^2}{6}$

CHAPTER 12 PRACTICE AND ADDITIONAL EXERCISES

1. $\int_1^{10}\int_0^{1/y} ye^{xy}\,dx\,dy = \int_1^{10}\left[e^{xy}\right]_0^{1/y}dy$

$= \int_1^{10}(e-1)\,dy = 9e-9$

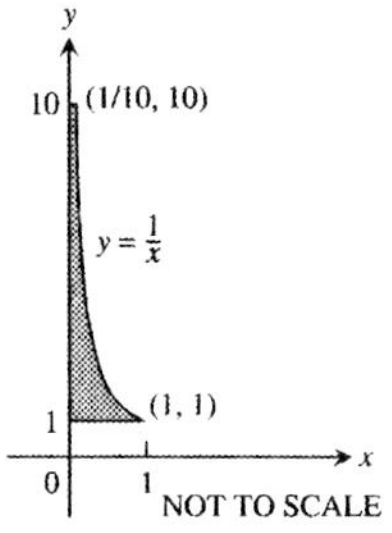

3. $\int_0^{3/2}\int_{-\sqrt{9-4t^2}}^{\sqrt{9-4t^2}} t\,ds\,dt = \int_0^{3/2}\left[ts\right]_{-\sqrt{9-4t^2}}^{\sqrt{9-4t^2}}dt$

$= \int_0^{3/2} 2t\sqrt{9-4t^2}\,dt = \left[-\frac{1}{6}(9-4t^2)^{3/2}\right]_0^{3/2}$

$= -\frac{1}{6}\left(0^{3/2}-9^{3/2}\right) = \frac{27}{6} = \frac{9}{2}$

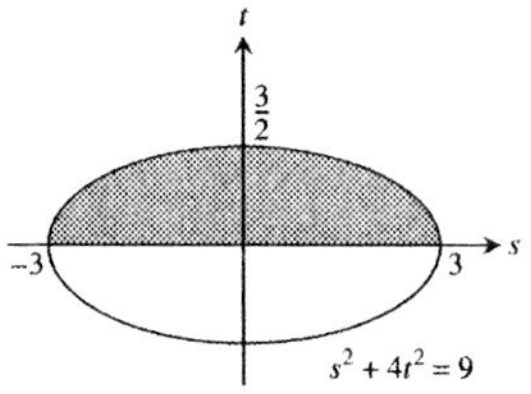

5. $\int_{-2}^{0}\int_{2x+4}^{4-x^2} dy\,dx = \int_{-2}^{0}(-x^2-2x)\,dx$

$= \left[-\frac{x^3}{3}-x^2\right]_{-2}^{0} = -\left(\frac{8}{3}-4\right) = \frac{4}{3}$

$\int_0^4\int_{-\sqrt{4-y}}^{(y-4)/2} dx\,dy = \int_0^4\left(\frac{y-4}{2}+\sqrt{4-y}\right)dy$

$= \left[\frac{y^2}{4}-2y-\frac{2}{3}(4-y)^{3/2}\right]_0^4 = 4-8+\frac{2}{3}\cdot 4^{3/2}$

$= -4+\frac{16}{3} = \frac{4}{3}$

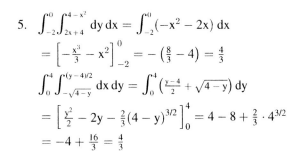

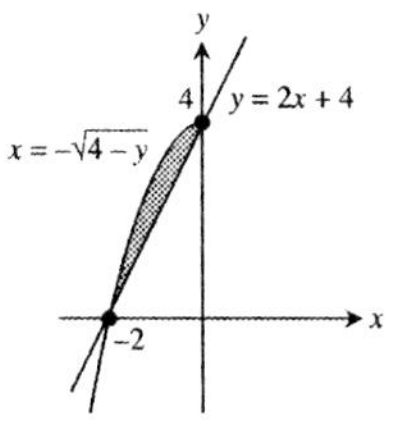

7. $\int_{-3}^{3} \int_0^{(1/2)\sqrt{9-x^2}} y\, dy\, dx = \int_{-3}^{3} \left[\frac{y^2}{2}\right]_0^{(1/2)\sqrt{9-x^2}} dx$
$= \int_{-3}^{3} \frac{1}{8}(9 - x^2)\, dx = \left[\frac{9x}{8} - \frac{x^3}{24}\right]_{-3}^{3}$
$= \left(\frac{27}{8} - \frac{27}{24}\right) - \left(-\frac{27}{8} + \frac{27}{24}\right) = \frac{27}{6} = \frac{9}{2}$
$\int_0^{3/2} \int_{-\sqrt{9-4y^2}}^{\sqrt{9-4y^2}} y\, dx\, dy = \int_0^{3/2} 2y\sqrt{9-4y^2}\, dy$
$= -\frac{1}{4} \cdot \frac{2}{3}(9-4y^2)^{3/2}\Big|_0^{3/2} = \frac{1}{6} \cdot 9^{3/2} = \frac{27}{6} = \frac{9}{2}$

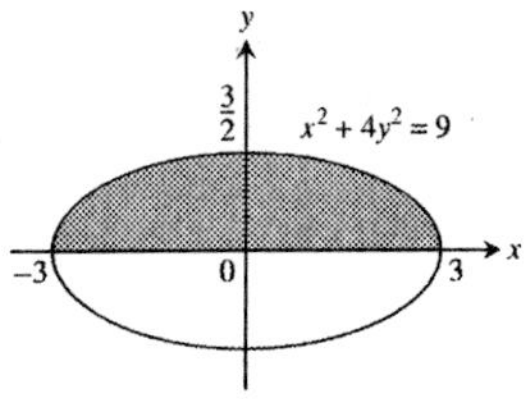

9. $\int_0^1 \int_{2y}^2 4\cos(x^2)\, dx\, dy = \int_0^2 \int_0^{x/2} 4\cos(x^2)\, dy\, dx = \int_0^2 2x\cos(x^2)\, dx = [\sin(x^2)]_0^2 = \sin 4$

11. $\int_0^8 \int_{\sqrt[3]{x}}^2 \frac{1}{y^4+1}\, dy\, dx = \int_0^2 \int_0^{y^3} \frac{1}{y^4+1}\, dx\, dy = \frac{1}{4} \int_0^2 \frac{4y^3}{y^4+1}\, dy = \frac{\ln 17}{4}$

13. $A = \int_{-2}^{0} \int_{2x+4}^{4-x^2} dy\, dx = \int_{-2}^{0} (-x^2 - 2x)\, dx = \frac{4}{3}$

15. $V = \int_0^1 \int_x^{2-x} (x^2 + y^2)\, dy\, dx = \int_0^1 \left[x^2 y + \frac{y^3}{3}\right]_x^{2-x} dx = \int_0^1 \left[2x^2 + \frac{(2-x)^3}{3} - \frac{7x^3}{3}\right] dx = \left[\frac{2x^3}{3} - \frac{(2-x)^4}{12} - \frac{7x^4}{12}\right]_0^1$
$= \left(\frac{2}{3} - \frac{1}{12} - \frac{7}{12}\right) + \frac{2^4}{12} = \frac{4}{3}$

17. $\text{average value} = \int_0^1 \int_0^1 xy\, dy\, dx = \int_0^1 \left[\frac{xy^2}{2}\right]_0^1 dx = \int_0^1 \frac{x}{2}\, dx = \frac{1}{4}$

19. $\int_{-1}^{1} \int_{-\sqrt{1-x^2}}^{\sqrt{1-x^2}} \frac{2}{(1+x^2+y^2)^2}\, dy\, dx = \int_0^{2\pi} \int_0^1 \frac{2r}{(1+r^2)^2}\, dr\, d\theta = \int_0^{2\pi} \left[-\frac{1}{1+r^2}\right]_0^1 d\theta = \frac{1}{2} \int_0^{2\pi} d\theta = \pi$

21. $(x^2+y^2)^2 - (x^2-y^2) = 0 \Rightarrow r^4 - r^2\cos 2\theta = 0 \Rightarrow r^2 = \cos 2\theta$ so the integral is $\int_{-\pi/4}^{\pi/4} \int_0^{\sqrt{\cos 2\theta}} \frac{r}{(1+r^2)^2}\, dr\, d\theta$
$= \int_{-\pi/4}^{\pi/4} \left[-\frac{1}{2(1+r^2)}\right]_0^{\sqrt{\cos 2\theta}} d\theta = \frac{1}{2} \int_{-\pi/4}^{\pi/4} \left(1 - \frac{1}{1+\cos 2\theta}\right) d\theta = \frac{1}{2} \int_{-\pi/4}^{\pi/4} \left(1 - \frac{1}{2\cos^2\theta}\right) d\theta$
$= \frac{1}{2} \int_{-\pi/4}^{\pi/4} \left(1 - \frac{\sec^2\theta}{2}\right) d\theta = \frac{1}{2}\left[\theta - \frac{\tan\theta}{2}\right]_{-\pi/4}^{\pi/4} = \frac{\pi-2}{4}$

23. $\int_0^{\pi} \int_0^{\pi} \int_0^{\pi} \cos(x+y+z)\, dx\, dy\, dz = \int_0^{\pi} \int_0^{\pi} [\sin(z+y+\pi) - \sin(z+y)]\, dy\, dz$
$= \int_0^{\pi} [-\cos(z+2\pi) + \cos(z+\pi) - \cos z + \cos(z+\pi)]\, dz = 0$

25. $\int_0^1 \int_0^{x^2} \int_0^{x+y} (2x - y - z)\, dz\, dy\, dx = \int_0^1 \int_0^{x^2} \left(\frac{3x^2}{2} - \frac{3y^2}{2}\right) dy\, dx = \int_0^1 \left(\frac{3x^4}{2} - \frac{x^6}{2}\right) dx = \frac{8}{35}$

27. $V = 2 \int_0^{\pi/2} \int_{-\cos y}^{0} \int_0^{-2x} dz\, dx\, dy = 2 \int_0^{\pi/2} \int_{-\cos y}^{0} -2x\, dx\, dy = 2 \int_0^{\pi/2} \cos^2 y\, dy = 2\left[\frac{y}{2} + \frac{\sin 2y}{4}\right]_0^{\pi/2} = \frac{\pi}{2}$

29. $\text{average} = \frac{1}{3} \int_0^1 \int_0^3 \int_0^1 30xz\sqrt{x^2+y}\, dz\, dy\, dx = \frac{1}{3} \int_0^1 \int_0^3 15x\sqrt{x^2+y}\, dy\, dx = \frac{1}{3} \int_0^3 \int_0^1 15x\sqrt{x^2+y}\, dx\, dy$
$= \frac{1}{3} \int_0^3 \left[5(x^2+y)^{3/2}\right]_0^1 dy = \frac{1}{3} \int_0^3 [5(1+y)^{3/2} - 5y^{3/2}]\, dy = \frac{1}{3}\left[2(1+y)^{5/2} - 2y^{5/2}\right]_0^3 = \frac{1}{3}\left[2(4)^{5/2} - 2(3)^{5/2} - 2\right]$
$= \frac{1}{3}\left[2\left(31 - 3^{5/2}\right)\right]$

31. (a) $\int_{-\sqrt{2}}^{\sqrt{2}}\int_{-\sqrt{2-y^2}}^{\sqrt{2-y^2}}\int_{\sqrt{x^2+y^2}}^{\sqrt{4-x^2-y^2}} 3\, dz\, dx\, dy$

(b) $\int_0^{2\pi}\int_0^{\pi/4}\int_0^{2} 3\rho^2 \sin\phi\, d\rho\, d\phi\, d\theta$

(c) $\int_0^{2\pi}\int_0^{\sqrt{2}}\int_r^{\sqrt{4-r^2}} 3\, dz\, r\, dr\, d\theta = 3\int_0^{2\pi}\int_0^{\sqrt{2}}\left[r\left(4-r^2\right)^{1/2} - r^2\right] dr\, d\theta = 3\int_0^{2\pi}\left[-\frac{1}{3}\left(4-r^2\right)^{3/2} - \frac{r^3}{3}\right]_0^{\sqrt{2}} d\theta$
$= \int_0^{2\pi}\left(-2^{3/2} - 2^{3/2} + 4^{3/2}\right) d\theta = \left(8 - 4\sqrt{2}\right)\int_0^{2\pi} d\theta = 2\pi\left(8 - 4\sqrt{2}\right)$

33. (a) $\int_0^{2\pi}\int_0^{\pi/4}\int_0^{\sec\phi} \rho^2 \sin\phi\, d\rho\, d\phi\, d\theta$

(b) $\int_0^{2\pi}\int_0^{\pi/4}\int_0^{\sec\phi} \rho^2 \sin\phi\, d\rho\, d\phi\, d\theta = \frac{1}{3}\int_0^{2\pi}\int_0^{\pi/4} (\sec\phi)(\sec\phi\tan\phi)\, d\phi\, d\theta = \frac{1}{3}\int_0^{2\pi}\left[\frac{1}{2}\tan^2\phi\right]_0^{\pi/4} d\theta = \frac{1}{6}\int_0^{2\pi} d\theta = \frac{\pi}{3}$

35. $\int_0^{1}\int_{\sqrt{1-x^2}}^{\sqrt{3-x^2}}\int_1^{\sqrt{4-x^2-y^2}} z^2yx\, dz\, dy\, dx + \int_1^{\sqrt{3}}\int_0^{\sqrt{3-x^2}}\int_1^{\sqrt{4-x^2-y^2}} z^2yx\, dz\, dy\, dx$

37. (a) $V = \int_0^{2\pi}\int_0^{2}\int_2^{\sqrt{8-r^2}} dz\, r\, dr\, d\theta = \int_0^{2\pi}\int_0^{2}\left(r\sqrt{8-r^2} - 2r\right) dr\, d\theta = \int_0^{2\pi}\left[-\frac{1}{3}\left(8-r^2\right)^{3/2} - r^2\right]_0^2 d\theta$
$= \int_0^{2\pi}\left[-\frac{1}{3}(4)^{3/2} - 4 + \frac{1}{3}(8)^{3/2}\right] d\theta = \int_0^{2\pi}\frac{4}{3}\left(-2 - 3 + 2\sqrt{8}\right) d\theta = \frac{4}{3}\left(4\sqrt{2} - 5\right)\int_0^{2\pi} d\theta = \frac{8\pi\left(4\sqrt{2}-5\right)}{3}$

(b) $V = \int_0^{2\pi}\int_0^{\pi/4}\int_{2\sec\phi}^{\sqrt{8}} \rho^2 \sin\phi\, d\rho\, d\phi\, d\theta = \frac{8}{3}\int_0^{2\pi}\int_0^{\pi/4}\left(2\sqrt{2}\sin\phi - \sec^3\phi\sin\phi\right) d\phi\, d\theta$
$= \frac{8}{3}\int_0^{2\pi}\int_0^{\pi/4}\left(2\sqrt{2}\sin\phi - \tan\phi\sec^2\phi\right) d\phi\, d\theta = \frac{8}{3}\int_0^{2\pi}\left[-2\sqrt{2}\cos\phi - \frac{1}{2}\tan^2\phi\right]_0^{\pi/4} d\theta$
$= \frac{8}{3}\int_0^{2\pi}\left(-2 - \frac{1}{2} + 2\sqrt{2}\right) d\theta = \frac{8}{3}\int_0^{2\pi}\left(\frac{-5+4\sqrt{2}}{2}\right) d\theta = \frac{8\pi\left(4\sqrt{2}-5\right)}{3}$

39. With the centers of the spheres at the origin, $I_z = \int_0^{2\pi}\int_0^{\pi}\int_a^{b} \delta(\rho\sin\phi)^2\left(\rho^2\sin\phi\right) d\rho\, d\phi\, d\theta$
$= \frac{\delta\left(b^5-a^5\right)}{5}\int_0^{2\pi}\int_0^{\pi}\sin^3\phi\, d\phi\, d\theta = \frac{\delta\left(b^5-a^5\right)}{5}\int_0^{2\pi}\int_0^{\pi}\left(\sin\phi - \cos^2\phi\sin\phi\right) d\phi\, d\theta$
$= \frac{\delta\left(b^5-a^5\right)}{5}\int_0^{2\pi}\left[-\cos\phi + \frac{\cos^3\phi}{3}\right]_0^{\pi} d\theta = \frac{4\delta\left(b^5-a^5\right)}{15}\int_0^{2\pi} d\theta = \frac{8\pi\delta\left(b^5-a^5\right)}{15}$

41. $M = \int_1^2\int_{2/x}^{2} dy\, dx = \int_1^2\left(2 - \frac{2}{x}\right) dx = 2 - \ln 4$; $M_y = \int_1^2\int_{2/x}^{2} x\, dy\, dx = \int_1^2 x\left(2 - \frac{2}{x}\right) dx = 1$;
$M_x = \int_1^2\int_{2/x}^{2} y\, dy\, dx = \int_1^2\left(2 - \frac{2}{x^2}\right) dx = 1 \Rightarrow \bar{x} = \bar{y} = \frac{1}{2 - \ln 4}$

43. $I_o = \int_0^2\int_{2x}^{4}\left(x^2 + y^2\right)(3)\, dy\, dx = 3\int_0^2\left(4x^2 + \frac{64}{3} - \frac{14x^3}{3}\right) dx = 104$

45. $M = \delta\int_0^3\int_0^{2x/3} dy\, dx = \delta\int_0^3 \frac{2x}{3}\, dx = 3\delta$; $I_x = \delta\int_0^3\int_0^{2x/3} y^2\, dy\, dx = \frac{8\delta}{81}\int_0^3 x^3\, dx = \left(\frac{8\delta}{81}\right)\left(\frac{3^4}{4}\right) = 2\delta$

47. $M = \int_{-1}^{1}\int_{-1}^{1}\left(x^2 + y^2 + \frac{1}{3}\right) dy\, dx = \int_{-1}^{1}\left(2x^2 + \frac{4}{3}\right) dx = 4$; $M_x = \int_{-1}^{1}\int_{-1}^{1} y\left(x^2 + y^2 + \frac{1}{3}\right) dy\, dx = \int_{-1}^{1} 0\, dx = 0$;
$M_y = \int_{-1}^{1}\int_{-1}^{1} x\left(x^2 + y^2 + \frac{1}{3}\right) dy\, dx = \int_{-1}^{1}\left(2x^3 + \frac{4}{3}x\right) dx = 0$

49. $M = \int_{-\pi/3}^{\pi/3}\int_0^{3} r\, dr\, d\theta = \frac{9}{2}\int_{-\pi/3}^{\pi/3} d\theta = 3\pi$; $M_y = \int_{-\pi/3}^{\pi/3}\int_0^{3} r^2\cos\theta\, dr\, d\theta = 9\int_{-\pi/3}^{\pi/3}\cos\theta\, d\theta = 9\sqrt{3} \Rightarrow \bar{x} = \frac{3\sqrt{3}}{\pi}$,
and $\bar{y} = 0$ by symmetry

51. (a) $M = 2\int_0^{\pi/2}\int_1^{1+\cos\theta} r\,dr\,d\theta$

$= \int_0^{\pi/2}\left(2\cos\theta + \frac{1+\cos 2\theta}{2}\right)d\theta = \frac{8+\pi}{4}$;

$M_y = \int_{-\pi/2}^{\pi/2}\int_1^{1+\cos\theta} (r\cos\theta)\,r\,dr\,d\theta$

$= \int_{-\pi/2}^{\pi/2}\left(\cos^2\theta + \cos^3\theta + \frac{\cos^4\theta}{3}\right)d\theta$

$= \frac{32+15\pi}{24} \Rightarrow \overline{x} = \frac{15\pi+32}{6\pi+48}$, and

$\overline{y} = 0$ by symmetry

(b)

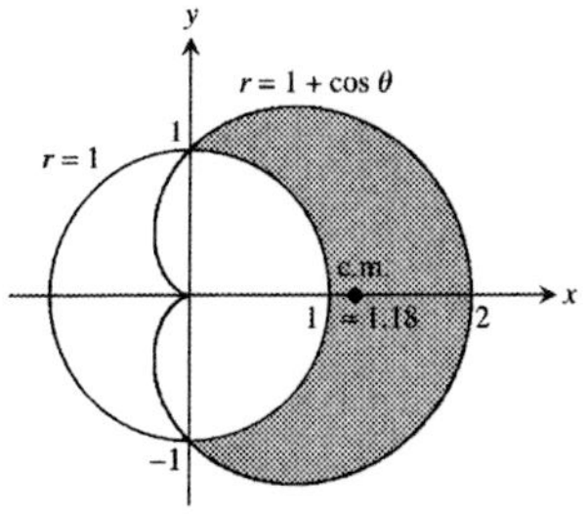

53. $x = u + y$ and $y = v \Rightarrow x = u + v$ and $y = v$

$\Rightarrow J(u, v) = \begin{vmatrix} 1 & 1 \\ 0 & 1 \end{vmatrix} = 1$; the boundary of the image G is obtained from the boundary of R as follows:

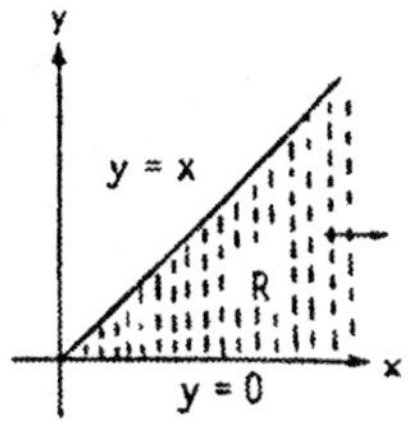

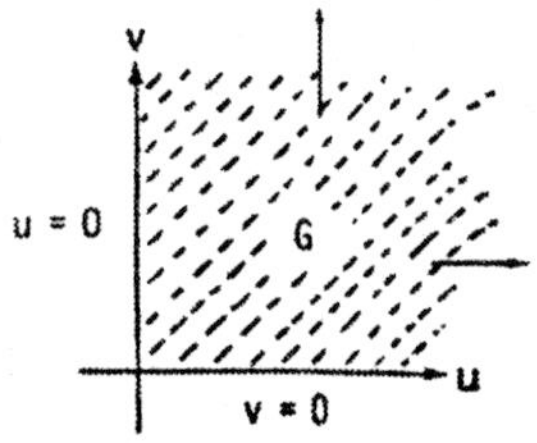

xy-equations for the boundary of R	Corresponding uv-equations for the boundary of G	Simplified uv-equations
$y = x$	$v = u + v$	$u = 0$
$y = 0$	$v = 0$	$v = 0$

$\Rightarrow \int_0^{\infty}\int_0^{x} e^{-sx} f(x-y, y)\,dy\,dx = \int_0^{\infty}\int_0^{\infty} e^{-s(u+v)} f(u, v)\,du\,dv$

NOTES:

CHAPTER 13 INTEGRATION IN VECTOR FIELDS

13.1 LINE INTEGRALS

1. $\mathbf{r} = t\mathbf{i} + (1-t)\mathbf{j} \Rightarrow x = t$ and $y = 1 - t \Rightarrow y = 1 - x \Rightarrow$ (c)

3. $\mathbf{r} = (2\cos t)\mathbf{i} + (2\sin t)\mathbf{j} \Rightarrow x = 2\cos t$ and $y = 2\sin t \Rightarrow x^2 + y^2 = 4 \Rightarrow$ (g)

5. $\mathbf{r} = t\mathbf{i} + t\mathbf{j} + t\mathbf{k} \Rightarrow x = t, y = t$, and $z = t \Rightarrow$ (d)

7. $\mathbf{r} = (t^2-1)\mathbf{j} + 2t\mathbf{k} \Rightarrow y = t^2 - 1$ and $z = 2t \Rightarrow y = \frac{z^2}{4} - 1 \Rightarrow$ (f)

9. $\mathbf{r}(t) = t\mathbf{i} + (1-t)\mathbf{j},\ 0 \le t \le 1 \Rightarrow \frac{d\mathbf{r}}{dt} = \mathbf{i} - \mathbf{j} \Rightarrow \left|\frac{d\mathbf{r}}{dt}\right| = \sqrt{2}\,\mathbf{j}$; $x = t$ and $y = 1 - t \Rightarrow x + y = t + (1-t) = 1$
$\Rightarrow \int_C f(x,y,z)\,ds = \int_0^1 f(t, 1-t, 0)\left|\frac{d\mathbf{r}}{dt}\right| dt = \int_0^1 (1)\left(\sqrt{2}\right) dt = \left[\sqrt{2}\,t\right]_0^1 = \sqrt{2}$

11. $\mathbf{r}(t) = 2t\mathbf{i} + t\mathbf{j} + (2-2t)\mathbf{k},\ 0 \le t \le 1 \Rightarrow \frac{d\mathbf{r}}{dt} = 2\mathbf{i} + \mathbf{j} - 2\mathbf{k} \Rightarrow \left|\frac{d\mathbf{r}}{dt}\right| = \sqrt{4+1+4} = 3$; $xy + y + z$
$= (2t)t + t + (2-2t) \Rightarrow \int_C f(x,y,z)\,ds = \int_0^1 (2t^2 - t + 2)\,3\,dt = 3\left[\frac{2}{3}t^3 - \frac{1}{2}t^2 + 2t\right]_0^1 = 3\left(\frac{2}{3} - \frac{1}{2} + 2\right) = \frac{13}{2}$

13. $\mathbf{r}(t) = (\mathbf{i} + 2\mathbf{j} + 3\mathbf{k}) + t(-\mathbf{i} - 3\mathbf{j} - 2\mathbf{k}) = (1-t)\mathbf{i} + (2-3t)\mathbf{j} + (3-2t)\mathbf{k},\ 0 \le t \le 1 \Rightarrow \frac{d\mathbf{r}}{dt} = -\mathbf{i} - 3\mathbf{j} - 2\mathbf{k}$
$\Rightarrow \left|\frac{d\mathbf{r}}{dt}\right| = \sqrt{1+9+4} = \sqrt{14}$; $x + y + z = (1-t) + (2-3t) + (3-2t) = 6 - 6t \Rightarrow \int_C f(x,y,z)\,ds$
$= \int_0^1 (6-6t)\sqrt{14}\,dt = 6\sqrt{14}\left[t - \frac{t^2}{2}\right]_0^1 = \left(6\sqrt{14}\right)\left(\frac{1}{2}\right) = 3\sqrt{14}$

15. C_1: $\mathbf{r}(t) = t\mathbf{i} + t^2\mathbf{j},\ 0 \le t \le 1 \Rightarrow \frac{d\mathbf{r}}{dt} = \mathbf{i} + 2t\mathbf{j} \Rightarrow \left|\frac{d\mathbf{r}}{dt}\right| = \sqrt{1+4t^2}$; $x + \sqrt{y} - z^2 = t + \sqrt{t^2} - 0 = t + |t| = 2t$
since $t \ge 0 \Rightarrow \int_{C_1} f(x,y,z)\,ds = \int_0^1 2t\sqrt{1+4t^2}\,dt = \left[\frac{1}{6}\left(1+4t^2\right)^{3/2}\right]_0^1 = \frac{1}{6}(5)^{3/2} - \frac{1}{6} = \frac{1}{6}\left(5\sqrt{5} - 1\right)$;
C_2: $\mathbf{r}(t) = \mathbf{i} + \mathbf{j} + t\mathbf{k},\ 0 \le t \le 1 \Rightarrow \frac{d\mathbf{r}}{dt} = \mathbf{k} \Rightarrow \left|\frac{d\mathbf{r}}{dt}\right| = 1$; $x + \sqrt{y} - z^2 = 1 + \sqrt{1} - t^2 = 2 - t^2$
$\Rightarrow \int_{C_2} f(x,y,z)\,ds = \int_0^1 (2-t^2)(1)\,dt = \left[2t - \frac{1}{3}t^3\right]_0^1 = 2 - \frac{1}{3} = \frac{5}{3}$; therefore $\int_C f(x,y,z)\,ds$
$= \int_{C_1} f(x,y,z)\,ds + \int_{C_2} f(x,y,z)\,ds = \frac{5}{6}\sqrt{5} + \frac{3}{2}$

17. $\mathbf{r}(t) = t\mathbf{i} + t\mathbf{j} + t\mathbf{k},\ 0 < a \le t \le b \Rightarrow \frac{d\mathbf{r}}{dt} = \mathbf{i} + \mathbf{j} + \mathbf{k} \Rightarrow \left|\frac{d\mathbf{r}}{dt}\right| = \sqrt{3}$; $\frac{x+y+z}{x^2+y^2+z^2} = \frac{t+t+t}{t^2+t^2+t^2} = \frac{1}{t}$
$\Rightarrow \int_C f(x,y,z)\,ds = \int_a^b \left(\frac{1}{t}\right)\sqrt{3}\,dt = \left[\sqrt{3}\ln|t|\right]_a^b = \sqrt{3}\ln\left(\frac{b}{a}\right)$, since $0 < a \le b$

19. $\mathbf{r}(x) = x\mathbf{i} + y\mathbf{j} = x\mathbf{i} + \frac{x^2}{2}\mathbf{j},\ 0 \le x \le 2 \Rightarrow \frac{d\mathbf{r}}{dx} = \mathbf{i} + x\mathbf{j} \Rightarrow \left|\frac{d\mathbf{r}}{dx}\right| = \sqrt{1+x^2}$; $f(x,y) = f\left(x, \frac{x^2}{2}\right) = \frac{x^3}{\left(\frac{x^2}{2}\right)} = 2x \Rightarrow \int_C f\,ds$
$= \int_0^2 (2x)\sqrt{1+x^2}\,dx = \left[\frac{2}{3}\left(1+x^2\right)^{3/2}\right]_0^2 = \frac{2}{3}\left(5^{3/2} - 1\right) = \frac{10\sqrt{5}-2}{3}$

21. $\mathbf{r}(t) = (2\cos t)\mathbf{i} + (2\sin t)\mathbf{j},\ 0 \le t \le \frac{\pi}{2} \Rightarrow \frac{d\mathbf{r}}{dt} = (-2\sin t)\mathbf{i} + (2\cos t)\mathbf{j} \Rightarrow \left|\frac{d\mathbf{r}}{dt}\right| = 2$; $f(x,y) = f(2\cos t, 2\sin t)$
$= 2\cos t + 2\sin t \Rightarrow \int_C f\,ds = \int_0^{\pi/2} (2\cos t + 2\sin t)(2)\,dt = \left[4\sin t - 4\cos t\right]_0^{\pi/2} = 4 - (-4) = 8$

23. $\mathbf{r}(t) = (t^2-1)\,\mathbf{j} + 2t\mathbf{k},\ 0 \le t \le 1 \Rightarrow \frac{d\mathbf{r}}{dt} = 2t\mathbf{j} + 2\mathbf{k} \Rightarrow \left|\frac{d\mathbf{r}}{dt}\right| = 2\sqrt{t^2+1};\ M = \int_C \delta(x,y,z)\,ds = \int_0^1 \delta(t)\left(2\sqrt{t^2+1}\right) dt$
$= \int_0^1 \left(\frac{3}{2}t\right)\left(2\sqrt{t^2+1}\right) dt = \left[(t^2+1)^{3/2}\right]_0^1 = 2^{3/2} - 1 = 2\sqrt{2} - 1$

25. $\mathbf{r}(t) = \sqrt{2}t\,\mathbf{i} + \sqrt{2}t\,\mathbf{j} + (4-t^2)\,\mathbf{k},\ 0 \le t \le 1 \Rightarrow \frac{d\mathbf{r}}{dt} = \sqrt{2}\mathbf{i} + \sqrt{2}\mathbf{j} - 2t\mathbf{k} \Rightarrow \left|\frac{d\mathbf{r}}{dt}\right| = \sqrt{2+2+4t^2} = 2\sqrt{1+t^2};$

(a) $M = \int_C \delta\,ds = \int_0^1 (3t)\left(2\sqrt{1+t^2}\right) dt = \left[2\left(1+t^2\right)^{3/2}\right]_0^1 = 2\left(2^{3/2}-1\right) = 4\sqrt{2} - 2$

(b) $M = \int_C \delta\,ds = \int_0^1 (1)\left(2\sqrt{1+t^2}\right) dt = \left[t\sqrt{1+t^2} + \ln\left(t + \sqrt{1+t^2}\right)\right]_0^1 = \left[\sqrt{2} + \ln\left(1+\sqrt{2}\right)\right] - (0 + \ln 1)$
$= \sqrt{2} + \ln\left(1+\sqrt{2}\right)$

27. Let $x = a\cos t$ and $y = a\sin t$, $0 \le t \le 2\pi$. Then $\frac{dx}{dt} = -a\sin t$, $\frac{dy}{dt} = a\cos t$, $\frac{dz}{dt} = 0$
$\Rightarrow \sqrt{\left(\frac{dx}{dt}\right)^2 + \left(\frac{dy}{dt}\right)^2 + \left(\frac{dz}{dt}\right)^2}\,dt = a\,dt;\ I_z = \int_C (x^2+y^2)\,\delta\,ds = \int_0^{2\pi} (a^2\sin^2 t + a^2\cos^2 t)\,a\delta\,dt$
$= \int_0^{2\pi} a^3\delta\,dt = 2\pi\delta a^3;\ M = \int_C \delta(x,y,z)\,ds = \int_0^{2\pi} \delta a\,dt = 2\pi\delta a\,.$

29. $\mathbf{r}(t) = (\cos t)\mathbf{i} + (\sin t)\mathbf{j} + t\mathbf{k},\ 0 \le t \le 2\pi \Rightarrow \frac{d\mathbf{r}}{dt} = (-\sin t)\mathbf{i} + (\cos t)\mathbf{j} + \mathbf{k} \Rightarrow \left|\frac{d\mathbf{r}}{dt}\right| = \sqrt{\sin^2 t + \cos^2 t + 1} = \sqrt{2};$

(a) $M = \int_C \delta\,ds = \int_0^{2\pi} \delta\sqrt{2}\,dt = 2\pi\delta\sqrt{2};\ I_z = \int_C (x^2+y^2)\,\delta\,ds = \int_0^{2\pi} (\cos^2 t + \sin^2 t)\,\delta\sqrt{2}\,dt = 2\pi\delta\sqrt{2}$

(b) $M = \int_C \delta(x,y,z)\,ds = \int_0^{4\pi} \delta\sqrt{2}\,dt = 4\pi\delta\sqrt{2}$ and $I_z = \int_C (x^2+y^2)\,\delta\,ds = \int_0^{4\pi} \delta\sqrt{2}\,dt = 4\pi\delta\sqrt{2}$

31. $\delta(x,y,z) = 2 - z$ and $\mathbf{r}(t) = (\cos t)\mathbf{j} + (\sin t)\mathbf{k},\ 0 \le t \le \pi \Rightarrow M = 2\pi - 2$ as found in Example 3 of the text;
also $\left|\frac{d\mathbf{r}}{dt}\right| = 1;\ I_x = \int_C (y^2+z^2)\,\delta\,ds = \int_0^{\pi} (\cos^2 t + \sin^2 t)(2 - \sin t)\,dt = \int_0^{\pi} (2 - \sin t)\,dt = 2\pi - 2$

13.2 VECTOR FIELDS, WORK, CIRCULATION, AND FLUX

1. $f(x,y,z) = \left(x^2+y^2+z^2\right)^{-1/2} \Rightarrow \frac{\partial f}{\partial x} = -\frac{1}{2}\left(x^2+y^2+z^2\right)^{-3/2}(2x) = -x\left(x^2+y^2+z^2\right)^{-3/2}$; similarly,
$\frac{\partial f}{\partial y} = -y\left(x^2+y^2+z^2\right)^{-3/2}$ and $\frac{\partial f}{\partial z} = -z\left(x^2+y^2+z^2\right)^{-3/2} \Rightarrow \nabla f = \frac{-x\mathbf{i} - y\mathbf{j} - z\mathbf{k}}{\left(x^2+y^2+z^2\right)^{3/2}}$

3. $g(x,y,z) = e^z - \ln(x^2+y^2) \Rightarrow \frac{\partial g}{\partial x} = -\frac{2x}{x^2+y^2},\ \frac{\partial g}{\partial y} = -\frac{2y}{x^2+y^2}$ and $\frac{\partial g}{\partial z} = e^z \Rightarrow \nabla g = \left(\frac{-2x}{x^2+y^2}\right)\mathbf{i} - \left(\frac{2y}{x^2+y^2}\right)\mathbf{j} + e^z\mathbf{k}$

5. $|\mathbf{F}|$ inversely proportional to the square of the distance from (x,y) to the origin $\Rightarrow \sqrt{(M(x,y))^2 + (N(x,y))^2}$
$= \frac{k}{x^2+y^2}$, $k > 0$; $\mathbf{F}$ points toward the origin $\Rightarrow$ $\mathbf{F}$ is in the direction of $\mathbf{n} = \frac{-x}{\sqrt{x^2+y^2}}\,\mathbf{i} - \frac{y}{\sqrt{x^2+y^2}}\,\mathbf{j}$
$\Rightarrow \mathbf{F} = a\mathbf{n}$, for some constant $a > 0$. Then $M(x,y) = \frac{-ax}{\sqrt{x^2+y^2}}$ and $N(x,y) = \frac{-ay}{\sqrt{x^2+y^2}} \Rightarrow \sqrt{(M(x,y))^2 + (N(x,y))^2} = a$
$\Rightarrow a = \frac{k}{x^2+y^2} \Rightarrow \mathbf{F} = \frac{-kx}{\left(x^2+y^2\right)^{3/2}}\,\mathbf{i} - \frac{ky}{\left(x^2+y^2\right)^{3/2}}\,\mathbf{j}$, for any constant $k > 0$

7. Substitute the parametric representations for $\mathbf{r}(t) = x(t)\mathbf{i} + y(t)\mathbf{j} + z(t)\mathbf{k}$ representing each path into the vector field $\mathbf{F}$, and calculate the work $W = \int_C \mathbf{F} \cdot \frac{d\mathbf{r}}{dt}$.

(a) $\mathbf{F} = 3t\mathbf{i} + 2t\mathbf{j} + 4t\mathbf{k}$ and $\frac{d\mathbf{r}}{dt} = \mathbf{i} + \mathbf{j} + \mathbf{k} \Rightarrow \mathbf{F} \cdot \frac{d\mathbf{r}}{dt} = 9t \Rightarrow W = \int_0^1 9t\,dt = \frac{9}{2}$

(b) $\mathbf{F} = 3t^2\mathbf{i} + 2t\mathbf{j} + 4t^4\mathbf{k}$ and $\frac{d\mathbf{r}}{dt} = \mathbf{i} + 2t\mathbf{j} + 4t^3\mathbf{k} \Rightarrow \mathbf{F} \cdot \frac{d\mathbf{r}}{dt} = 7t^2 + 16t^7 \Rightarrow W = \int_0^1 (7t^2 + 16t^7)\,dt = \left[\frac{7}{3}t^3 + 2t^8\right]_0^1$
$= \frac{7}{3} + 2 = \frac{13}{3}$

(c) $\mathbf{r}_1 = t\mathbf{i} + t\mathbf{j}$ and $\mathbf{r}_2 = \mathbf{i} + \mathbf{j} + t\mathbf{k}$; $\mathbf{F}_1 = 3t\mathbf{i} + 2t\mathbf{j}$ and $\frac{d\mathbf{r}_1}{dt} = \mathbf{i} + \mathbf{j} \Rightarrow \mathbf{F}_1 \cdot \frac{d\mathbf{r}_1}{dt} = 5t \Rightarrow W_1 = \int_0^1 5t\, dt = \frac{5}{2}$;
$\mathbf{F}_2 = 3\mathbf{i} + 2\mathbf{j} + 4t\mathbf{k}$ and $\frac{d\mathbf{r}_2}{dt} = \mathbf{k} \Rightarrow \mathbf{F}_2 \cdot \frac{d\mathbf{r}_2}{dt} = 4t \Rightarrow W_2 = \int_0^1 4t\, dt = 2 \Rightarrow W = W_1 + W_2 = \frac{9}{2}$

9. Substitute the parametric representation for $\mathbf{r}(t) = x(t)\mathbf{i} + y(t)\mathbf{j} + z(t)\mathbf{k}$ representing each path into the vector field $\mathbf{F}$, and calculate the work $W = \int_C \mathbf{F} \cdot \frac{d\mathbf{r}}{dt}$.

(a) $\mathbf{F} = \sqrt{t}\mathbf{i} - 2t\mathbf{j} + \sqrt{t}\mathbf{k}$ and $\frac{d\mathbf{r}}{dt} = \mathbf{i} + \mathbf{j} + \mathbf{k} \Rightarrow \mathbf{F} \cdot \frac{d\mathbf{r}}{dt} = 2\sqrt{t} - 2t \Rightarrow W = \int_0^1 \left(2\sqrt{t} - 2t\right) dt = \left[\frac{4}{3}t^{3/2} - t^2\right]_0^1 = \frac{1}{3}$

(b) $\mathbf{F} = t^2\mathbf{i} - 2t\mathbf{j} + t\mathbf{k}$ and $\frac{d\mathbf{r}}{dt} = \mathbf{i} + 2t\mathbf{j} + 4t^3\mathbf{k} \Rightarrow \mathbf{F} \cdot \frac{d\mathbf{r}}{dt} = 4t^4 - 3t^2 \Rightarrow W = \int_0^1 (4t^4 - 3t^2)\, dt = \left[\frac{4}{5}t^5 - t^3\right]_0^1 = -\frac{1}{5}$

(c) $\mathbf{r}_1 = t\mathbf{i} + t\mathbf{j}$ and $\mathbf{r}_2 = \mathbf{i} + \mathbf{j} + t\mathbf{k}$; $\mathbf{F}_1 = -2t\mathbf{j} + \sqrt{t}\mathbf{k}$ and $\frac{d\mathbf{r}_1}{dt} = \mathbf{i} + \mathbf{j} \Rightarrow \mathbf{F}_1 \cdot \frac{d\mathbf{r}_1}{dt} = -2t \Rightarrow W_1 = \int_0^1 -2t\, dt$
$= -1$; $\mathbf{F}_2 = \sqrt{t}\mathbf{i} - 2\mathbf{j} + \mathbf{k}$ and $\frac{d\mathbf{r}_2}{dt} = \mathbf{k} \Rightarrow \mathbf{F}_2 \cdot \frac{d\mathbf{r}_2}{dt} = 1 \Rightarrow W_2 = \int_0^1 dt = 1 \Rightarrow W = W_1 + W_2 = 0$

11. Substitute the parametric representation for $\mathbf{r}(t) = x(t)\mathbf{i} + y(t)\mathbf{j} + z(t)\mathbf{k}$ representing each path into the vector field $\mathbf{F}$, and calculate the work $W = \int_C \mathbf{F} \cdot \frac{d\mathbf{r}}{dt}$.

(a) $\mathbf{F} = (3t^2 - 3t)\mathbf{i} + 3t\mathbf{j} + \mathbf{k}$ and $\frac{d\mathbf{r}}{dt} = \mathbf{i} + \mathbf{j} + \mathbf{k} \Rightarrow \mathbf{F} \cdot \frac{d\mathbf{r}}{dt} = 3t^2 + 1 \Rightarrow W = \int_0^1 (3t^2 + 1)\, dt = \left[t^3 + t\right]_0^1 = 2$

(b) $\mathbf{F} = (3t^2 - 3t)\mathbf{i} + 3t^4\mathbf{j} + \mathbf{k}$ and $\frac{d\mathbf{r}}{dt} = \mathbf{i} + 2t\mathbf{j} + 4t^3\mathbf{k} \Rightarrow \mathbf{F} \cdot \frac{d\mathbf{r}}{dt} = 6t^5 + 4t^3 + 3t^2 - 3t$
$\Rightarrow W = \int_0^1 (6t^5 + 4t^3 + 3t^2 - 3t)\, dt = \left[t^6 + t^4 + t^3 - \frac{3}{2}t^2\right]_0^1 = \frac{3}{2}$

(c) $\mathbf{r}_1 = t\mathbf{i} + t\mathbf{j}$ and $\mathbf{r}_2 = \mathbf{i} + \mathbf{j} + t\mathbf{k}$; $\mathbf{F}_1 = (3t^2 - 3t)\mathbf{i} + \mathbf{k}$ and $\frac{d\mathbf{r}_1}{dt} = \mathbf{i} + \mathbf{j} \Rightarrow \mathbf{F}_1 \cdot \frac{d\mathbf{r}_1}{dt} = 3t^2 - 3t$
$\Rightarrow W_1 = \int_0^1 (3t^2 - 3t)\, dt = \left[t^3 - \frac{3}{2}t^2\right]_0^1 = -\frac{1}{2}$; $\mathbf{F}_2 = 3t\mathbf{j} + \mathbf{k}$ and $\frac{d\mathbf{r}_2}{dt} = \mathbf{k} \Rightarrow \mathbf{F}_2 \cdot \frac{d\mathbf{r}_2}{dt} = 1 \Rightarrow W_2 = \int_0^1 dt = 1$
$\Rightarrow W = W_1 + W_2 = \frac{1}{2}$

13. $\mathbf{r} = t\mathbf{i} + t^2\mathbf{j} + t\mathbf{k}$, $0 \le t \le 1$, and $\mathbf{F} = xy\mathbf{i} + y\mathbf{j} - yz\mathbf{k} \Rightarrow \mathbf{F} = t^3\mathbf{i} + t^2\mathbf{j} - t^3\mathbf{k}$ and $\frac{d\mathbf{r}}{dt} = \mathbf{i} + 2t\mathbf{j} + \mathbf{k} \Rightarrow \mathbf{F} \cdot \frac{d\mathbf{r}}{dt} = 2t^3$
$\Rightarrow$ work $= \int_0^1 2t^3\, dt = \frac{1}{2}$

15. $\mathbf{r} = (\sin t)\mathbf{i} + (\cos t)\mathbf{j} + t\mathbf{k}$, $0 \le t \le 2\pi$, and $\mathbf{F} = z\mathbf{i} + x\mathbf{j} + y\mathbf{k} \Rightarrow \mathbf{F} = t\mathbf{i} + (\sin t)\mathbf{j} + (\cos t)\mathbf{k}$ and
$\frac{d\mathbf{r}}{dt} = (\cos t)\mathbf{i} - (\sin t)\mathbf{j} + \mathbf{k} \Rightarrow \mathbf{F} \cdot \frac{d\mathbf{r}}{dt} = t\cos t - \sin^2 t + \cos t \Rightarrow$ work $= \int_0^{2\pi} (t\cos t - \sin^2 t + \cos t)\, dt$
$= \left[\cos t + t\sin t - \frac{t}{2} + \frac{\sin 2t}{4} + \sin t\right]_0^{2\pi} = -\pi$

17. $x = t$ and $y = x^2 = t^2 \Rightarrow \mathbf{r} = t\mathbf{i} + t^2\mathbf{j}$, $-1 \le t \le 2$, and $\mathbf{F} = xy\mathbf{i} + (x + y)\mathbf{j} \Rightarrow \mathbf{F} = t^3\mathbf{i} + (t + t^2)\mathbf{j}$ and $\frac{d\mathbf{r}}{dt} = \mathbf{i} + 2t\mathbf{j}$
$\Rightarrow \mathbf{F} \cdot \frac{d\mathbf{r}}{dt} = t^3 + (2t^2 + 2t^3) = 3t^3 + 2t^2 \Rightarrow \int_C xy\, dx + (x + y)\, dy = \int_C \mathbf{F} \cdot \frac{d\mathbf{r}}{dt}\, dt = \int_{-1}^2 (3t^3 + 2t^2)\, dt = \left[\frac{3}{4}t^4 + \frac{2}{3}t^3\right]_{-1}^2$
$= \left(12 + \frac{16}{3}\right) - \left(\frac{3}{4} - \frac{2}{3}\right) = \frac{45}{4} + \frac{18}{3} = \frac{69}{4}$

19. $\mathbf{r} = x\mathbf{i} + y\mathbf{j} = y^2\mathbf{i} + y\mathbf{j}$, $2 \ge y \ge -1$, and $\mathbf{F} = x^2\mathbf{i} - y\mathbf{j} = y^4\mathbf{i} - y\mathbf{j} \Rightarrow \frac{d\mathbf{r}}{dy} = 2y\mathbf{i} + \mathbf{j}$ and $\mathbf{F} \cdot \frac{d\mathbf{r}}{dy} = 2y^5 - y$
$\Rightarrow \int_C \mathbf{F} \cdot \mathbf{T}\, ds = \int_2^{-1} \mathbf{F} \cdot \frac{d\mathbf{r}}{dy}\, dy = \int_2^{-1} (2y^5 - y)\, dy = \left[\frac{1}{3}y^6 - \frac{1}{2}y^2\right]_2^{-1} = \left(\frac{1}{3} - \frac{1}{2}\right) - \left(\frac{64}{3} - \frac{4}{2}\right) = \frac{3}{2} - \frac{63}{3} = -\frac{39}{2}$

21. $\mathbf{r} = (\mathbf{i} + \mathbf{j}) + t(\mathbf{i} + 2\mathbf{j}) = (1 + t)\mathbf{i} + (1 + 2t)\mathbf{j}$, $0 \le t \le 1$, and $\mathbf{F} = xy\mathbf{i} + (y - x)\mathbf{j} \Rightarrow \mathbf{F} = (1 + 3t + 2t^2)\mathbf{i} + t\mathbf{j}$ and
$\frac{d\mathbf{r}}{dt} = \mathbf{i} + 2\mathbf{j} \Rightarrow \mathbf{F} \cdot \frac{d\mathbf{r}}{dt} = 1 + 5t + 2t^2 \Rightarrow$ work $= \int_C \mathbf{F} \cdot \frac{d\mathbf{r}}{dt}\, dt = \int_0^1 (1 + 5t + 2t^2)\, dt = \left[t + \frac{5}{2}t^2 + \frac{2}{3}t^3\right]_0^1 = \frac{25}{6}$

23. (a) $\mathbf{r} = (\cos t)\mathbf{i} + (\sin t)\mathbf{j}$, $0 \le t \le 2\pi$, $\mathbf{F}_1 = x\mathbf{i} + y\mathbf{j}$, and $\mathbf{F}_2 = -y\mathbf{i} + x\mathbf{j} \Rightarrow \frac{d\mathbf{r}}{dt} = (-\sin t)\mathbf{i} + (\cos t)\mathbf{j}$,
$\mathbf{F}_1 = (\cos t)\mathbf{i} + (\sin t)\mathbf{j}$, and $\mathbf{F}_2 = (-\sin t)\mathbf{i} + (\cos t)\mathbf{j} \Rightarrow \mathbf{F}_1 \cdot \frac{d\mathbf{r}}{dt} = 0$ and $\mathbf{F}_2 \cdot \frac{d\mathbf{r}}{dt} = \sin^2 t + \cos^2 t = 1$

$\Rightarrow \text{Circ}_1 = \int_0^{2\pi} 0\, dt = 0$ and $\text{Circ}_2 = \int_0^{2\pi} dt = 2\pi$; $\mathbf{n} = (\cos t)\mathbf{i} + (\sin t)\mathbf{j} \Rightarrow \mathbf{F}_1 \cdot \mathbf{n} = \cos^2 t + \sin^2 t = 1$ and $\mathbf{F}_2 \cdot \mathbf{n} = 0 \Rightarrow \text{Flux}_1 = \int_0^{2\pi} dt = 2\pi$ and $\text{Flux}_2 = \int_0^{2\pi} 0\, dt = 0$

(b) $\mathbf{r} = (\cos t)\mathbf{i} + (4\sin t)\mathbf{j},\ 0 \le t \le 2\pi \Rightarrow \frac{d\mathbf{r}}{dt} = (-\sin t)\mathbf{i} + (4\cos t)\mathbf{j}$, $\mathbf{F}_1 = (\cos t)\mathbf{i} + (4\sin t)\mathbf{j}$, and $\mathbf{F}_2 = (-4\sin t)\mathbf{i} + (\cos t)\mathbf{j} \Rightarrow \mathbf{F}_1 \cdot \frac{d\mathbf{r}}{dt} = 15\sin t\cos t$ and $\mathbf{F}_2 \cdot \frac{d\mathbf{r}}{dt} = 4 \Rightarrow \text{Circ}_1 = \int_0^{2\pi} 15\sin t\cos t\, dt$ $= \left[\frac{15}{2}\sin^2 t\right]_0^{2\pi} = 0$ and $\text{Circ}_2 = \int_0^{2\pi} 4\, dt = 8\pi$; $\mathbf{n} = \left(\frac{4}{\sqrt{17}}\cos t\right)\mathbf{i} + \left(\frac{1}{\sqrt{17}}\sin t\right)\mathbf{j} \Rightarrow \mathbf{F}_1 \cdot \mathbf{n}$ $= \frac{4}{\sqrt{17}}\cos^2 t + \frac{4}{\sqrt{17}}\sin^2 t$ and $\mathbf{F}_2 \cdot \mathbf{n} = -\frac{15}{\sqrt{17}}\sin t\cos t \Rightarrow \text{Flux}_1 = \int_0^{2\pi} (\mathbf{F}_1 \cdot \mathbf{n})\,|\mathbf{v}|\, dt = \int_0^{2\pi} \left(\frac{4}{\sqrt{17}}\right)\sqrt{17}\, dt$ $= 8\pi$ and $\text{Flux}_2 = \int_0^{2\pi} (\mathbf{F}_2 \cdot \mathbf{n})\,|\mathbf{v}|\, dt = \int_0^{2\pi} \left(-\frac{15}{\sqrt{17}}\sin t\cos t\right)\sqrt{17}\, dt = \left[-\frac{15}{2}\sin^2 t\right]_0^{2\pi} = 0$

25. $\mathbf{F}_1 = (a\cos t)\mathbf{i} + (a\sin t)\mathbf{j}$, $\frac{d\mathbf{r}_1}{dt} = (-a\sin t)\mathbf{i} + (a\cos t)\mathbf{j} \Rightarrow \mathbf{F}_1 \cdot \frac{d\mathbf{r}_1}{dt} = 0 \Rightarrow \text{Circ}_1 = 0$; $M_1 = a\cos t$, $N_1 = a\sin t$, $dx = -a\sin t\, dt$, $dy = a\cos t\, dt \Rightarrow \text{Flux}_1 = \int_C M_1\, dy - N_1\, dx = \int_0^{\pi} (a^2\cos^2 t + a^2\sin^2 t)\, dt$ $= \int_0^{\pi} a^2\, dt = a^2\pi$;

$\mathbf{F}_2 = t\mathbf{i}$, $\frac{d\mathbf{r}_2}{dt} = \mathbf{i} \Rightarrow \mathbf{F}_2 \cdot \frac{d\mathbf{r}_2}{dt} = t \Rightarrow \text{Circ}_2 = \int_{-a}^{a} t\, dt = 0$; $M_2 = t$, $N_2 = 0$, $dx = dt$, $dy = 0 \Rightarrow \text{Flux}_2$ $= \int_C M_2\, dy - N_2\, dx = \int_{-a}^{a} 0\, dt = 0$; therefore, $\text{Circ} = \text{Circ}_1 + \text{Circ}_2 = 0$ and $\text{Flux} = \text{Flux}_1 + \text{Flux}_2 = a^2\pi$

27. $\mathbf{F}_1 = (-a\sin t)\mathbf{i} + (a\cos t)\mathbf{j}$, $\frac{d\mathbf{r}_1}{dt} = (-a\sin t)\mathbf{i} + (a\cos t)\mathbf{j} \Rightarrow \mathbf{F}_1 \cdot \frac{d\mathbf{r}_1}{dt} = a^2\sin^2 t + a^2\cos^2 t = a^2$ $\Rightarrow \text{Circ}_1 = \int_0^{\pi} a^2\, dt = a^2\pi$; $M_1 = -a\sin t$, $N_1 = a\cos t$, $dx = -a\sin t\, dt$, $dy = a\cos t\, dt$ $\Rightarrow \text{Flux}_1 = \int_C M_1\, dy - N_1\, dx = \int_0^{\pi} (-a^2\sin t\cos t + a^2\sin t\cos t)\, dt = 0$; $\mathbf{F}_2 = t\mathbf{j}$, $\frac{d\mathbf{r}_2}{dt} = \mathbf{i} \Rightarrow \mathbf{F}_2 \cdot \frac{d\mathbf{r}_2}{dt} = 0$ $\Rightarrow \text{Circ}_2 = 0$; $M_2 = 0$, $N_2 = t$, $dx = dt$, $dy = 0 \Rightarrow \text{Flux}_2 = \int_C M_2\, dy - N_2\, dx = \int_{-a}^{a} -t\, dt = 0$; therefore, $\text{Circ} = \text{Circ}_1 + \text{Circ}_2 = a^2\pi$ and $\text{Flux} = \text{Flux}_1 + \text{Flux}_2 = 0$

29. (a) $\mathbf{r} = (\cos t)\mathbf{i} + (\sin t)\mathbf{j},\ 0 \le t \le \pi$, and $\mathbf{F} = (x+y)\mathbf{i} - (x^2+y^2)\mathbf{j} \Rightarrow \frac{d\mathbf{r}}{dt} = (-\sin t)\mathbf{i} + (\cos t)\mathbf{j}$ and $\mathbf{F} = (\cos t + \sin t)\mathbf{i} - (\cos^2 t + \sin^2 t)\mathbf{j} \Rightarrow \mathbf{F} \cdot \frac{d\mathbf{r}}{dt} = -\sin t\cos t - \sin^2 t - \cos t \Rightarrow \int_C \mathbf{F} \cdot \mathbf{T}\, ds$ $= \int_0^{\pi} (-\sin t\cos t - \sin^2 t - \cos t)\, dt = \left[-\frac{1}{2}\sin^2 t - \frac{t}{2} + \frac{\sin 2t}{4} - \sin t\right]_0^{\pi} = -\frac{\pi}{2}$

(b) $\mathbf{r} = (1-2t)\mathbf{i},\ 0 \le t \le 1$, and $\mathbf{F} = (x+y)\mathbf{i} - (x^2+y^2)\mathbf{j} \Rightarrow \frac{d\mathbf{r}}{dt} = -2\mathbf{i}$ and $\mathbf{F} = (1-2t)\mathbf{i} - (1-2t)^2\mathbf{j} \Rightarrow$ $\mathbf{F} \cdot \frac{d\mathbf{r}}{dt} = 4t - 2 \Rightarrow \int_C \mathbf{F} \cdot \mathbf{T}\, ds = \int_0^1 (4t-2)\, dt = [2t^2 - 2t]_0^1 = 0$

(c) $\mathbf{r}_1 = (1-t)\mathbf{i} - t\mathbf{j},\ 0 \le t \le 1$, and $\mathbf{F} = (x+y)\mathbf{i} - (x^2+y^2)\mathbf{j} \Rightarrow \frac{d\mathbf{r}_1}{dt} = -\mathbf{i} - \mathbf{j}$ and $\mathbf{F} = (1-2t)\mathbf{i} - (1 - 2t + 2t^2)\mathbf{j}$ $\Rightarrow \mathbf{F} \cdot \frac{d\mathbf{r}_1}{dt} = (2t-1) + (1 - 2t + 2t^2) = 2t^2 \Rightarrow \text{Flow}_1 = \int_{C_1} \mathbf{F} \cdot \frac{d\mathbf{r}_1}{dt} = \int_0^1 2t^2\, dt = \frac{2}{3}$; $\mathbf{r}_2 = -t\mathbf{i} + (t-1)\mathbf{j}$, $0 \le t \le 1$, and $\mathbf{F} = (x+y)\mathbf{i} - (x^2+y^2)\mathbf{j} \Rightarrow \frac{d\mathbf{r}_2}{dt} = -\mathbf{i} + \mathbf{j}$ and $\mathbf{F} = -\mathbf{i} - (t^2 + t^2 - 2t + 1)\mathbf{j}$ $= -\mathbf{i} - (2t^2 - 2t + 1)\mathbf{j} \Rightarrow \mathbf{F} \cdot \frac{d\mathbf{r}_2}{dt} = 1 - (2t^2 - 2t + 1) = 2t - 2t^2 \Rightarrow \text{Flow}_2 = \int_{C_2} \mathbf{F} \cdot \frac{d\mathbf{r}_2}{dt} = \int_0^1 (2t - 2t^2)\, dt$ $= \left[t^2 - \frac{2}{3}t^3\right]_0^1 = \frac{1}{3} \Rightarrow \text{Flow} = \text{Flow}_1 + \text{Flow}_2 = \frac{2}{3} + \frac{1}{3} = 1$

31. $\mathbf{F} = -\frac{y}{\sqrt{x^2+y^2}}\mathbf{i} + \frac{x}{\sqrt{x^2+y^2}}\mathbf{j}$ on $x^2 + y^2 = 4$; at $(2,0)$, $\mathbf{F} = \mathbf{j}$; at $(0,2)$, $\mathbf{F} = -\mathbf{i}$; at $(-2,0)$, $\mathbf{F} = -\mathbf{j}$; at $(0,-2)$, $\mathbf{F} = \mathbf{i}$; at $\left(\sqrt{2}, \sqrt{2}\right)$, $\mathbf{F} = -\frac{\sqrt{3}}{2}\mathbf{i} + \frac{1}{2}\mathbf{j}$; at $\left(\sqrt{2}, -\sqrt{2}\right)$, $\mathbf{F} = \frac{\sqrt{3}}{2}\mathbf{i} + \frac{1}{2}\mathbf{j}$; at $\left(-\sqrt{2}, \sqrt{2}\right)$, $\mathbf{F} = -\frac{\sqrt{3}}{2}\mathbf{i} - \frac{1}{2}\mathbf{j}$; at $\left(-\sqrt{2}, -\sqrt{2}\right)$, $\mathbf{F} = \frac{\sqrt{3}}{2}\mathbf{i} - \frac{1}{2}\mathbf{j}$

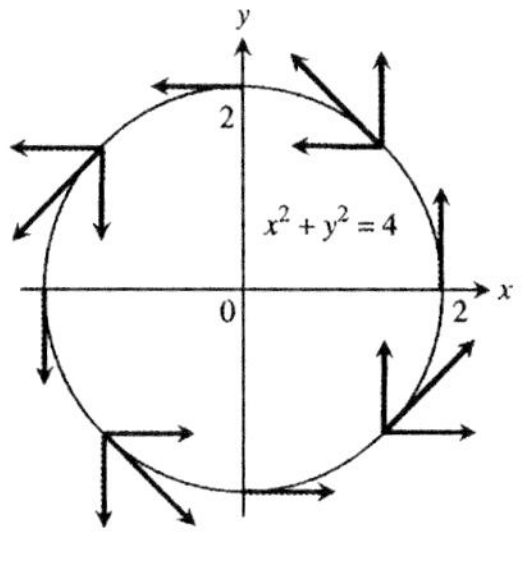

33. (a) $\mathbf{G} = P(x,y)\mathbf{i} + Q(x,y)\mathbf{j}$ is to have a magnitude $\sqrt{a^2+b^2}$ and to be tangent to $x^2+y^2=a^2+b^2$ in a counterclockwise direction. Thus $x^2+y^2=a^2+b^2 \Rightarrow 2x+2yy'=0 \Rightarrow y' = -\frac{x}{y}$ is the slope of the tangent line at any point on the circle $\Rightarrow y' = -\frac{a}{b}$ at (a,b). Let $\mathbf{v} = -b\mathbf{i} + a\mathbf{j} \Rightarrow |\mathbf{v}| = \sqrt{a^2+b^2}$, with $\mathbf{v}$ in a counterclockwise direction and tangent to the circle. Then let $P(x,y) = -y$ and $Q(x,y) = x$ $\Rightarrow \mathbf{G} = -y\mathbf{i} + x\mathbf{j} \Rightarrow$ for (a,b) on $x^2+y^2=a^2+b^2$ we have $\mathbf{G} = -b\mathbf{i} + a\mathbf{j}$ and $|\mathbf{G}| = \sqrt{a^2+b^2}$.

(b) $\mathbf{G} = \left(\sqrt{x^2+y^2}\right)\mathbf{F} = \left(\sqrt{a^2+b^2}\right)\mathbf{F}$.

35. The slope of the line through (x,y) and the origin is $\frac{y}{x} \Rightarrow \mathbf{v} = x\mathbf{i} + y\mathbf{j}$ is a vector parallel to that line and pointing away from the origin $\Rightarrow \mathbf{F} = -\frac{x\mathbf{i}+y\mathbf{j}}{\sqrt{x^2+y^2}}$ is the unit vector pointing toward the origin.

37. $\mathbf{F} = -4t^3\mathbf{i} + 8t^2\mathbf{j} + 2\mathbf{k}$ and $\frac{d\mathbf{r}}{dt} = \mathbf{i} + 2t\mathbf{j} \Rightarrow \mathbf{F}\cdot\frac{d\mathbf{r}}{dt} = 12t^3 \Rightarrow \text{Flow} = \int_0^2 12t^3\,dt = \left[3t^4\right]_0^2 = 48$

39. $\mathbf{F} = (\cos t - \sin t)\mathbf{i} + (\cos t)\mathbf{k}$ and $\frac{d\mathbf{r}}{dt} = (-\sin t)\mathbf{i} + (\cos t)\mathbf{k} \Rightarrow \mathbf{F}\cdot\frac{d\mathbf{r}}{dt} = -\sin t\cos t + 1$

$\Rightarrow \text{Flow} = \int_0^{\pi} (-\sin t\cos t + 1)\,dt = \left[\frac{1}{2}\cos^2 t + t\right]_0^{\pi} = \left(\frac{1}{2}+\pi\right) - \left(\frac{1}{2}+0\right) = \pi$

41. C_1: $\mathbf{r} = (\cos t)\mathbf{i} + (\sin t)\mathbf{j} + t\mathbf{k},\ 0 \le t \le \frac{\pi}{2} \Rightarrow \mathbf{F} = (2\cos t)\mathbf{i} + 2t\mathbf{j} + (2\sin t)\mathbf{k}$ and $\frac{d\mathbf{r}}{dt} = (-\sin t)\mathbf{i} + (\cos t)\mathbf{j} + \mathbf{k}$

$\Rightarrow \mathbf{F}\cdot\frac{d\mathbf{r}}{dt} = -2\cos t\sin t + 2t\cos t + 2\sin t = -\sin 2t + 2t\cos t + 2\sin t$

$\Rightarrow \text{Flow}_1 = \int_0^{\pi/2} (-\sin 2t + 2t\cos t + 2\sin t)\,dt = \left[\frac{1}{2}\cos 2t + 2t\sin t + 2\cos t - 2\cos t\right]_0^{\pi/2} = -1 + \pi;$

C_2: $\mathbf{r} = \mathbf{j} + \frac{\pi}{2}(1-t)\mathbf{k},\ 0 \le t \le 1 \Rightarrow \mathbf{F} = \pi(1-t)\mathbf{j} + 2\mathbf{k}$ and $\frac{d\mathbf{r}}{dt} = -\frac{\pi}{2}\mathbf{k} \Rightarrow \mathbf{F}\cdot\frac{d\mathbf{r}}{dt} = -\pi$

$\Rightarrow \text{Flow}_2 = \int_0^1 -\pi\,dt = \left[-\pi t\right]_0^1 = -\pi;$

C_3: $\mathbf{r} = t\mathbf{i} + (1-t)\mathbf{j},\ 0 \le t \le 1 \Rightarrow \mathbf{F} = 2t\mathbf{i} + 2(1-t)\mathbf{k}$ and $\frac{d\mathbf{r}}{dt} = \mathbf{i} - \mathbf{j} \Rightarrow \mathbf{F}\cdot\frac{d\mathbf{r}}{dt} = 2t$

$\Rightarrow \text{Flow}_3 = \int_0^1 2t\,dt = \left[t^2\right]_0^1 = 1 \Rightarrow \text{Circulation} = (-1+\pi) - \pi + 1 = 0$

43. Let $x = t$ be the parameter $\Rightarrow y = x^2 = t^2$ and $z = x = t \Rightarrow \mathbf{r} = t\mathbf{i} + t^2\mathbf{j} + t\mathbf{k},\ 0 \le t \le 1$ from $(0,0,0)$ to $(1,1,1)$ $\Rightarrow \frac{d\mathbf{r}}{dt} = \mathbf{i} + 2t\mathbf{j} + \mathbf{k}$ and $\mathbf{F} = xy\mathbf{i} + y\mathbf{j} - yz\mathbf{k} = t^3\mathbf{i} + t^2\mathbf{j} - t^3\mathbf{k} \Rightarrow \mathbf{F}\cdot\frac{d\mathbf{r}}{dt} = t^3 + 2t^3 - t^3 = 2t^3 \Rightarrow \text{Flow} = \int_0^1 2t^3\,dt$ $= \frac{1}{2}$

45. Yes. The work and area have the same numerical value because work $= \int_C \mathbf{F}\cdot d\mathbf{r} = \int_C y\mathbf{i}\cdot d\mathbf{r}$

$= \int_b^a [f(t)\mathbf{i}]\cdot\left[\mathbf{i} + \frac{df}{dt}\mathbf{j}\right]dt$ [On the path, y equals f(t)]

$= \int_a^b f(t)\,dt = \text{Area under the curve}$ [because $f(t) > 0$]

13.3 PATH INDEPENDENCE, POTENTIAL FUNCTIONS, AND CONSERVATIVE FIELDS

1. $\frac{\partial P}{\partial y} = x = \frac{\partial N}{\partial z},\ \frac{\partial M}{\partial z} = y = \frac{\partial P}{\partial x},\ \frac{\partial N}{\partial x} = z = \frac{\partial M}{\partial y} \Rightarrow$ Conservative

3. $\frac{\partial P}{\partial y} = -1 \ne 1 = \frac{\partial N}{\partial z} \Rightarrow$ Not Conservative

5. $\frac{\partial N}{\partial x} = 0 \ne 1 = \frac{\partial M}{\partial y} \Rightarrow$ Not Conservative

7. $\frac{\partial f}{\partial x} = 2x \Rightarrow f(x,y,z) = x^2 + g(y,z) \Rightarrow \frac{\partial f}{\partial y} = \frac{\partial g}{\partial y} = 3y \Rightarrow g(y,z) = \frac{3y^2}{2} + h(z) \Rightarrow f(x,y,z) = x^2 + \frac{3y^2}{2} + h(z)$ $\Rightarrow \frac{\partial f}{\partial z} = h'(z) = 4z \Rightarrow h(z) = 2z^2 + C \Rightarrow f(x,y,z) = x^2 + \frac{3y^2}{2} + 2z^2 + C$

9. $\frac{\partial f}{\partial x} = e^{y+2z} \Rightarrow f(x,y,z) = xe^{y+2z} + g(y,z) \Rightarrow \frac{\partial f}{\partial y} = xe^{y+2z} + \frac{\partial g}{\partial y} = xe^{y+2z} \Rightarrow \frac{\partial g}{\partial y} = 0 \Rightarrow f(x,y,z)$
$= xe^{y+2z} + h(z) \Rightarrow \frac{\partial f}{\partial z} = 2xe^{y+2z} + h'(z) = 2xe^{y+2z} \Rightarrow h'(z) = 0 \Rightarrow h(z) = C \Rightarrow f(x,y,z) = xe^{y+2z} + C$

11. $\frac{\partial f}{\partial z} = \frac{z}{y^2+z^2} \Rightarrow f(x,y,z) = \frac{1}{2}\ln(y^2+z^2) + g(x,y) \Rightarrow \frac{\partial f}{\partial x} = \frac{\partial g}{\partial x} = \ln x + \sec^2(x+y) \Rightarrow g(x,y)$
$= (x\ln x - x) + \tan(x+y) + h(y) \Rightarrow f(x,y,z) = \frac{1}{2}\ln(y^2+z^2) + (x\ln x - x) + \tan(x+y) + h(y)$
$\Rightarrow \frac{\partial f}{\partial y} = \frac{y}{y^2+z^2} + \sec^2(x+y) + h'(y) = \sec^2(x+y) + \frac{y}{y^2+z^2} \Rightarrow h'(y) = 0 \Rightarrow h(y) = C \Rightarrow f(x,y,z)$
$= \frac{1}{2}\ln(y^2+z^2) + (x\ln x - x) + \tan(x+y) + C$

13. Let $\mathbf{F}(x,y,z) = 2x\mathbf{i} + 2y\mathbf{j} + 2z\mathbf{k} \Rightarrow \frac{\partial P}{\partial y} = 0 = \frac{\partial N}{\partial z}, \frac{\partial M}{\partial z} = 0 = \frac{\partial P}{\partial x}, \frac{\partial N}{\partial x} = 0 = \frac{\partial M}{\partial y} \Rightarrow M\,dx + N\,dy + P\,dz$ is exact; $\frac{\partial f}{\partial x} = 2x \Rightarrow f(x,y,z) = x^2 + g(y,z) \Rightarrow \frac{\partial f}{\partial y} = \frac{\partial g}{\partial y} = 2y \Rightarrow g(y,z) = y^2 + h(z) \Rightarrow f(x,y,z) = x^2 + y^2 = h(z)$
$\Rightarrow \frac{\partial f}{\partial z} = h'(z) = 2z \Rightarrow h(z) = z^2 + C \Rightarrow f(x,y,z) = x^2 + y^2 + z^2 + C \Rightarrow \int_{(0,0,0)}^{(2,3,-6)} 2x\,dx + 2y\,dy + 2z\,dz$
$= f(2,3,-6) - f(0,0,0) = 2^2 + 3^2 + (-6)^2 = 49$

15. Let $\mathbf{F}(x,y,z) = 2xy\mathbf{i} + (x^2 - z^2)\mathbf{j} - 2yz\mathbf{k} \Rightarrow \frac{\partial P}{\partial y} = -2z = \frac{\partial N}{\partial z}, \frac{\partial M}{\partial z} = 0 = \frac{\partial P}{\partial x}, \frac{\partial N}{\partial x} = 2x = \frac{\partial M}{\partial y}$
$\Rightarrow M\,dx + N\,dy + P\,dz$ is exact; $\frac{\partial f}{\partial x} = 2xy \Rightarrow f(x,y,z) = x^2y + g(y,z) \Rightarrow \frac{\partial f}{\partial y} = x^2 + \frac{\partial g}{\partial y} = x^2 - z^2 \Rightarrow \frac{\partial g}{\partial y} = -z^2$
$\Rightarrow g(y,z) = -yz^2 + h(z) \Rightarrow f(x,y,z) = x^2y - yz^2 + h(z) \Rightarrow \frac{\partial f}{\partial z} = -2yz + h'(z) = -2yz \Rightarrow h'(z) = 0 \Rightarrow h(z) = C$
$\Rightarrow f(x,y,z) = x^2y - yz^2 + C \Rightarrow \int_{(0,0,0)}^{(1,2,3)} 2xy\,dx + (x^2 - z^2)\,dy - 2yz\,dz = f(1,2,3) - f(0,0,0) = 2 - 2(3)^2 = -16$

17. Let $\mathbf{F}(x,y,z) = (\sin y \cos x)\mathbf{i} + (\cos y \sin x)\mathbf{j} + \mathbf{k} \Rightarrow \frac{\partial P}{\partial y} = 0 = \frac{\partial N}{\partial z}, \frac{\partial M}{\partial z} = 0 = \frac{\partial P}{\partial x}, \frac{\partial N}{\partial x} = \cos y \cos x = \frac{\partial M}{\partial y}$
$\Rightarrow M\,dx + N\,dy + P\,dz$ is exact; $\frac{\partial f}{\partial x} = \sin y \cos x \Rightarrow f(x,y,z) = \sin y \sin x + g(y,z) \Rightarrow \frac{\partial f}{\partial y} = \cos y \sin x + \frac{\partial g}{\partial y}$
$= \cos y \sin x \Rightarrow \frac{\partial g}{\partial y} = 0 \Rightarrow g(y,z) = h(z) \Rightarrow f(x,y,z) = \sin y \sin x + h(z) \Rightarrow \frac{\partial f}{\partial z} = h'(z) = 1 \Rightarrow h(z) = z + C$
$\Rightarrow f(x,y,z) = \sin y \sin x + z + C \Rightarrow \int_{(1,0,0)}^{(0,1,1)} \sin y \cos x\,dx + \cos y \sin x\,dy + dz = f(0,1,1) - f(1,0,0)$
$= (0+1) - (0+0) = 1$

19. Let $\mathbf{F}(x,y,z) = 3x^2\mathbf{i} + \left(\frac{z^2}{y}\right)\mathbf{j} + (2z\ln y)\mathbf{k} \Rightarrow \frac{\partial P}{\partial y} = \frac{2z}{y} = \frac{\partial N}{\partial z}, \frac{\partial M}{\partial z} = 0 = \frac{\partial P}{\partial x}, \frac{\partial N}{\partial x} = 0 = \frac{\partial M}{\partial y}$
$\Rightarrow M\,dx + N\,dy + P\,dz$ is exact; $\frac{\partial f}{\partial x} = 3x^2 \Rightarrow f(x,y,z) = x^3 + g(y,z) \Rightarrow \frac{\partial f}{\partial y} = \frac{\partial g}{\partial y} = \frac{z^2}{y} \Rightarrow g(y,z) = z^2\ln y + h(z)$
$\Rightarrow f(x,y,z) = x^3 + z^2\ln y + h(z) \Rightarrow \frac{\partial f}{\partial z} = 2z\ln y + h'(z) = 2z\ln y \Rightarrow h'(z) = 0 \Rightarrow h(z) = C \Rightarrow f(x,y,z)$
$= x^3 + z^2\ln y + C \Rightarrow \int_{(1,1,1)}^{(1,2,3)} 3x^2\,dx + \frac{z^2}{y}\,dy + 2z\ln y\,dz = f(1,2,3) - f(1,1,1)$
$= (1 + 9\ln 2 + C) - (1 + 0 + C) = 9\ln 2$

21. Let $\mathbf{F}(x,y,z) = \left(\frac{1}{y}\right)\mathbf{i} + \left(\frac{1}{z} - \frac{x}{y^2}\right)\mathbf{j} - \left(\frac{y}{z^2}\right)\mathbf{k} \Rightarrow \frac{\partial P}{\partial y} = -\frac{1}{z^2} = \frac{\partial N}{\partial z}, \frac{\partial M}{\partial z} = 0 = \frac{\partial P}{\partial x}, \frac{\partial N}{\partial x} = -\frac{1}{y^2} = \frac{\partial M}{\partial y}$
$\Rightarrow M\,dx + N\,dy + P\,dz$ is exact; $\frac{\partial f}{\partial x} = \frac{1}{y} \Rightarrow f(x,y,z) = \frac{x}{y} + g(y,z) \Rightarrow \frac{\partial f}{\partial y} = -\frac{x}{y^2} + \frac{\partial g}{\partial y} = \frac{1}{z} - \frac{x}{y^2}$
$\Rightarrow \frac{\partial g}{\partial y} = \frac{1}{z} \Rightarrow g(y,z) = \frac{y}{z} + h(z) \Rightarrow f(x,y,z) = \frac{x}{y} + \frac{y}{z} + h(z) \Rightarrow \frac{\partial f}{\partial z} = -\frac{y}{z^2} + h'(z) = -\frac{y}{z^2} \Rightarrow h'(z) = 0 \Rightarrow h(z) = C$
$\Rightarrow f(x,y,z) = \frac{x}{y} + \frac{y}{z} + C \Rightarrow \int_{(1,1,1)}^{(2,2,2)} \frac{1}{y}\,dx + \left(\frac{1}{z} - \frac{x}{y^2}\right)dy - \frac{y}{z^2}\,dz = f(2,2,2) - f(1,1,1) = \left(\frac{2}{2} + \frac{2}{2} + C\right) - \left(\frac{1}{1} + \frac{1}{1} + C\right)$
$= 0$

23. $\mathbf{r} = (\mathbf{i} + \mathbf{j} + \mathbf{k}) + t(\mathbf{i} + 2\mathbf{j} - 2\mathbf{k}) = (1+t)\mathbf{i} + (1+2t)\mathbf{j} + (1-2t)\mathbf{k},\ 0 \le t \le 1 \Rightarrow dx = dt,\ dy = 2\,dt,\ dz = -2\,dt$
$\Rightarrow \int_{(1,1,1)}^{(2,3,-1)} y\,dx + x\,dy + 4\,dz = \int_0^1 (2t+1)\,dt + (t+1)(2\,dt) + 4(-2)\,dt = \int_0^1 (4t-5)\,dt = \left[2t^2 - 5t\right]_0^1 = -3$

25. $\frac{\partial P}{\partial y} = 0 = \frac{\partial N}{\partial z}, \frac{\partial M}{\partial z} = 2z = \frac{\partial P}{\partial x}, \frac{\partial N}{\partial x} = 0 = \frac{\partial M}{\partial y} \Rightarrow M\,dx + N\,dy + P\,dz$ is exact $\Rightarrow$ **F** is conservative $\Rightarrow$ path independence

27. $\frac{\partial P}{\partial y} = 0 = \frac{\partial N}{\partial z}, \frac{\partial M}{\partial z} = 0 = \frac{\partial P}{\partial x}, \frac{\partial N}{\partial x} = -\frac{2x}{y^2} = \frac{\partial M}{\partial y} \Rightarrow$ **F** is conservative $\Rightarrow$ there exists an f so that $\mathbf{F} = \nabla f$;
$\frac{\partial f}{\partial x} = \frac{2x}{y} \Rightarrow f(x,y) = \frac{x^2}{y} + g(y) \Rightarrow \frac{\partial f}{\partial y} = -\frac{x^2}{y^2} + g'(y) = \frac{1-x^2}{y^2} \Rightarrow g'(y) = \frac{1}{y^2} \Rightarrow g(y) = -\frac{1}{y} + C$
$\Rightarrow f(x,y) = \frac{x^2}{y} - \frac{1}{y} + C \Rightarrow \mathbf{F} = \nabla\left(\frac{x^2-1}{y}\right)$

29. $\frac{\partial P}{\partial y} = 0 = \frac{\partial N}{\partial z}, \frac{\partial M}{\partial z} = 0 = \frac{\partial P}{\partial x}, \frac{\partial N}{\partial x} = 1 = \frac{\partial M}{\partial y} \Rightarrow$ **F** is conservative $\Rightarrow$ there exists an f so that $\mathbf{F} = \nabla f$;
$\frac{\partial f}{\partial x} = x^2 + y \Rightarrow f(x,y,z) = \frac{1}{3}x^3 + xy + g(y,z) \Rightarrow \frac{\partial f}{\partial y} = x + \frac{\partial g}{\partial y} = y^2 + x \Rightarrow \frac{\partial g}{\partial y} = y^2 \Rightarrow g(y,z) = \frac{1}{3}y^3 + h(z)$
$\Rightarrow f(x,y,z) = \frac{1}{3}x^3 + xy + \frac{1}{3}y^3 + h(z) \Rightarrow \frac{\partial f}{\partial z} = h'(z) = ze^z \Rightarrow h(z) = ze^z - e^z + C \Rightarrow f(x,y,z)$
$= \frac{1}{3}x^3 + xy + \frac{1}{3}y^3 + ze^z - e^z + C \Rightarrow \mathbf{F} = \nabla\left(\frac{1}{3}x^3 + xy + \frac{1}{3}y^3 + ze^z - e^z\right)$

(a) work $= \int_A^B \mathbf{F} \cdot \frac{d\mathbf{r}}{dt}\,dt = \int_A^B \mathbf{F} \cdot d\mathbf{r} = \left[\frac{1}{3}x^3 + xy + \frac{1}{3}y^3 + ze^z - e^z\right]_{(1,0,0)}^{(1,0,1)} = \left(\frac{1}{3} + 0 + 0 + e - e\right) - \left(\frac{1}{3} + 0 + 0 - 1\right)$
$= 1$

(b) work $= \int_A^B \mathbf{F} \cdot d\mathbf{r} = \left[\frac{1}{3}x^3 + xy + \frac{1}{3}y^3 + ze^z - e^z\right]_{(1,0,0)}^{(1,0,1)} = 1$

(c) work $= \int_A^B \mathbf{F} \cdot d\mathbf{r} = \left[\frac{1}{3}x^3 + xy + \frac{1}{3}y^3 + ze^z - e^z\right]_{(1,0,0)}^{(1,0,1)} = 1$

Note: Since **F** is conservative, $\int_A^B \mathbf{F} \cdot d\mathbf{r}$ is independent of the path from (1, 0, 0) to (1, 0, 1).

31. (a) $\mathbf{F} = \nabla(x^3y^2) \Rightarrow \mathbf{F} = 3x^2y^2\mathbf{i} + 2x^3y\mathbf{j}$; let C_1 be the path from $(-1,1)$ to $(0,0) \Rightarrow x = t-1$ and $y = -t+1, 0 \le t \le 1 \Rightarrow \mathbf{F} = 3(t-1)^2(-t+1)^2\mathbf{i} + 2(t-1)^3(-t+1)\mathbf{j} = 3(t-1)^4\mathbf{i} - 2(t-1)^4\mathbf{j}$ and $\mathbf{r}_1 = (t-1)\mathbf{i} + (-t+1)\mathbf{j} \Rightarrow d\mathbf{r}_1 = dt\,\mathbf{i} - dt\,\mathbf{j} \Rightarrow \int_{C_1} \mathbf{F} \cdot d\mathbf{r}_1 = \int_0^1 [3(t-1)^4 + 2(t-1)^4]\,dt$
$= \int_0^1 5(t-1)^4\,dt = [(t-1)^5]_0^1 = 1$; let C_2 be the path from $(0,0)$ to $(1,1) \Rightarrow x = t$ and $y = t$,
$0 \le t \le 1 \Rightarrow \mathbf{F} = 3t^4\mathbf{i} + 2t^4\mathbf{j}$ and $\mathbf{r}_2 = t\mathbf{i} + t\mathbf{j} \Rightarrow d\mathbf{r}_2 = dt\,\mathbf{i} + dt\,\mathbf{j} \Rightarrow \int_{C_2} \mathbf{F} \cdot d\mathbf{r}_2 = \int_0^1 (3t^4 + 2t^4)\,dt$
$= \int_0^1 5t^4\,dt = 1 \Rightarrow \int_C \mathbf{F} \cdot d\mathbf{r} = \int_{C_1} \mathbf{F} \cdot d\mathbf{r}_1 + \int_{C_2} \mathbf{F} \cdot d\mathbf{r}_2 = 2$

(b) Since $f(x,y) = x^3y^2$ is a potential function for **F**, $\int_{(-1,1)}^{(1,1)} \mathbf{F} \cdot d\mathbf{r} = f(1,1) - f(-1,1) = 2$

33. (a) If the differential form is exact, then $\frac{\partial P}{\partial y} = \frac{\partial N}{\partial z} \Rightarrow 2ay = cy$ for all $y \Rightarrow 2a = c$, $\frac{\partial M}{\partial z} = \frac{\partial P}{\partial x} \Rightarrow 2cx = 2cx$ for all x, and $\frac{\partial N}{\partial x} = \frac{\partial M}{\partial y} \Rightarrow by = 2ay$ for all $y \Rightarrow b = 2a$ and $c = 2a$

(b) $\mathbf{F} = \nabla f \Rightarrow$ the differential form with $a = 1$ in part (a) is exact $\Rightarrow b = 2$ and $c = 2$

35. The path will not matter; the work along any path will be the same because the field is conservative.

37. Let the coordinates of points A and B be (x_A, y_A, z_A) and (x_B, y_B, z_B), respectively. The force $\mathbf{F} = a\mathbf{i} + b\mathbf{j} + c\mathbf{k}$ is conservative because all the partial derivatives of M, N, and P are zero. Therefore, the potential function is $f(x,y,z) = ax + by + cz + C$, and the work done by the force in moving a particle along any path from A to B is $f(B) - f(A) = f(x_B, y_B, z_B) - f(x_A, y_A, z_A) = (ax_B + by_B + cz_B + C) - (ax_A + by_A + cz_A + C)$
$= a(x_B - x_A) + b(y_B - y_A) + c(z_B - z_A) = \mathbf{F} \cdot \overrightarrow{AB}$

13.4 GREEN'S THEOREM IN THE PLANE

1. $M = -y = -a \sin t$, $N = x = a \cos t$, $dx = -a \sin t\, dt$, $dy = a \cos t\, dt \Rightarrow \frac{\partial M}{\partial x} = 0$, $\frac{\partial M}{\partial y} = -1$, $\frac{\partial N}{\partial x} = 1$, and $\frac{\partial N}{\partial y} = 0$;

Equation (3): $\oint_C M\, dy - N\, dx = \int_0^{2\pi} [(-a \sin t)(a \cos t) - (a \cos t)(-a \sin t)]\, dt = \int_0^{2\pi} 0\, dt = 0$;
$\iint_R \left(\frac{\partial M}{\partial x} + \frac{\partial N}{\partial y}\right) dx\, dy = \iint_R 0\, dx\, dy = 0$, Flux

Equation (4): $\oint_C M\, dx + N\, dy = \int_0^{2\pi} [(-a \sin t)(-a \sin t) - (a \cos t)(a \cos t)]\, dt = \int_0^{2\pi} a^2\, dt = 2\pi a^2$;
$\iint_R \left(\frac{\partial N}{\partial x} - \frac{\partial M}{\partial y}\right) dx\, dy = \int_{-a}^{a} \int_{-c}^{\sqrt{a^2 - x^2}} 2\, dy\, dx = \int_{-a}^{a} 4\sqrt{a^2 - x^2}\, dx = 4\left[\frac{x}{2}\sqrt{a^2 - x^2} + \frac{a^2}{2} \sin^{-1} \frac{x}{a}\right]_{-a}^{a}$
$= 2a^2\left(\frac{\pi}{2} + \frac{\pi}{2}\right) = 2a^2\pi$, Circulation

3. $M = 2x = 2a \cos t$, $N = -3y = -3a \sin t$, $dx = -a \sin t\, dt$, $dy = a \cos t\, dt \Rightarrow \frac{\partial M}{\partial x} = 2$, $\frac{\partial M}{\partial y} = 0$, $\frac{\partial N}{\partial x} = 0$, and $\frac{\partial N}{\partial y} = -3$;

Equation (3): $\oint_C M\, dy - N\, dx = \int_0^{2\pi} [(2a \cos t)(a \cos t) + (3a \sin t)(-a \sin t)]\, dt$
$= \int_0^{2\pi} (2a^2 \cos^2 t - 3a^2 \sin^2 t)\, dt = 2a^2\left[\frac{t}{2} + \frac{\sin 2t}{4}\right]_0^{2\pi} - 3a^2\left[\frac{t}{2} - \frac{\sin 2t}{4}\right]_0^{2\pi} = 2\pi a^2 - 3\pi a^2 = -\pi a^2$;
$\iint_R \left(\frac{\partial M}{\partial x} + \frac{\partial N}{\partial y}\right) = \iint_R -1\, dx\, dy = \int_0^{2\pi} \int_0^{a} -r\, dr\, d\theta = \int_0^{2\pi} -\frac{a^2}{2}\, d\theta = -\pi a^2$, Flux

Equation (4): $\oint_C M\, dx + N\, dy = \int_0^{2\pi} [(2a \cos t)(-a \sin t) + (-3a \sin t)(a \cos t)]\, dt$
$= \int_0^{2\pi} (-2a^2 \sin t \cos t - 3a^2 \sin t \cos t)\, dt = -5a^2\left[\frac{1}{2} \sin^2 t\right]_0^{2\pi} = 0$; $\iint_R 0\, dx\, dy = 0$, Circulation

5. $M = x - y$, $N = y - x \Rightarrow \frac{\partial M}{\partial x} = 1$, $\frac{\partial M}{\partial y} = -1$, $\frac{\partial N}{\partial x} = -1$, $\frac{\partial N}{\partial y} = 1 \Rightarrow$ Flux $= \iint_R 2\, dx\, dy = \int_0^1 \int_0^1 2\, dx\, dy = 2$;
Circ $= \iint_R [-1 - (-1)]\, dx\, dy = 0$

7. $M = y^2 - x^2$, $N = x^2 + y^2 \Rightarrow \frac{\partial M}{\partial x} = -2x$, $\frac{\partial M}{\partial y} = 2y$, $\frac{\partial N}{\partial x} = 2x$, $\frac{\partial N}{\partial y} = 2y \Rightarrow$ Flux $= \iint_R (-2x + 2y)\, dx\, dy$
$= \int_0^3 \int_0^x (-2x + 2y)\, dy\, dx = \int_0^3 (-2x^2 + x^2)\, dx = \left[-\frac{1}{3}x^3\right]_0^3 = -9$; Circ $= \iint_R (2x - 2y)\, dx\, dy$
$= \int_0^3 \int_0^x (2x - 2y)\, dy\, dx = \int_0^3 x^2\, dx = 9$

9. $M = x + e^x \sin y$, $N = x + e^x \cos y \Rightarrow \frac{\partial M}{\partial x} = 1 + e^x \sin y$, $\frac{\partial M}{\partial y} = e^x \cos y$, $\frac{\partial N}{\partial x} = 1 + e^x \cos y$, $\frac{\partial N}{\partial y} = -e^x \sin y$
$\Rightarrow$ Flux $= \iint_R dx\, dy = \int_{-\pi/4}^{\pi/4} \int_0^{\sqrt{\cos 2\theta}} r\, dr\, d\theta = \int_{-\pi/4}^{\pi/4} \left(\frac{1}{2} \cos 2\theta\right) d\theta = \left[\frac{1}{4} \sin 2\theta\right]_{-\pi/4}^{\pi/4} = \frac{1}{2}$;
Circ $= \iint_R (1 + e^x \cos y - e^x \cos y)\, dx\, dy = \iint_R dx\, dy = \int_{-\pi/4}^{\pi/4} \int_0^{\sqrt{\cos 2\theta}} r\, dr\, d\theta = \int_{-\pi/4}^{\pi/4} \left(\frac{1}{2} \cos 2\theta\right) d\theta = \frac{1}{2}$

11. $M = xy$, $N = y^2 \Rightarrow \frac{\partial M}{\partial x} = y$, $\frac{\partial M}{\partial y} = x$, $\frac{\partial N}{\partial x} = 0$, $\frac{\partial N}{\partial y} = 2y \Rightarrow$ Flux $= \iint_R (y + 2y)\, dy\, dx = \int_0^1 \int_{x^2}^{x} 3y\, dy\, dx$
$= \int_0^1 \left(\frac{3x^2}{2} - \frac{3x^4}{2}\right) dx = \frac{1}{5}$; Circ $= \iint_R -x\, dy\, dx = \int_0^1 \int_{x^2}^{x} -x\, dy\, dx = \int_0^1 (-x^2 + x^3)\, dx = -\frac{1}{12}$

13. $M = 3xy - \frac{x}{1+y^2}$, $N = e^x + \tan^{-1} y \Rightarrow \frac{\partial M}{\partial x} = 3y - \frac{1}{1+y^2}$, $\frac{\partial N}{\partial y} = \frac{1}{1+y^2}$

$\Rightarrow \text{Flux} = \iint_R \left(3y - \frac{1}{1+y^2} + \frac{1}{1+y^2}\right) dx\,dy = \iint_R 3y\,dx\,dy = \int_0^{2\pi}\int_0^{a(1+\cos\theta)} (3r\sin\theta)\,r\,dr\,d\theta$

$= \int_0^{2\pi} a^3(1+\cos\theta)^3(\sin\theta)\,d\theta = \left[-\frac{a^3}{4}(1+\cos\theta)^4\right]_0^{2\pi} = -4a^3 - (-4a^3) = 0$

15. $M = 2xy^3$, $N = 4x^2y^2 \Rightarrow \frac{\partial M}{\partial y} = 6xy^2$, $\frac{\partial N}{\partial x} = 8xy^2 \Rightarrow \text{work} = \oint_C 2xy^3\,dx + 4x^2y^2\,dy = \iint_R (8xy^2 - 6xy^2)\,dx\,dy$

$= \int_0^1\int_0^{x^3} 2xy^2\,dy\,dx = \int_0^1 \frac{2}{3}x^{10}\,dx = \frac{2}{33}$

17. $M = y^2$, $N = x^2 \Rightarrow \frac{\partial M}{\partial y} = 2y$, $\frac{\partial N}{\partial x} = 2x \Rightarrow \oint_C y^2\,dx + x^2\,dy = \iint_R (2x - 2y)\,dy\,dx$

$= \int_0^1\int_0^{1-x} (2x-2y)\,dy\,dx = \int_0^1 (-3x^2 + 4x - 1)\,dx = \left[-x^3 + 2x^2 - x\right]_0^1 = -1 + 2 - 1 = 0$

19. $M = 6y + x$, $N = y + 2x \Rightarrow \frac{\partial M}{\partial y} = 6$, $\frac{\partial N}{\partial x} = 2 \Rightarrow \oint_C (6y+x)\,dx + (y+2x)\,dy = \iint_R (2-6)\,dy\,dx$

$= -4(\text{Area of the circle}) = -16\pi$

21. $M = x = a\cos t$, $N = y = a\sin t \Rightarrow dx = -a\sin t\,dt$, $dy = a\cos t\,dt \Rightarrow \text{Area} = \frac{1}{2}\oint_C x\,dy - y\,dx$

$= \frac{1}{2}\int_0^{2\pi}(a^2\cos^2 t + a^2\sin^2 t)\,dt = \frac{1}{2}\int_0^{2\pi} a^2\,dt = \pi a^2$

23. $M = x = \cos^3 t$, $N = y = \sin^3 t \Rightarrow dx = -3\cos^2 t\sin t\,dt$, $dy = 3\sin^2 t\cos t\,dt \Rightarrow \text{Area} = \frac{1}{2}\oint_C x\,dy - y\,dx$

$= \frac{1}{2}\int_0^{2\pi}(3\sin^2 t\cos^2 t)(\cos^2 t + \sin^2 t)\,dt = \frac{1}{2}\int_0^{2\pi}(3\sin^2 t\cos^2 t)\,dt = \frac{3}{8}\int_0^{2\pi}\sin^2 2t\,dt = \frac{3}{16}\int_0^{4\pi}\sin^2 u\,du$

$= \frac{3}{16}\left[\frac{u}{2} - \frac{\sin 2u}{4}\right]_0^{4\pi} = \frac{3}{8}\pi$

25. (a) $M = f(x)$, $N = g(y) \Rightarrow \frac{\partial M}{\partial y} = 0$, $\frac{\partial N}{\partial x} = 0 \Rightarrow \oint_C f(x)\,dx + g(y)\,dy = \iint_R \left(\frac{\partial N}{\partial x} - \frac{\partial M}{\partial y}\right) dx\,dy$

$= \iint_R 0\,dx\,dy = 0$

(b) $M = ky$, $N = hx \Rightarrow \frac{\partial M}{\partial y} = k$, $\frac{\partial N}{\partial x} = h \Rightarrow \oint_C ky\,dx + hx\,dy = \iint_R \left(\frac{\partial N}{\partial x} - \frac{\partial M}{\partial y}\right) dx\,dy$

$= \iint_R (h-k)\,dx\,dy = (h-k)(\text{Area of the region})$

27. Let $M = x$ and $N = 0 \Rightarrow \frac{\partial M}{\partial x} = 1$ and $\frac{\partial N}{\partial y} = 0 \Rightarrow \oint_C M\,dy - N\,dx = \iint_R \left(\frac{\partial M}{\partial x} + \frac{\partial N}{\partial y}\right) dx\,dy \Rightarrow \oint_C x\,dy$

$= \iint_R (1+0)\,dx\,dy \Rightarrow \text{Area of R} = \iint_R dx\,dy = \oint_C x\,dy$; similarly, $M = y$ and $N = 0 \Rightarrow \frac{\partial M}{\partial y} = 1$ and

$\frac{\partial N}{\partial x} = 0 \Rightarrow \oint_C M\,dx + N\,dy = \iint_R \left(\frac{\partial N}{\partial x} + \frac{\partial M}{\partial y}\right) dy\,dx \Rightarrow \oint_C y\,dx = \iint_R (0-1)\,dy\,dx \Rightarrow -\oint_C y\,dx$

$= \iint_R dx\,dy = \text{Area of R}$

29. Let $\delta(x,y) = 1 \Rightarrow \bar{x} = \frac{M_y}{M} = \frac{\iint_R x\,\delta(x,y)\,dA}{\iint_R \delta(x,y)\,dA} = \frac{\iint_R x\,dA}{\iint_R dA} = \frac{\iint_R x\,dA}{A} \Rightarrow A\bar{x} = \iint_R x\,dA = \iint_R (x+0)\,dx\,dy$

$= \oint_C \frac{x^2}{2}\,dy$, $A\bar{x} = \iint_R x\,dA = \iint_R (0+x)\,dx\,dy = -\oint_C xy\,dx$, and $A\bar{x} = \iint_R x\,dA = \iint_R \left(\frac{2}{3}x + \frac{1}{3}x\right) dx\,dy$

$= \oint_C \frac{1}{3}x^2\,dy - \frac{1}{3}xy\,dx \Rightarrow \frac{1}{2}\oint_C x^2\,dy = -\oint_C xy\,dx = \frac{1}{3}\oint_C x^2\,dy - xy\,dx = A\bar{x}$

31. (a) $\nabla f = \left(\frac{2x}{x^2+y^2}\right)\mathbf{i} + \left(\frac{2y}{x^2+y^2}\right)\mathbf{j} \Rightarrow M = \frac{2x}{x^2+y^2}, N = \frac{2y}{x^2+y^2}$; since M, N are discontinuous at (0, 0), we compute $\int_C \nabla f \cdot \mathbf{n}\, ds$ directly since Green's Theorem does not apply. Let $x = a\cos t$, $y = a\sin t \Rightarrow dx = -a\sin t\, dt$, $dy = a\cos t\, dt$, $M = \frac{2}{a}\cos t$, $N = \frac{2}{a}\sin t$, $0 \le t \le 2\pi$, so $\int_C \nabla f \cdot \mathbf{n}\, ds = \int_C M\, dy - N\, dx$
$= \int_0^{2\pi} \left[\left(\frac{2}{a}\cos t\right)(a\cos t) - \left(\frac{2}{a}\sin t\right)(-a\sin t)\right]dt = \int_0^{2\pi} 2(\cos^2 t + \sin^2 t)dt = 4\pi$. Note that this holds for any $a > 0$, so $\int_C \nabla f \cdot \mathbf{n}\, ds = 4\pi$ for any circle C centered at (0, 0) traversed counterclockwise and $\int_C \nabla f \cdot \mathbf{n}\, ds = -4\pi$ if C is traversed clockwise.

(b) If K does not enclose the point (0, 0) we may apply Green's Theorem: $\int_C \nabla f \cdot \mathbf{n}\, ds = \int_C M\, dy - N\, dx$
$= \iint_R \left(\frac{\partial M}{\partial x} + \frac{\partial N}{\partial y}\right) dx\, dy = \iint_R \left(\frac{2(y^2-x^2)}{(x^2+y^2)^2} + \frac{2(x^2-y^2)}{(x^2+y^2)^2}\right) dx\, dy = \iint_R 0\, dx\, dy = 0$. If K does enclose the point (0, 0) we proceed as in Example 6:
Choose a small enough so that the circle C centered at (0, 0) of radius a lies entirely within K. Green's Theorem applies to the region R that lies between K and C. Thus, as before, $0 = \iint_R \left(\frac{\partial M}{\partial x} + \frac{\partial N}{\partial y}\right) dx\, dy$
$= \int_K M\, dy - N\, dx + \int_C M\, dy - N\, dx$ where K is traversed counterclockwise and C is traversed clockwise.
Hence by part (a) $0 = \left[\int_K M\, dy - N\, dx\right] - 4\pi \Rightarrow 4\pi = \int_K M\, dy - N\, dx = \int_K \nabla f \cdot \mathbf{n}\, ds$. We have shown:
$$\int_K \nabla f \cdot \mathbf{n}\, ds = \begin{cases} 0 & \text{if } (0,0) \text{ lies inside K} \\ 4\pi & \text{if } (0,0) \text{ lies outside K} \end{cases}$$

13.5 SURFACES AND AREA

1. In cylindrical coordinates, let $x = r\cos\theta$, $y = r\sin\theta$, $z = \left(\sqrt{x^2+y^2}\right)^2 = r^2$. Then $\mathbf{r}(r,\theta) = (r\cos\theta)\mathbf{i} + (r\sin\theta)\mathbf{j} + r^2\mathbf{k}$, $0 \le r \le 2$, $0 \le \theta \le 2\pi$.

3. In cylindrical coordinates, let $x = r\cos\theta$, $y = r\sin\theta$, $z = \frac{\sqrt{x^2+y^2}}{2} \Rightarrow z = \frac{r}{2}$. Then $\mathbf{r}(r,\theta) = (r\cos\theta)\mathbf{i} + (r\sin\theta)\mathbf{j} + \left(\frac{r}{2}\right)\mathbf{k}$. For $0 \le z \le 3$, $0 \le \frac{r}{2} \le 3 \Rightarrow 0 \le r \le 6$; to get only the first octant, let $0 \le \theta \le \frac{\pi}{2}$.

5. In cylindrical coordinates, let $x = r\cos\theta$, $y = r\sin\theta$; since $x^2+y^2 = r^2 \Rightarrow z^2 = 9 - (x^2+y^2) = 9 - r^2$ $\Rightarrow z = \sqrt{9-r^2}$, $z \ge 0$. Then $\mathbf{r}(r,\theta) = (r\cos\theta)\mathbf{i} + (r\sin\theta)\mathbf{j} + \sqrt{9-r^2}\mathbf{k}$. Let $0 \le \theta \le 2\pi$. For the domain of r: $z = \sqrt{x^2+y^2}$ and $x^2+y^2+z^2 = 9 \Rightarrow x^2+y^2+\left(\sqrt{x^2+y^2}\right)^2 = 9 \Rightarrow 2(x^2+y^2) = 9 \Rightarrow 2r^2 = 9$ $\Rightarrow r = \frac{3}{\sqrt{2}} \Rightarrow 0 \le r \le \frac{3}{\sqrt{2}}$.

7. In spherical coordinates, $x = \rho\sin\phi\cos\theta$, $y = \rho\sin\phi\sin\theta$, $\rho = \sqrt{x^2+y^2+z^2} \Rightarrow \rho^2 = 3 \Rightarrow \rho = \sqrt{3}$ $\Rightarrow z = \sqrt{3}\cos\phi$ for the sphere; $z = \frac{\sqrt{3}}{2} = \sqrt{3}\cos\phi \Rightarrow \cos\phi = \frac{1}{2} \Rightarrow \phi = \frac{\pi}{3}$; $z = -\frac{\sqrt{3}}{2} \Rightarrow -\frac{\sqrt{3}}{2} = \sqrt{3}\cos\phi$ $\Rightarrow \cos\phi = -\frac{1}{2} \Rightarrow \phi = \frac{2\pi}{3}$. Then $\mathbf{r}(\phi,\theta) = \left(\sqrt{3}\sin\phi\cos\theta\right)\mathbf{i} + \left(\sqrt{3}\sin\phi\sin\theta\right)\mathbf{j} + \left(\sqrt{3}\cos\phi\right)\mathbf{k}$, $\frac{\pi}{3} \le \phi \le \frac{2\pi}{3}$ and $0 \le \theta \le 2\pi$.

9. Since $z = 4 - y^2$, we can let $\mathbf{r}$ be a function of x and y $\Rightarrow \mathbf{r}(x,y) = x\mathbf{i} + y\mathbf{j} + (4-y^2)\mathbf{k}$. Then $z = 0$ $\Rightarrow 0 = 4 - y^2 \Rightarrow y = \pm 2$. Thus, let $-2 \le y \le 2$ and $0 \le x \le 2$.

11. When $x = 0$, let $y^2 + z^2 = 9$ be the circular section in the yz-plane. Use polar coordinates in the yz-plane $\Rightarrow y = 3\cos\theta$ and $z = 3\sin\theta$. Thus let $x = u$ and $\theta = v \Rightarrow \mathbf{r}(u,v) = u\mathbf{i} + (3\cos v)\mathbf{j} + (3\sin v)\mathbf{k}$ where $0 \le u \le 3$, and $0 \le v \le 2\pi$.

13. (a) $x+y+z=1 \Rightarrow z=1-x-y$. In cylindrical coordinates, let $x=r\cos\theta$ and $y=r\sin\theta$ $\Rightarrow z=1-r\cos\theta-r\sin\theta \Rightarrow \mathbf{r}(r,\theta)=(r\cos\theta)\mathbf{i}+(r\sin\theta)\mathbf{j}+(1-r\cos\theta-r\sin\theta)\mathbf{k}$, $0\le\theta\le 2\pi$ and $0\le r\le 3$.

(b) In a fashion similar to cylindrical coordinates, but working in the yz-plane instead of the xy-plane, let $y=u\cos v$, $z=u\sin v$ where $u=\sqrt{y^2+z^2}$ and v is the angle formed by (x, y, z), (x, 0, 0), and (x, y, 0) with (x, 0, 0) as vertex. Since $x+y+z=1 \Rightarrow x=1-y-z \Rightarrow x=1-u\cos v-u\sin v$, then $\mathbf{r}$ is a function of u and v $\Rightarrow \mathbf{r}(u,v)=(1-u\cos v-u\sin v)\mathbf{i}+(u\cos v)\mathbf{j}+(u\sin v)\mathbf{k}$, $0\le u\le 3$ and $0\le v\le 2\pi$.

15. Let $x=w\cos v$ and $z=w\sin v$. Then $(x-2)^2+z^2=4 \Rightarrow x^2-4x+z^2=0 \Rightarrow w^2\cos^2 v-4w\cos v+w^2\sin^2 v=0 \Rightarrow w^2-4w\cos v=0 \Rightarrow w=0$ or $w-4\cos v=0 \Rightarrow w=0$ or $w=4\cos v$. Now $w=0 \Rightarrow x=0$ and $y=0$, which is a line not a cylinder. Therefore, let $w=4\cos v \Rightarrow x=(4\cos v)(\cos v)=4\cos^2 v$ and $z=4\cos v\sin v$. Finally, let $y=u$. Then $\mathbf{r}(u,v)=(4\cos^2 v)\,\mathbf{i}+u\mathbf{j}+(4\cos v\sin v)\mathbf{k}$, $-\frac{\pi}{2}\le v\le\frac{\pi}{2}$ and $0\le u\le 3$.

17. Let $x=r\cos\theta$ and $y=r\sin\theta$. Then $\mathbf{r}(r,\theta)=(r\cos\theta)\mathbf{i}+(r\sin\theta)\mathbf{j}+\left(\frac{2-r\sin\theta}{2}\right)\mathbf{k}$, $0\le r\le 1$ and $0\le\theta\le 2\pi$ $\Rightarrow \mathbf{r}_r=(\cos\theta)\mathbf{i}+(\sin\theta)\mathbf{j}-\left(\frac{\sin\theta}{2}\right)\mathbf{k}$ and $\mathbf{r}_\theta=(-r\sin\theta)\mathbf{i}+(r\cos\theta)\mathbf{j}-\left(\frac{r\cos\theta}{2}\right)\mathbf{k}$

$$\Rightarrow \mathbf{r}_r\times\mathbf{r}_\theta=\begin{vmatrix}\mathbf{i} & \mathbf{j} & \mathbf{k}\\ \cos\theta & \sin\theta & -\frac{\sin\theta}{2}\\ -r\sin\theta & r\cos\theta & -\frac{r\cos\theta}{2}\end{vmatrix}$$

$$=\left(\frac{-r\sin\theta\cos\theta}{2}+\frac{(\sin\theta)(r\cos\theta)}{2}\right)\mathbf{i}+\left(\frac{r\sin^2\theta}{2}+\frac{r\cos^2\theta}{2}\right)\mathbf{j}+\left(r\cos^2\theta+r\sin^2\theta\right)\mathbf{k}=\frac{r}{2}\mathbf{j}+r\mathbf{k}$$

$$\Rightarrow |\mathbf{r}_r\times\mathbf{r}_\theta|=\sqrt{\frac{r^2}{4}+r^2}=\frac{\sqrt{5}r}{2} \Rightarrow A=\int_0^{2\pi}\int_0^1\frac{\sqrt{5}r}{2}\,dr\,d\theta=\int_0^{2\pi}\left[\frac{\sqrt{5}r^2}{4}\right]_0^1 d\theta=\int_0^{2\pi}d\theta=\frac{\pi\sqrt{5}}{2}$$

19. Let $x=r\cos\theta$ and $y=r\sin\theta \Rightarrow z=2\sqrt{x^2+y^2}=2r$, $1\le r\le 3$ and $0\le\theta\le 2\pi$. Then $\mathbf{r}(r,\theta)=(r\cos\theta)\mathbf{i}+(r\sin\theta)\mathbf{j}+2r\mathbf{k} \Rightarrow \mathbf{r}_r=(\cos\theta)\mathbf{i}+(\sin\theta)\mathbf{j}+2\mathbf{k}$ and $\mathbf{r}_\theta=(-r\sin\theta)\mathbf{i}+(r\cos\theta)\mathbf{j}$

$$\Rightarrow \mathbf{r}_r\times\mathbf{r}_\theta=\begin{vmatrix}\mathbf{i} & \mathbf{j} & \mathbf{k}\\ \cos\theta & \sin\theta & 2\\ -r\sin\theta & r\cos\theta & 0\end{vmatrix}=(-2r\cos\theta)\mathbf{i}-(2r\sin\theta)\mathbf{j}+\left(r\cos^2\theta+r\sin^2\theta\right)\mathbf{k}$$

$$=(-2r\cos\theta)\mathbf{i}-(2r\sin\theta)\mathbf{j}+r\mathbf{k} \Rightarrow |\mathbf{r}_r\times\mathbf{r}_\theta|=\sqrt{4r^2\cos^2\theta+4r^2\sin^2\theta+r^2}=\sqrt{5r^2}=r\sqrt{5}$$

$$\Rightarrow A=\int_0^{2\pi}\int_1^3 r\sqrt{5}\,dr\,d\theta=\int_0^{2\pi}\left[\frac{r^2\sqrt{5}}{2}\right]_1^3 d\theta=\int_0^{2\pi}4\sqrt{5}\,d\theta=8\pi\sqrt{5}$$

21. Let $x=r\cos\theta$ and $y=r\sin\theta \Rightarrow r^2=x^2+y^2=1$, $1\le z\le 4$ and $0\le\theta\le 2\pi$. Then

$$\mathbf{r}(z,\theta)=(\cos\theta)\mathbf{i}+(\sin\theta)\mathbf{j}+z\mathbf{k} \Rightarrow \mathbf{r}_z=\mathbf{k} \text{ and } \mathbf{r}_\theta=(-\sin\theta)\mathbf{i}+(\cos\theta)\mathbf{j} \Rightarrow \mathbf{r}_\theta\times\mathbf{r}_z=\begin{vmatrix}\mathbf{i} & \mathbf{j} & \mathbf{k}\\ -\sin\theta & \cos\theta & 0\\ 0 & 0 & 1\end{vmatrix}$$

$$=(\cos\theta)\mathbf{i}+(\sin\theta)\,\mathbf{j} \Rightarrow |\mathbf{r}_\theta\times\mathbf{r}_z|=\sqrt{\cos^2\theta+\sin^2\theta}=1 \Rightarrow A=\int_0^{2\pi}\int_1^4 1\,dr\,d\theta=\int_0^{2\pi}3\,d\theta=6\pi$$

23. $z=2-x^2-y^2$ and $z=\sqrt{x^2+y^2} \Rightarrow z=2-z^2 \Rightarrow z^2+z-2=0 \Rightarrow z=-2$ or $z=1$. Since $z=\sqrt{x^2+y^2}\ge 0$, we get $z=1$ where the cone intersects the paraboloid. When $x=0$ and $y=0$, $z=2 \Rightarrow$ the vertex of the paraboloid is $(0,0,2)$. Therefore, z ranges from 1 to 2 on the "cap" $\Rightarrow$ r ranges from 1 (when $x^2+y^2=1$) to 0 (when $x=0$ and $y=0$ at the vertex). Let $x=r\cos\theta$, $y=r\sin\theta$, and $z=2-r^2$. Then $\mathbf{r}(r,\theta)=(r\cos\theta)\mathbf{i}+(r\sin\theta)\mathbf{j}+(2-r^2)\,\mathbf{k}$, $0\le r\le 1$,

$$0\le\theta\le 2\pi \Rightarrow \mathbf{r}_r=(\cos\theta)\mathbf{i}+(\sin\theta)\mathbf{j}-2r\mathbf{k} \text{ and } \mathbf{r}_\theta=(-r\sin\theta)\mathbf{i}+(r\cos\theta)\mathbf{j} \Rightarrow \mathbf{r}_r\times\mathbf{r}_\theta=\begin{vmatrix}\mathbf{i} & \mathbf{j} & \mathbf{k}\\ \cos\theta & \sin\theta & -2r\\ -r\sin\theta & r\cos\theta & 0\end{vmatrix}$$

$$=(2r^2\cos\theta)\,\mathbf{i}+(2r^2\sin\theta)\,\mathbf{j}+r\mathbf{k} \Rightarrow |\mathbf{r}_r\times\mathbf{r}_\theta|=\sqrt{4r^4\cos^2\theta+4r^4\sin^2\theta+r^2}=r\sqrt{4r^2+1}$$

$$\Rightarrow A=\int_0^{2\pi}\int_0^1 r\sqrt{4r^2+1}\,dr\,d\theta=\int_0^{2\pi}\left[\frac{1}{12}(4r^2+1)^{3/2}\right]_0^1 d\theta=\int_0^{2\pi}\left(\frac{5\sqrt{5}-1}{12}\right)d\theta=\frac{\pi}{6}\left(5\sqrt{5}-1\right)$$

25. Let $x = \rho \sin\phi \cos\theta$, $y = \rho \sin\phi \sin\theta$, and $z = \rho \cos\phi \Rightarrow \rho = \sqrt{x^2+y^2+z^2} = \sqrt{2}$ on the sphere. Next, $x^2+y^2+z^2 = 2$ and $z = \sqrt{x^2+y^2} \Rightarrow z^2+z^2 = 2 \Rightarrow z^2 = 1 \Rightarrow z = 1$ since $z \ge 0 \Rightarrow \phi = \frac{\pi}{4}$. For the lower portion of the sphere cut by the cone, we get $\phi = \pi$. Then

$$\mathbf{r}(\phi,\theta) = \left(\sqrt{2}\sin\phi\cos\theta\right)\mathbf{i} + \left(\sqrt{2}\sin\phi\sin\theta\right)\mathbf{j} + \left(\sqrt{2}\cos\phi\right)\mathbf{k},\ \tfrac{\pi}{4} \le \phi \le \pi,\ 0 \le \theta \le 2\pi$$

$$\Rightarrow \mathbf{r}_\phi = \left(\sqrt{2}\cos\phi\cos\theta\right)\mathbf{i} + \left(\sqrt{2}\cos\phi\sin\theta\right)\mathbf{j} - \left(\sqrt{2}\sin\phi\right)\mathbf{k} \text{ and } \mathbf{r}_\theta = \left(-\sqrt{2}\sin\phi\sin\theta\right)\mathbf{i} + \left(\sqrt{2}\sin\phi\cos\theta\right)\mathbf{j}$$

$$\Rightarrow \mathbf{r}_\phi \times \mathbf{r}_\theta = \begin{vmatrix} \mathbf{i} & \mathbf{j} & \mathbf{k} \\ \sqrt{2}\cos\phi\cos\theta & \sqrt{2}\cos\phi\sin\theta & -\sqrt{2}\sin\phi \\ -\sqrt{2}\sin\phi\sin\theta & \sqrt{2}\sin\phi\cos\theta & 0 \end{vmatrix}$$

$$= \left(2\sin^2\phi\cos\theta\right)\mathbf{i} + \left(2\sin^2\phi\sin\theta\right)\mathbf{j} + (2\sin\phi\cos\phi)\mathbf{k}$$

$$\Rightarrow |\mathbf{r}_\phi \times \mathbf{r}_\theta| = \sqrt{4\sin^4\phi\cos^2\theta + 4\sin^4\phi\sin^2\theta + 4\sin^2\phi\cos^2\phi} = \sqrt{4\sin^2\phi} = 2|\sin\phi| = 2\sin\phi$$

$$\Rightarrow A = \int_0^{2\pi}\int_{\pi/4}^{\pi} 2\sin\phi\, d\phi\, d\theta = \int_0^{2\pi}\left(2+\sqrt{2}\right)d\theta = \left(4+2\sqrt{2}\right)\pi$$

27. The parametrization $\mathbf{r}(r,\theta) = (r\cos\theta)\mathbf{i} + (r\sin\theta)\mathbf{j} + r\mathbf{k}$ at $P_0 = \left(\sqrt{2}, \sqrt{2}, 2\right) \Rightarrow \theta = \frac{\pi}{4}, r = 2$,

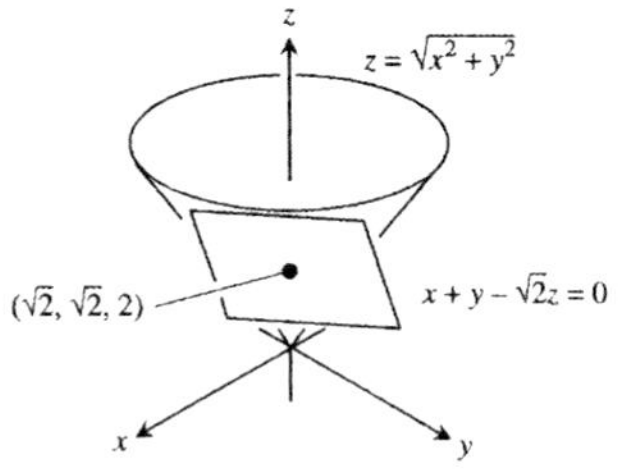

$\mathbf{r}_r = (\cos\theta)\mathbf{i} + (\sin\theta)\mathbf{j} + \mathbf{k} = \frac{\sqrt{2}}{2}\mathbf{i} + \frac{\sqrt{2}}{2}\mathbf{j} + \mathbf{k}$ and

$\mathbf{r}_\theta = (-r\sin\theta)\mathbf{i} + (r\cos\theta)\mathbf{j} = -\sqrt{2}\mathbf{i} + \sqrt{2}\mathbf{j}$

$$\Rightarrow \mathbf{r}_r \times \mathbf{r}_\theta = \begin{vmatrix} \mathbf{i} & \mathbf{j} & \mathbf{k} \\ \sqrt{2}/2 & \sqrt{2}/2 & 1 \\ -\sqrt{2} & \sqrt{2} & 0 \end{vmatrix}$$

$= -\sqrt{2}\mathbf{i} - \sqrt{2}\mathbf{j} + 2\mathbf{k} \Rightarrow$ the tangent plane is

$0 = \left(-\sqrt{2}\mathbf{i} - \sqrt{2}\mathbf{j} + 2\mathbf{k}\right) \cdot \left[\left(x-\sqrt{2}\right)\mathbf{i} + \left(y-\sqrt{2}\right)\mathbf{j} + (z-2)\mathbf{k}\right] \Rightarrow \sqrt{2}x + \sqrt{2}y - 2z = 0$, or $x + y - \sqrt{2}z = 0$.

The parametrization $\mathbf{r}(r,\theta) \Rightarrow x = r\cos\theta$, $y = r\sin\theta$ and $z = r \Rightarrow x^2+y^2 = r^2 = z^2 \Rightarrow$ the surface is $z = \sqrt{x^2+y^2}$.

29. The parametrization $\mathbf{r}(\theta, z) = (3\sin 2\theta)\mathbf{i} + \left(6\sin^2\theta\right)\mathbf{j} + z\mathbf{k}$ at $P_0 = \left(\frac{3\sqrt{3}}{2}, \frac{9}{2}, 0\right) \Rightarrow \theta = \frac{\pi}{3}$ and $z = 0$. Then

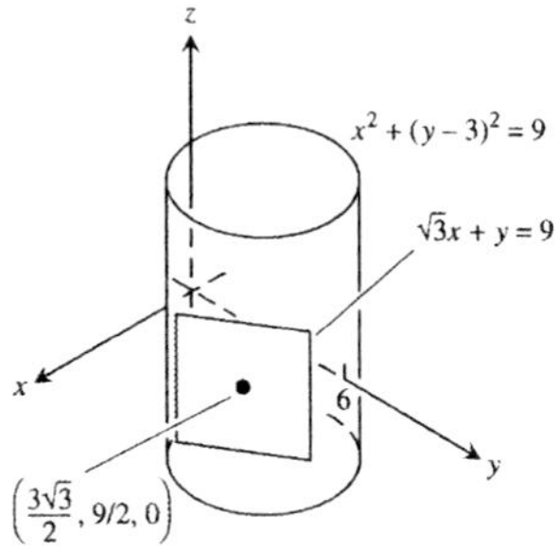

$\mathbf{r}_\theta = (6\cos 2\theta)\mathbf{i} + (12\sin\theta\cos\theta)\mathbf{j}$

$= -3\mathbf{i} + 3\sqrt{3}\mathbf{j}$ and $\mathbf{r}_z = \mathbf{k}$ at P_0

$$\Rightarrow \mathbf{r}_\theta \times \mathbf{r}_z = \begin{vmatrix} \mathbf{i} & \mathbf{j} & \mathbf{k} \\ -3 & 3\sqrt{3} & 0 \\ 0 & 0 & 1 \end{vmatrix} = 3\sqrt{3}\mathbf{i} + 3\mathbf{j}$$

$\Rightarrow$ the tangent plane is

$$\left(3\sqrt{3}\mathbf{i} + 3\mathbf{j}\right) \cdot \left[\left(x - \tfrac{3\sqrt{3}}{2}\right)\mathbf{i} + \left(y - \tfrac{9}{2}\right)\mathbf{j} + (z-0)\mathbf{k}\right] = 0$$

$\Rightarrow \sqrt{3}x + y = 9$. The parametrization $\Rightarrow x = 3\sin 2\theta$ and $y = 6\sin^2\theta \Rightarrow x^2+y^2 = 9\sin^2 2\theta + \left(6\sin^2\theta\right)^2$

$= 9\left(4\sin^2\theta\cos^2\theta\right) + 36\sin^4\theta = 6\left(6\sin^2\theta\right) = 6y \Rightarrow x^2+y^2-6y+9 = 9 \Rightarrow x^2+(y-3)^2 = 9$

31. (a) An arbitrary point on the circle C is $(x,z) = (R + r\cos u, r\sin u) \Rightarrow (x,y,z)$ is on the torus with $x = (R + r\cos u)\cos v$, $y = (R + r\cos u)\sin v$, and $z = r\sin u$, $0 \le u \le 2\pi$, $0 \le v \le 2\pi$

(b) $\mathbf{r}_u = (-r\sin u\cos v)\mathbf{i} - (r\sin u\sin v)\mathbf{j} + (r\cos u)\mathbf{k}$ and $\mathbf{r}_v = (-(R + r\cos u)\sin v)\mathbf{i} + ((R + r\cos u)\cos v)\mathbf{j}$

$$\Rightarrow \mathbf{r}_u \times \mathbf{r}_v = \begin{vmatrix} \mathbf{i} & \mathbf{j} & \mathbf{k} \\ -r\sin u\cos v & -r\sin u\sin v & r\cos u \\ -(R + r\cos u)\sin v & (R + r\cos u)\cos v & 0 \end{vmatrix}$$

$$= -(R + r\cos u)(r\cos v\cos u)\mathbf{i} - (R + r\cos u)(r\sin v\cos u)\mathbf{j} + (-r\sin u)(R + r\cos u)\mathbf{k}$$

$\Rightarrow\ |\mathbf{r}_u \times \mathbf{r}_v|^2 = (R + r\cos u)^2\,(r^2\cos^2 v\cos^2 u + r^2\sin^2 v\cos^2 u + r^2\sin^2 u)\ \Rightarrow\ |\mathbf{r}_u \times \mathbf{r}_v| = r(R + r\cos u)$

$\Rightarrow\ A = \int_0^{2\pi}\int_0^{2\pi} (rR + r^2\cos u)\,du\,dv = \int_0^{2\pi} 2\pi rR\,dv = 4\pi^2 rR$

33. $\mathbf{r}(\theta, u) = (5\cosh u\cos\theta)\mathbf{i} + (5\cosh u\sin\theta)\mathbf{j} + (5\sinh u)\mathbf{k}\ \Rightarrow\ \mathbf{r}_\theta = (-5\cosh u\sin\theta)\mathbf{i} + (5\cosh u\cos\theta)\mathbf{j}$ and $\mathbf{r}_u = (5\sinh u\cos\theta)\mathbf{i} + (5\sinh u\sin\theta)\mathbf{j} + (5\cosh u)\mathbf{k}$

$\Rightarrow\ \mathbf{r}_\theta \times \mathbf{r}_u = \begin{vmatrix} \mathbf{i} & \mathbf{j} & \mathbf{k} \\ -5\cosh u\sin\theta & 5\cosh u\cos\theta & 0 \\ 5\sinh u\cos\theta & 5\sinh u\sin\theta & 5\cosh u \end{vmatrix}$

$= (25\cosh^2 u\cos\theta)\,\mathbf{i} + (25\cosh^2 u\sin\theta)\,\mathbf{j} - (25\cosh u\sinh u)\mathbf{k}$. At the point $(x_0, y_0, 0)$, where $x_0^2 + y_0^2 = 25$ we have $5\sinh u = 0\ \Rightarrow\ u = 0$ and $x_0 = 25\cos\theta$, $y_0 = 25\sin\theta\ \Rightarrow$ the tangent plane is $5(x_0\mathbf{i} + y_0\mathbf{j})\cdot[(x - x_0)\mathbf{i} + (y - y_0)\mathbf{j} + z\mathbf{k}] = 0\ \Rightarrow\ x_0x - x_0^2 + y_0y - y_0^2 = 0\ \Rightarrow\ x_0x + y_0y = 25$

35. $\mathbf{p} = \mathbf{k},\ \nabla f = 2x\mathbf{i} + 2y\mathbf{j} - \mathbf{k}\ \Rightarrow\ |\nabla f| = \sqrt{(2x)^2 + (2y)^2 + (-1)^2} = \sqrt{4x^2 + 4y^2 + 1}$ and $|\nabla f\cdot\mathbf{p}| = 1$; $z = 2\ \Rightarrow\ x^2 + y^2 = 2$; thus $S = \iint_R \frac{|\nabla f|}{|\nabla f\cdot\mathbf{p}|}\,dA = \iint_R \sqrt{4x^2 + 4y^2 + 1}\,dx\,dy$

$= \iint_R \sqrt{4r^2\cos^2\theta + 4r^2\sin^2\theta + 1}\,r\,dr\,d\theta = \int_0^{2\pi}\int_0^{\sqrt{2}} \sqrt{4r^2 + 1}\,r\,dr\,d\theta = \int_0^{2\pi}\left[\frac{1}{12}(4r^2 + 1)^{3/2}\right]_0^{\sqrt{2}} d\theta$

$= \int_0^{2\pi} \frac{13}{6}\,d\theta = \frac{13}{3}\pi$

37. $\mathbf{p} = \mathbf{k},\ \nabla f = \mathbf{i} + 2\mathbf{j} + 2\mathbf{k}\ \Rightarrow\ |\nabla f| = 3$ and $|\nabla f\cdot\mathbf{p}| = 2$; $x = y^2$ and $x = 2 - y^2$ intersect at $(1, 1)$ and $(1, -1)$

$\Rightarrow\ S = \iint_R \frac{|\nabla f|}{|\nabla f\cdot\mathbf{p}|}\,dA = \iint_R \frac{3}{2}\,dx\,dy = \int_{-1}^{1}\int_{y^2}^{2-y^2} \frac{3}{2}\,dx\,dy = \int_{-1}^{1}(3 - 3y^2)\,dy = 4$

39. $\mathbf{p} = \mathbf{k},\ \nabla f = 2x\mathbf{i} - 2\mathbf{j} - 2\mathbf{k}\ \Rightarrow\ |\nabla f| = \sqrt{(2x)^2 + (-2)^2 + (-2)^2} = \sqrt{4x^2 + 8} = 2\sqrt{x^2 + 2}$ and $|\nabla f\cdot\mathbf{p}| = 2$

$\Rightarrow\ S = \iint_R \frac{|\nabla f|}{|\nabla f\cdot\mathbf{p}|}\,dA = \iint_R \frac{2\sqrt{x^2+2}}{2}\,dx\,dy = \int_0^2\int_0^{3x} \sqrt{x^2 + 2}\,dy\,dx = \int_0^2 3x\sqrt{x^2 + 2}\,dx = \left[(x^2 + 2)^{3/2}\right]_0^2$

$= 6\sqrt{6} - 2\sqrt{2}$

41. $\mathbf{p} = \mathbf{k},\ \nabla f = c\mathbf{i} - \mathbf{k}\ \Rightarrow\ |\nabla f| = \sqrt{c^2 + 1}$ and $|\nabla f\cdot\mathbf{p}| = 1\ \Rightarrow\ S = \iint_R \frac{|\nabla f|}{|\nabla f\cdot\mathbf{p}|}\,dA = \iint_R \sqrt{c^2 + 1}\,dx\,dy$

$= \int_0^{2\pi}\int_0^1 \sqrt{c^2 + 1}\,r\,dr\,d\theta = \int_0^{2\pi} \frac{\sqrt{c^2+1}}{2}\,d\theta = \pi\sqrt{c^2 + 1}$

43. $\mathbf{p} = \mathbf{i},\ \nabla f = \mathbf{i} + 2y\mathbf{j} + 2z\mathbf{k}\ \Rightarrow\ |\nabla f| = \sqrt{1^2 + (2y)^2 + (2z)^2} = \sqrt{1 + 4y^2 + 4z^2}$ and $|\nabla f\cdot\mathbf{p}| = 1$; $1 \le y^2 + z^2 \le 4$

$\Rightarrow\ S = \iint_R \frac{|\nabla f|}{|\nabla f\cdot\mathbf{p}|}\,dA = \iint_R \sqrt{1 + 4y^2 + 4z^2}\,dy\,dz = \int_0^{2\pi}\int_1^2 \sqrt{1 + 4r^2\cos^2\theta + 4r^2\sin^2\theta}\,r\,dr\,d\theta$

$= \int_0^{2\pi}\int_1^2 \sqrt{1 + 4r^2}\,r\,dr\,d\theta = \int_0^{2\pi}\left[\frac{1}{12}(1 + 4r^2)^{3/2}\right]_1^2 d\theta = \int_0^{2\pi} \frac{1}{12}\left(17\sqrt{17} - 5\sqrt{5}\right)d\theta = \frac{\pi}{6}\left(17\sqrt{17} - 5\sqrt{5}\right)$

45. $\mathbf{p} = \mathbf{k},\ \nabla f = \left(2x - \frac{2}{x}\right)\mathbf{i} + \sqrt{15}\,\mathbf{j} - \mathbf{k} \Rightarrow |\nabla f| = \sqrt{\left(2x - \frac{2}{x}\right)^2 + \left(\sqrt{15}\right)^2 + (-1)^2} = \sqrt{4x^2 + 8 + \frac{4}{x^2}} = \sqrt{\left(2x + \frac{2}{x}\right)^2}$

$= 2x + \frac{2}{x}$, on $1 \le x \le 2$ and $|\nabla f\cdot\mathbf{p}| = 1\ \Rightarrow\ S = \iint_R \frac{|\nabla f|}{|\nabla f\cdot\mathbf{p}|}\,dA = \iint_R (2x + 2x^{-1})\,dx\,dy$

$= \int_0^1\int_1^2 (2x + 2x^{-1})\,dx\,dy = \int_0^1 \left[x^2 + 2\ln x\right]_1^2 dy = \int_0^1 (3 + 2\ln 2)\,dy = 3 + 2\ln 2$

47. $f_x(x, y) = 2x,\ f_y(x, y) = 2y\ \Rightarrow\ \sqrt{f_x^2 + f_y^2 + 1} = \sqrt{4x^2 + 4y^2 + 1}\ \Rightarrow\ \text{Area} = \iint_R \sqrt{4x^2 + 4y^2 + 1}\,dx\,dy$

$= \int_0^{2\pi}\int_0^{\sqrt{3}} \sqrt{4r^2 + 1}\,r\,dr\,d\theta = \frac{\pi}{6}\left(13\sqrt{13} - 1\right)$

49. $f_x(x,y) = \frac{x}{\sqrt{x^2+y^2}}$, $f_y(x,y) = \frac{y}{\sqrt{x^2+y^2}} \Rightarrow \sqrt{f_x^2+f_y^2+1} = \sqrt{\frac{x^2}{x^2+y^2}+\frac{y^2}{x^2+y^2}+1} = \sqrt{2}$
$\Rightarrow$ Area $= \iint\limits_{R_{xy}} \sqrt{2}\,dx\,dy = \sqrt{2}$(Area between the ellipse and the circle) $= \sqrt{2}(6\pi - \pi) = 5\pi\sqrt{2}$

51. $y = \frac{2}{3}z^{3/2} \Rightarrow f_x(x,z) = 0,\ f_z(x,z) = z^{1/2} \Rightarrow \sqrt{f_x^2+f_z^2+1} = \sqrt{z+1}$; $y = \frac{16}{3} \Rightarrow \frac{16}{3} = \frac{2}{3}z^{3/2} \Rightarrow z = 4$
$\Rightarrow$ Area $= \int_0^4\int_0^1 \sqrt{z+1}\,dx\,dz = \int_0^4 \sqrt{z+1}\,dz = \frac{2}{3}\left(5\sqrt{5}-1\right)$

53. $\mathbf{r}(x,y) = x\mathbf{i} + y\mathbf{j} + f(x,y)\mathbf{k} \Rightarrow \mathbf{r}_x(x,y) = \mathbf{i} + f_x(x,y)\mathbf{k},\ \mathbf{r}_y(x,y) = \mathbf{j} + f_y(x,y)\mathbf{k}$
$$\Rightarrow \mathbf{r}_x \times \mathbf{r}_y = \begin{vmatrix} \mathbf{i} & \mathbf{j} & \mathbf{k} \\ 1 & 0 & f_x(x,y) \\ 0 & 1 & f_y(x,y) \end{vmatrix} = -f_x(x,y)\mathbf{i} - f_y(x,y)\mathbf{j} + \mathbf{k}$$
$$\Rightarrow |\mathbf{r}_x \times \mathbf{r}_y| = \sqrt{(-f_x(x,y))^2 + (-f_y(x,y))^2 + 1^2} = \sqrt{f_x(x,y)^2 + f_y(x,y)^2 + 1}$$
$$\Rightarrow d\sigma = \sqrt{f_x(x,y)^2 + f_y(x,y)^2 + 1}\,dA$$

13.6 SURFACE INTEGRALS AND FLUX

1. Let the parametrization be $\mathbf{r}(x,z) = x\mathbf{i} + x^2\mathbf{j} + z\mathbf{k} \Rightarrow \mathbf{r}_x = \mathbf{i} + 2x\mathbf{j}$ and $\mathbf{r}_z = \mathbf{k} \Rightarrow \mathbf{r}_x \times \mathbf{r}_z = \begin{vmatrix} \mathbf{i} & \mathbf{j} & \mathbf{k} \\ 1 & 2x & 0 \\ 0 & 0 & 1 \end{vmatrix}$
$= 2x\mathbf{i} + \mathbf{j} \Rightarrow |\mathbf{r}_x \times \mathbf{r}_z| = \sqrt{4x^2+1} \Rightarrow \iint\limits_S G(x,y,z)\,d\sigma = \int_0^3\int_0^2 x\sqrt{4x^2+1}\,dx\,dz = \int_0^3 \left[\frac{1}{12}\left(4x^2+1\right)^{3/2}\right]_0^2 dz$
$= \int_0^3 \frac{1}{12}\left(17\sqrt{17}-1\right)dz = \frac{17\sqrt{17}-1}{4}$

3. Let the parametrization be $\mathbf{r}(\phi,\theta) = (\sin\phi\cos\theta)\mathbf{i} + (\sin\phi\sin\theta)\mathbf{j} + (\cos\phi)\mathbf{k}$ (spherical coordinates with $\rho = 1$ on the sphere), $0 \le \phi \le \pi$, $0 \le \theta \le 2\pi \Rightarrow \mathbf{r}_\phi = (\cos\phi\cos\theta)\mathbf{i} + (\cos\phi\sin\theta)\mathbf{j} - (\sin\phi)\mathbf{k}$ and
$\mathbf{r}_\theta = (-\sin\phi\sin\theta)\mathbf{i} + (\sin\phi\cos\theta)\mathbf{j} \Rightarrow \mathbf{r}_\phi \times \mathbf{r}_\theta = \begin{vmatrix} \mathbf{i} & \mathbf{j} & \mathbf{k} \\ \cos\phi\cos\theta & \cos\phi\sin\theta & -\sin\phi \\ -\sin\phi\sin\theta & \sin\phi\cos\theta & 0 \end{vmatrix}$
$= (\sin^2\phi\cos\theta)\,\mathbf{i} + (\sin^2\phi\sin\theta)\,\mathbf{j} + (\sin\phi\cos\phi)\mathbf{k} \Rightarrow |\mathbf{r}_\phi \times \mathbf{r}_\theta| = \sqrt{\sin^4\phi\cos^2\theta + \sin^4\phi\sin^2\theta + \sin^2\phi\cos^2\phi}$
$= \sin\phi$; $x = \sin\phi\cos\theta \Rightarrow G(x,y,z) = \cos^2\theta\sin^2\phi \Rightarrow \iint\limits_S G(x,y,z)\,d\sigma = \int_0^{2\pi}\int_0^{\pi} (\cos^2\theta\sin^2\phi)(\sin\phi)\,d\phi\,d\theta$
$= \int_0^{2\pi}\int_0^{\pi} (\cos^2\theta)(1-\cos^2\phi)(\sin\phi)\,d\phi\,d\theta$; $\begin{bmatrix} u = \cos\phi \\ du = -\sin\phi\,d\phi \end{bmatrix} \to \int_0^{2\pi}\int_1^{-1} (\cos^2\theta)(u^2-1)\,du\,d\theta$
$= \int_0^{2\pi} (\cos^2\theta)\left[\frac{u^3}{3} - u\right]_1^{-1} d\theta = \frac{4}{3}\int_0^{2\pi} \cos^2\theta\,d\theta = \frac{4}{3}\left[\frac{\theta}{2} + \frac{\sin 2\theta}{4}\right]_0^{2\pi} = \frac{4\pi}{3}$

5. Let the parametrization be $\mathbf{r}(x,y) = x\mathbf{i} + y\mathbf{j} + (4-x-y)\mathbf{k} \Rightarrow \mathbf{r}_x = \mathbf{i} - \mathbf{k}$ and $\mathbf{r}_y = \mathbf{j} - \mathbf{k}$
$\Rightarrow \mathbf{r}_x \times \mathbf{r}_y = \begin{vmatrix} \mathbf{i} & \mathbf{j} & \mathbf{k} \\ 1 & 0 & -1 \\ 0 & 1 & -1 \end{vmatrix} = \mathbf{i} + \mathbf{j} + \mathbf{k} \Rightarrow |\mathbf{r}_x \times \mathbf{r}_y| = \sqrt{3} \Rightarrow \iint\limits_S F(x,y,z)\,d\sigma = \int_0^1\int_0^1 (4-x-y)\sqrt{3}\,dy\,dx$
$= \int_0^1 \sqrt{3}\left[4y - xy - \frac{y^2}{2}\right]_0^1 dx = \int_0^1 \sqrt{3}\left(\frac{7}{2} - x\right)dx = \sqrt{3}\left[\frac{7}{2}x - \frac{x^2}{2}\right]_0^1 = 3\sqrt{3}$

7. Let the parametrization be $\mathbf{r}(r,\theta) = (r\cos\theta)\mathbf{i} + (r\sin\theta)\mathbf{j} + (1-r^2)\mathbf{k}$, $0 \le r \le 1$ (since $0 \le z \le 1$) and $0 \le \theta \le 2\pi$
$\Rightarrow \mathbf{r}_r = (\cos\theta)\mathbf{i} + (\sin\theta)\mathbf{j} - 2r\mathbf{k}$ and $\mathbf{r}_\theta = (-r\sin\theta)\mathbf{i} + (r\cos\theta)\mathbf{j} \Rightarrow \mathbf{r}_r \times \mathbf{r}_\theta = \begin{vmatrix} \mathbf{i} & \mathbf{j} & \mathbf{k} \\ \cos\theta & \sin\theta & -2r \\ -r\sin\theta & r\cos\theta & 0 \end{vmatrix}$

$= (2r^2 \cos\theta)\,\mathbf{i} + (2r^2 \sin\theta)\,\mathbf{j} + r\mathbf{k} \Rightarrow |\mathbf{r}_r \times \mathbf{r}_\theta| = \sqrt{(2r^2\cos\theta)^2 + (2r^2\sin\theta) + r^2} = r\sqrt{1+4r^2}$; $z = 1 - r^2$ and $x = r\cos\theta \Rightarrow H(x,y,z) = (r^2\cos^2\theta)\sqrt{1+4r^2} \Rightarrow \iint_S H(x,y,z)\,d\sigma$

$= \int_0^{2\pi}\int_0^1 (r^2\cos^2\theta)\left(\sqrt{1+4r^2}\right)\left(r\sqrt{1+4r^2}\right) dr\,d\theta = \int_0^{2\pi}\int_0^1 r^3(1+4r^2)\cos^2\theta\,dr\,d\theta = \frac{11\pi}{12}$

9. The bottom face S of the cube is in the xy-plane $\Rightarrow z = 0 \Rightarrow g(x,y,0) = x + y$ and $f(x,y,z) = z = 0 \Rightarrow \mathbf{p} = \mathbf{k}$ and $\nabla f = \mathbf{k} \Rightarrow |\nabla f| = 1$ and $|\nabla f \cdot \mathbf{p}| = 1 \Rightarrow d\sigma = dx\,dy \Rightarrow \iint_S g\,d\sigma = \iint_R (x+y)\,dx\,dy$

$= \int_0^a\int_0^a (x+y)\,dx\,dy = \int_0^a\left(\frac{a^2}{2} + ay\right)dy = a^3$. Because of symmetry, we also get a^3 over the face of the cube in the xz-plane and a^3 over the face of the cube in the yz-plane. Next, on the top of the cube, $g(x,y,z) = g(x,y,a) = x + y + a$ and $f(x,y,z) = z = a \Rightarrow \mathbf{p} = \mathbf{k}$ and $\nabla f = \mathbf{k} \Rightarrow |\nabla f| = 1$ and $|\nabla f \cdot \mathbf{p}| = 1 \Rightarrow d\sigma = dx\,dy$

$\iint_S g\,d\sigma = \iint_R (x+y+a)\,dx\,dy = \int_0^a\int_0^a (x+y+a)\,dx\,dy = \int_0^a\int_0^a (x+y)\,dx\,dy + \int_0^a\int_0^a a\,dx\,dy = 2a^3$.

Because of symmetry, the integral is also $2a^3$ over each of the other two faces. Therefore,

$\iint_{\text{cube}} (x+y+z)\,d\sigma = 3(a^3 + 2a^3) = 9a^3$.

11. On the faces in the coordinate planes, $g(x,y,z) = 0 \Rightarrow$ the integral over these faces is 0.

On the face $x = a$, we have $f(x,y,z) = x = a$ and $g(x,y,z) = g(a,y,z) = ayz \Rightarrow \mathbf{p} = \mathbf{i}$ and $\nabla f = \mathbf{i} \Rightarrow |\nabla f| = 1$ and $|\nabla f \cdot \mathbf{p}| = 1 \Rightarrow d\sigma = dy\,dz \Rightarrow \iint_S g\,d\sigma = \iint_S ayz\,d\sigma = \int_0^c\int_0^b ayz\,dy\,dz = \frac{ab^2c^2}{4}$.

On the face $y = b$, we have $f(x,y,z) = y = b$ and $g(x,y,z) = g(x,b,z) = bxz \Rightarrow \mathbf{p} = \mathbf{j}$ and $\nabla f = \mathbf{j} \Rightarrow |\nabla f| = 1$ and $|\nabla f \cdot \mathbf{p}| = 1 \Rightarrow d\sigma = dx\,dz \Rightarrow \iint_S g\,d\sigma = \iint_S bxz\,d\sigma = \int_0^c\int_0^a bxz\,dx\,dz = \frac{a^2bc^2}{4}$.

On the face $z = c$, we have $f(x,y,z) = z = c$ and $g(x,y,z) = g(x,y,c) = cxy \Rightarrow \mathbf{p} = \mathbf{k}$ and $\nabla f = \mathbf{k} \Rightarrow |\nabla f| = 1$ and $|\nabla f \cdot \mathbf{p}| = 1 \Rightarrow d\sigma = dy\,dx \Rightarrow \iint_S g\,d\sigma = \iint_S cxy\,d\sigma = \int_0^b\int_0^a cxy\,dx\,dy = \frac{a^2b^2c}{4}$. Therefore,

$\iint_S g(x,y,z)\,d\sigma = \frac{abc(ab+ac+bc)}{4}$.

13. $f(x,y,z) = 2x + 2y + z = 2 \Rightarrow \nabla f = 2\mathbf{i} + 2\mathbf{j} + \mathbf{k}$ and $g(x,y,z) = x + y + (2 - 2x - 2y) = 2 - x - y \Rightarrow \mathbf{p} = \mathbf{k}$, $|\nabla f| = 3$ and $|\nabla f \cdot \mathbf{p}| = 1 \Rightarrow d\sigma = 3\,dy\,dx$; $z = 0 \Rightarrow 2x + 2y = 2 \Rightarrow y = 1 - x \Rightarrow \iint_S g\,d\sigma = \iint_S (2-x-y)\,d\sigma$

$= 3\int_0^1\int_0^{1-x} (2-x-y)\,dy\,dx = 3\int_0^1 \left[(2-x)(1-x) - \frac{1}{2}(1-x)^2\right] dx = 3\int_0^1\left(\frac{3}{2} - 2x + \frac{x^2}{2}\right) dx = 2$

15. Let the parametrization be $\mathbf{r}(x,y) = x\mathbf{i} + y\mathbf{j} + (4-y^2)\mathbf{k}$, $0 \le x \le 1$, $-2 \le y \le 2$; $z = 0 \Rightarrow 0 = 4 - y^2$

$\Rightarrow y = \pm 2$; $\mathbf{r}_x = \mathbf{i}$ and $\mathbf{r}_y = \mathbf{j} - 2y\mathbf{k} \Rightarrow \mathbf{r}_x \times \mathbf{r}_y = \begin{vmatrix} \mathbf{i} & \mathbf{j} & \mathbf{k} \\ 1 & 0 & 0 \\ 0 & 1 & -2y \end{vmatrix} = 2y\mathbf{j} + \mathbf{k} \Rightarrow \mathbf{F}\cdot\mathbf{n}\,d\sigma$

$= \mathbf{F}\cdot\frac{\mathbf{r}_x \times \mathbf{r}_y}{|\mathbf{r}_x \times \mathbf{r}_y|}\,|\mathbf{r}_x \times \mathbf{r}_y|\,dy\,dx = (2xy - 3z)\,dy\,dx = [2xy - 3(4-y^2)]\,dy\,dx \Rightarrow \iint_S \mathbf{F}\cdot\mathbf{n}\,d\sigma$

$= \int_0^1\int_{-2}^2 (2xy + 3y^2 - 12)\,dy\,dx = \int_0^1 \left[xy^2 + y^3 - 12y\right]_{-2}^2 dx = \int_0^1 -32\,dx = -32$

17. Let the parametrization be $\mathbf{r}(\phi,\theta) = (a\sin\phi\cos\theta)\mathbf{i} + (a\sin\phi\sin\theta)\mathbf{j} + (a\cos\phi)\mathbf{k}$ (spherical coordinates with $\rho = a$, $a \ge 0$, on the sphere), $0 \le \phi \le \frac{\pi}{2}$ (for the first octant), $0 \le \theta \le \frac{\pi}{2}$ (for the first octant)

$\Rightarrow \mathbf{r}_\phi = (a\cos\phi\cos\theta)\mathbf{i} + (a\cos\phi\sin\theta)\mathbf{j} - (a\sin\phi)\mathbf{k}$ and $\mathbf{r}_\theta = (-a\sin\phi\sin\theta)\mathbf{i} + (a\sin\phi\cos\theta)\mathbf{j}$

$\Rightarrow \mathbf{r}_\phi \times \mathbf{r}_\theta = \begin{vmatrix} \mathbf{i} & \mathbf{j} & \mathbf{k} \\ a\cos\phi\cos\theta & a\cos\phi\sin\theta & -a\sin\phi \\ -a\sin\phi\sin\theta & a\sin\phi\cos\theta & 0 \end{vmatrix}$

$= (a^2 \sin^2\phi \cos\theta)\,\mathbf{i} + (a^2 \sin^2\phi \sin\theta)\,\mathbf{j} + (a^2 \sin\phi \cos\phi)\,\mathbf{k} \Rightarrow \mathbf{F}\cdot\mathbf{n}\,d\sigma = \mathbf{F}\cdot\frac{\mathbf{r}_\phi\times\mathbf{r}_\theta}{|\mathbf{r}_\phi\times\mathbf{r}_\theta|}\,|\mathbf{r}_\phi\times\mathbf{r}_\theta|\,d\theta\,d\phi$

$= a^3\cos^2\phi\sin\phi\,d\theta\,d\phi$ since $\mathbf{F} = z\mathbf{k} = (a\cos\phi)\mathbf{k} \Rightarrow \iint_S \mathbf{F}\cdot\mathbf{n}\,d\sigma = \int_0^{\pi/2}\int_0^{\pi/2} a^3\cos^2\phi\sin\phi\,d\phi\,d\theta = \frac{\pi a^3}{6}$

19. Let the parametrization be $\mathbf{r}(x,y) = x\mathbf{i} + y\mathbf{j} + (2a - x - y)\mathbf{k}$, $0 \le x \le a$, $0 \le y \le a \Rightarrow \mathbf{r}_x = \mathbf{i} - \mathbf{k}$ and $\mathbf{r}_y = \mathbf{j} - \mathbf{k}$

$\Rightarrow \mathbf{r}_x\times\mathbf{r}_y = \begin{vmatrix} \mathbf{i} & \mathbf{j} & \mathbf{k} \\ 1 & 0 & -1 \\ 0 & 1 & -1 \end{vmatrix} = \mathbf{i} + \mathbf{j} + \mathbf{k} \Rightarrow \mathbf{F}\cdot\mathbf{n}\,d\sigma = \mathbf{F}\cdot\frac{\mathbf{r}_x\times\mathbf{r}_y}{|\mathbf{r}_x\times\mathbf{r}_y|}\,|\mathbf{r}_x\times\mathbf{r}_y|\,dy\,dx$

$= [2xy + 2y(2a - x - y) + 2x(2a - x - y)]\,dy\,dx$ since $\mathbf{F} = 2xy\mathbf{i} + 2yz\mathbf{j} + 2xz\mathbf{k}$

$= 2xy\mathbf{i} + 2y(2a - x - y)\mathbf{j} + 2x(2a - x - y)\mathbf{k} \Rightarrow \iint_S \mathbf{F}\cdot\mathbf{n}\,d\sigma$

$= \int_0^a\int_0^a [2xy + 2y(2a - x - y) + 2x(2a - x - y)]\,dy\,dx = \int_0^a\int_0^a (4ay - 2y^2 + 4ax - 2x^2 - 2xy)\,dy\,dx$

$= \int_0^a \left(\frac{4}{3}a^3 + 3a^2x - 2ax^2\right)dx = \left(\frac{4}{3} + \frac{3}{2} - \frac{2}{3}\right)a^4 = \frac{13a^4}{6}$

21. Let the parametrization be $\mathbf{r}(r,\theta) = (r\cos\theta)\mathbf{i} + (r\sin\theta)\mathbf{j} + r\mathbf{k}$, $0 \le r \le 1$ (since $0 \le z \le 1$) and $0 \le \theta \le 2\pi$

$\Rightarrow \mathbf{r}_r = (\cos\theta)\mathbf{i} + (\sin\theta)\mathbf{j} + \mathbf{k}$ and $\mathbf{r}_\theta = (-r\sin\theta)\mathbf{i} + (r\cos\theta)\mathbf{j} \Rightarrow \mathbf{r}_\theta\times\mathbf{r}_r = \begin{vmatrix} \mathbf{i} & \mathbf{j} & \mathbf{k} \\ -r\sin\theta & r\cos\theta & 0 \\ \cos\theta & \sin\theta & 1 \end{vmatrix}$

$= (r\cos\theta)\mathbf{i} + (r\sin\theta)\mathbf{j} - r\mathbf{k} \Rightarrow \mathbf{F}\cdot\mathbf{n}\,d\sigma = \mathbf{F}\cdot\frac{\mathbf{r}_\theta\times\mathbf{r}_r}{|\mathbf{r}_\theta\times\mathbf{r}_r|}\,|\mathbf{r}_\theta\times\mathbf{r}_r|\,d\theta\,dr = (r^3\sin\theta\cos^2\theta + r^2)\,d\theta\,dr$ since

$\mathbf{F} = (r^2\sin\theta\cos\theta)\,\mathbf{i} - r\mathbf{k} \Rightarrow \iint_S \mathbf{F}\cdot\mathbf{n}\,d\sigma = \int_0^{2\pi}\int_0^1 (r^3\sin\theta\cos^2\theta + r^2)\,dr\,d\theta = \int_0^{2\pi}\left(\frac{1}{4}\sin\theta\cos^2\theta + \frac{1}{3}\right)d\theta$

$= \left[-\frac{1}{12}\cos^3\theta + \frac{\theta}{3}\right]_0^{2\pi} = \frac{2\pi}{3}$

23. Let the parametrization be $\mathbf{r}(r,\theta) = (r\cos\theta)\mathbf{i} + (r\sin\theta)\mathbf{j} + r\mathbf{k}$, $1 \le r \le 2$ (since $1 \le z \le 2$) and $0 \le \theta \le 2\pi$

$\Rightarrow \mathbf{r}_r = (\cos\theta)\mathbf{i} + (\sin\theta)\mathbf{j} + \mathbf{k}$ and $\mathbf{r}_\theta = (-r\sin\theta)\mathbf{i} + (r\cos\theta)\mathbf{j} \Rightarrow \mathbf{r}_\theta\times\mathbf{r}_r = \begin{vmatrix} \mathbf{i} & \mathbf{j} & \mathbf{k} \\ -r\sin\theta & r\cos\theta & 0 \\ \cos\theta & \sin\theta & 1 \end{vmatrix}$

$= (r\cos\theta)\mathbf{i} + (r\sin\theta)\mathbf{j} - r\mathbf{k} \Rightarrow \mathbf{F}\cdot\mathbf{n}\,d\sigma = \mathbf{F}\cdot\frac{\mathbf{r}_\theta\times\mathbf{r}_r}{|\mathbf{r}_\theta\times\mathbf{r}_r|}\,|\mathbf{r}_\theta\times\mathbf{r}_r|\,d\theta\,dr = (-r^2\cos^2\theta - r^2\sin^2\theta - r^3)\,d\theta\,dr$

$= (-r^2 - r^3)\,d\theta\,dr$ since $\mathbf{F} = (-r\cos\theta)\mathbf{i} - (r\sin\theta)\mathbf{j} + r^2\mathbf{k} \Rightarrow \iint_S \mathbf{F}\cdot\mathbf{n}\,d\sigma = \int_0^{2\pi}\int_1^2 (-r^2 - r^3)\,dr\,d\theta = -\frac{73\pi}{6}$

25. $g(x,y,z) = z$, $\mathbf{p} = \mathbf{k} \Rightarrow \nabla g = \mathbf{k} \Rightarrow |\nabla g| = 1$ and $|\nabla g\cdot\mathbf{p}| = 1 \Rightarrow \text{Flux} = \iint_S \mathbf{F}\cdot\mathbf{n}\,d\sigma = \iint_R (\mathbf{F}\cdot\mathbf{k})\,dA$

$= \int_0^2\int_0^3 3\,dy\,dx = 18$

27. $\nabla g = 2x\mathbf{i} + 2y\mathbf{j} + 2z\mathbf{k} \Rightarrow |\nabla g| = \sqrt{4x^2 + 4y^2 + 4z^2} = 2a$; $\mathbf{n} = \frac{2x\mathbf{i} + 2y\mathbf{j} + 2z\mathbf{k}}{2\sqrt{x^2 + y^2 + z^2}} = \frac{x\mathbf{i} + y\mathbf{j} + z\mathbf{k}}{a} \Rightarrow \mathbf{F}\cdot\mathbf{n} = \frac{z^2}{a}$;

$|\nabla g\cdot\mathbf{k}| = 2z \Rightarrow d\sigma = \frac{2a}{2z}\,dA \Rightarrow \text{Flux} = \iint_R \left(\frac{z^2}{a}\right)\left(\frac{a}{z}\right)dA = \iint_R z\,dA = \iint_R \sqrt{a^2 - (x^2 + y^2)}\,dx\,dy$

$= \int_0^{\pi/2}\int_0^a \sqrt{a^2 - r^2}\,r\,dr\,d\theta = \frac{\pi a^3}{6}$

29. From Exercise 27, $\mathbf{n} = \frac{x\mathbf{i} + y\mathbf{j} + z\mathbf{k}}{a}$ and $d\sigma = \frac{a}{z}\,dA \Rightarrow \mathbf{F}\cdot\mathbf{n} = \frac{xy}{a} - \frac{xy}{a} + \frac{z}{a} = \frac{z}{a} \Rightarrow \text{Flux} = \iint_R \left(\frac{z}{a}\right)\left(\frac{a}{z}\right)dA$

$= \iint_R 1\,dA = \frac{\pi a^2}{4}$

31. From Exercise 27, $\mathbf{n} = \frac{x\mathbf{i}+y\mathbf{j}+z\mathbf{k}}{a}$ and $d\sigma = \frac{a}{z}\,dA \Rightarrow \mathbf{F}\cdot\mathbf{n} = \frac{x^2}{a} + \frac{y^2}{a} + \frac{z^2}{a} = a \Rightarrow$ Flux
$= \iint\limits_R a\left(\frac{a}{z}\right) dA = \iint\limits_R \frac{a^2}{z}\,dA = \iint\limits_R \frac{a^2}{\sqrt{a^2-(x^2+y^2)}}\,dA = \int_0^{\pi/2}\int_0^a \frac{a^2}{\sqrt{a^2-r^2}}\,r\,dr\,d\theta$
$= \int_0^{\pi/2} a^2\left[-\sqrt{a^2-r^2}\right]_0^a d\theta = \frac{\pi a^3}{2}$

33. $g(x,y,z) = y^2 + z = 4 \Rightarrow \nabla g = 2y\mathbf{j} + \mathbf{k} \Rightarrow |\nabla g| = \sqrt{4y^2+1} \Rightarrow \mathbf{n} = \frac{2y\mathbf{j}+\mathbf{k}}{\sqrt{4y^2+1}}$
$\Rightarrow \mathbf{F}\cdot\mathbf{n} = \frac{2xy-3z}{\sqrt{4y^2+1}}$; $\mathbf{p} = \mathbf{k} \Rightarrow |\nabla g\cdot\mathbf{p}| = 1 \Rightarrow d\sigma = \sqrt{4y^2+1}\,dA \Rightarrow$ Flux
$= \iint\limits_R \left(\frac{2xy-3z}{\sqrt{4y^2+1}}\right)\sqrt{4y^2+1}\,dA = \iint\limits_R (2xy-3z)\,dA$; $z = 0$ and $z = 4 - y^2 \Rightarrow y^2 = 4$
$\Rightarrow$ Flux $= \iint\limits_R [2xy - 3(4-y^2)]\,dA = \int_0^1\int_{-2}^2 (2xy - 12 + 3y^2)\,dy\,dx = \int_0^1 \left[xy^2 - 12y + y^3\right]_{-2}^2 dx$
$= \int_0^1 -32\,dx = -32$

35. $g(x,y,z) = y - e^x = 0 \Rightarrow \nabla g = -e^x\mathbf{i} + \mathbf{j} \Rightarrow |\nabla g| = \sqrt{e^{2x}+1} \Rightarrow \mathbf{n} = \frac{e^x\mathbf{i}-\mathbf{j}}{\sqrt{e^{2x}+1}} \Rightarrow \mathbf{F}\cdot\mathbf{n} = \frac{-2e^x-2y}{\sqrt{e^{2x}+1}}$; $\mathbf{p} = \mathbf{i}$
$\Rightarrow |\nabla g\cdot\mathbf{p}| = e^x \Rightarrow d\sigma = \frac{\sqrt{e^{2x}+1}}{e^x}\,dA \Rightarrow$ Flux $= \iint\limits_R \left(\frac{-2e^x-2y}{\sqrt{e^{2x}+1}}\right)\left(\frac{\sqrt{e^{2x}+1}}{e^x}\right) dA = \iint\limits_R \frac{-2e^x-2e^x}{e^x}\,dA$
$= \iint\limits_R -4\,dA = \int_0^1\int_1^2 -4\,dy\,dz = -4$

37. On the face $z = a$: $g(x,y,z) = z \Rightarrow \nabla g = \mathbf{k} \Rightarrow |\nabla g| = 1$; $\mathbf{n} = \mathbf{k} \Rightarrow \mathbf{F}\cdot\mathbf{n} = 2xz = 2ax$ since $z = a$;
$d\sigma = dx\,dy \Rightarrow$ Flux $= \iint\limits_R 2ax\,dx\,dy = \int_0^a\int_0^a 2ax\,dx\,dy = a^4$.
On the face $z = 0$: $g(x,y,z) = z \Rightarrow \nabla g = \mathbf{k} \Rightarrow |\nabla g| = 1$; $\mathbf{n} = -\mathbf{k} \Rightarrow \mathbf{F}\cdot\mathbf{n} = -2xz = 0$ since $z = 0$;
$d\sigma = dx\,dy \Rightarrow$ Flux $= \iint\limits_R 0\,dx\,dy = 0$.
On the face $x = a$: $g(x,y,z) = x \Rightarrow \nabla g = \mathbf{i} \Rightarrow |\nabla g| = 1$; $\mathbf{n} = \mathbf{i} \Rightarrow \mathbf{F}\cdot\mathbf{n} = 2xy = 2ay$ since $x = a$;
$d\sigma = dy\,dz \Rightarrow$ Flux $= \int_0^a\int_0^a 2ay\,dy\,dz = a^4$.
On the face $x = 0$: $g(x,y,z) = x \Rightarrow \nabla g = \mathbf{i} \Rightarrow |\nabla g| = 1$; $\mathbf{n} = -\mathbf{i} \Rightarrow \mathbf{F}\cdot\mathbf{n} = -2xy = 0$ since $x = 0$
$\Rightarrow$ Flux $= 0$.
On the face $y = a$: $g(x,y,z) = y \Rightarrow \nabla g = \mathbf{j} \Rightarrow |\nabla g| = 1$; $\mathbf{n} = \mathbf{j} \Rightarrow \mathbf{F}\cdot\mathbf{n} = 2yz = 2az$ since $y = a$;
$d\sigma = dz\,dx \Rightarrow$ Flux $= \int_0^a\int_0^a 2az\,dz\,dx = a^4$.
On the face $y = 0$: $g(x,y,z) = y \Rightarrow \nabla g = \mathbf{j} \Rightarrow |\nabla g| = 1$; $\mathbf{n} = -\mathbf{j} \Rightarrow \mathbf{F}\cdot\mathbf{n} = -2yz = 0$ since $y = 0$
$\Rightarrow$ Flux $= 0$. Therefore, Total Flux $= 3a^4$.

39. $\nabla f = 2x\mathbf{i} + 2y\mathbf{j} + 2z\mathbf{k} \Rightarrow |\nabla f| = \sqrt{4x^2+4y^2+4z^2} = 2a$; $\mathbf{p} = \mathbf{k} \Rightarrow |\nabla f\cdot\mathbf{p}| = 2z$ since $z \ge 0 \Rightarrow d\sigma = \frac{2a}{2z}\,dA$
$= \frac{a}{z}\,dA$; $M = \iint\limits_S \delta\,d\sigma = \frac{\delta}{8}$ (surface area of sphere) $= \frac{\delta\pi a^2}{2}$; $M_{xy} = \iint\limits_S z\delta\,d\sigma = \delta\iint\limits_R z\left(\frac{a}{z}\right) dA$
$= a\delta\iint\limits_R dA = a\delta\int_0^{\pi/2}\int_0^a r\,dr\,d\theta = \frac{\delta\pi a^3}{4} \Rightarrow \overline{z} = \frac{M_{xy}}{M} = \left(\frac{\delta\pi a^3}{4}\right)\left(\frac{2}{\delta\pi a^2}\right) = \frac{a}{2}$. Because of symmetry, $\overline{x} = \overline{y}$
$= \frac{a}{2} \Rightarrow$ the centroid is $\left(\frac{a}{2}, \frac{a}{2}, \frac{a}{2}\right)$.

41. Let the diameter lie on the z-axis and let $f(x,y,z) = x^2 + y^2 + z^2 = a^2$, $z \ge 0$ be the upper hemisphere
$\Rightarrow \nabla f = 2x\mathbf{i} + 2y\mathbf{j} + 2z\mathbf{k} \Rightarrow |\nabla f| = \sqrt{4x^2+4y^2+4z^2} = 2a$, $a > 0$; $\mathbf{p} = \mathbf{k} \Rightarrow |\nabla f\cdot\mathbf{p}| = 2z$ since $z \ge 0$
$\Rightarrow d\sigma = \frac{a}{z}\,dA \Rightarrow I_z = \iint\limits_S \delta(x^2+y^2)\left(\frac{a}{z}\right) d\sigma = a\delta\iint\limits_R \frac{x^2+y^2}{\sqrt{a^2-(x^2+y^2)}}\,dA = a\delta\int_0^{2\pi}\int_0^a \frac{r^2}{\sqrt{a^2-r^2}}\,r\,dr\,d\theta$
$= a\delta\int_0^{2\pi}\left[-r^2\sqrt{a^2-r^2} - \frac{2}{3}(a^2-r^2)^{3/2}\right]_0^a d\theta = a\delta\int_0^{2\pi}\frac{2}{3}a^3\,d\theta = \frac{4\pi}{3}a^4\delta \Rightarrow$ the moment of inertia is $\frac{8\pi}{3}a^4\delta$ for the whole sphere

13.7 STOKES' THEOREM

1. $\text{curl}\,\mathbf{F} = \nabla \times \mathbf{F} = \begin{vmatrix} \mathbf{i} & \mathbf{j} & \mathbf{k} \\ \frac{\partial}{\partial x} & \frac{\partial}{\partial y} & \frac{\partial}{\partial z} \\ x^2 & 2x & z^2 \end{vmatrix} = 0\mathbf{i} + 0\mathbf{j} + (2-0)\mathbf{k} = 2\mathbf{k}$ and $\mathbf{n} = \mathbf{k} \Rightarrow \text{curl}\,\mathbf{F}\cdot\mathbf{n} = 2 \Rightarrow d\sigma = dx\,dy$

$\Rightarrow \oint_C \mathbf{F}\cdot d\mathbf{r} = \iint_R 2\,dA = 2(\text{Area of the ellipse}) = 4\pi$

3. $\text{curl}\,\mathbf{F} = \nabla \times \mathbf{F} = \begin{vmatrix} \mathbf{i} & \mathbf{j} & \mathbf{k} \\ \frac{\partial}{\partial x} & \frac{\partial}{\partial y} & \frac{\partial}{\partial z} \\ y & xz & x^2 \end{vmatrix} = -x\mathbf{i} - 2x\mathbf{j} + (z-1)\mathbf{k}$ and $\mathbf{n} = \frac{\mathbf{i}+\mathbf{j}+\mathbf{k}}{\sqrt{3}} \Rightarrow \text{curl}\,\mathbf{F}\cdot\mathbf{n}$

$= \frac{1}{\sqrt{3}}(-x-2x+z-1) \Rightarrow d\sigma = \frac{\sqrt{3}}{1}\,dA \Rightarrow \oint_C \mathbf{F}\cdot d\mathbf{r} = \iint_R \frac{1}{\sqrt{3}}(-3x+z-1)\sqrt{3}\,dA$

$= \int_0^1 \int_0^{1-x} [-3x + (1-x-y) - 1]\,dy\,dx = \int_0^1 \int_0^{1-x} (-4x-y)\,dy\,dx = \int_0^1 -\left[4x(1-x) + \frac{1}{2}(1-x)^2\right] dx$

$= -\int_0^1 \left(\frac{1}{2} + 3x - \frac{7}{2}x^2\right) dx = -\frac{5}{6}$

5. $\text{curl}\,\mathbf{F} = \nabla \times \mathbf{F} = \begin{vmatrix} \mathbf{i} & \mathbf{j} & \mathbf{k} \\ \frac{\partial}{\partial x} & \frac{\partial}{\partial y} & \frac{\partial}{\partial z} \\ y^2+z^2 & x^2+y^2 & x^2+y^2 \end{vmatrix} = 2y\mathbf{i} + (2z-2x)\mathbf{j} + (2x-2y)\mathbf{k}$ and $\mathbf{n} = \mathbf{k}$

$\Rightarrow \text{curl}\,\mathbf{F}\cdot\mathbf{n} = 2x - 2y \Rightarrow d\sigma = dx\,dy \Rightarrow \oint_C \mathbf{F}\cdot d\mathbf{r} = \int_{-1}^1 \int_{-1}^1 (2x-2y)\,dx\,dy = \int_{-1}^1 \left[x^2 - 2xy\right]_{-1}^1 dy$

$= \int_{-1}^1 -4y\,dy = 0$

7. $x = 3\cos t$ and $y = 2\sin t \Rightarrow \mathbf{F} = (2\sin t)\mathbf{i} + (9\cos^2 t)\mathbf{j} + (9\cos^2 t + 16\sin^4 t)\sin e^{\sqrt{(6\sin t\cos t)(0)}}\mathbf{k}$ at the base of the shell; $\mathbf{r} = (3\cos t)\mathbf{i} + (2\sin t)\mathbf{j} \Rightarrow d\mathbf{r} = (-3\sin t)\mathbf{i} + (2\cos t)\mathbf{j} \Rightarrow \mathbf{F}\cdot\frac{d\mathbf{r}}{dt} = -6\sin^2 t + 18\cos^3 t$

$\Rightarrow \iint_S \nabla \times \mathbf{F}\cdot\mathbf{n}\,d\sigma = \int_0^{2\pi} (-6\sin^2 t + 18\cos^3 t)\,dt = \left[-3t + \frac{3}{2}\sin 2t + 6(\sin t)(\cos^2 t + 2)\right]_0^{2\pi} = -6\pi$

9. Flux of $\nabla \times \mathbf{F} = \iint_S \nabla \times \mathbf{F}\cdot\mathbf{n}\,d\sigma = \oint_C \mathbf{F}\cdot d\mathbf{r}$, so let C be parametrized by $\mathbf{r} = (a\cos t)\mathbf{i} + (a\sin t)\mathbf{j}$, $0 \le t \le 2\pi \Rightarrow \frac{d\mathbf{r}}{dt} = (-a\sin t)\mathbf{i} + (a\cos t)\mathbf{j} \Rightarrow \mathbf{F}\cdot\frac{d\mathbf{r}}{dt} = ay\sin t + ax\cos t = a^2\sin^2 t + a^2\cos^2 t = a^2$

$\Rightarrow$ Flux of $\nabla \times \mathbf{F} = \oint_C \mathbf{F}\cdot d\mathbf{r} = \int_0^{2\pi} a^2\,dt = 2\pi a^2$

11. Let S_1 and S_2 be oriented surfaces that span C and that induce the same positive direction on C. Then $\iint_{S_1} \nabla \times \mathbf{F}\cdot\mathbf{n}_1\,d\sigma_1 = \oint_C \mathbf{F}\cdot d\mathbf{r} = \iint_{S_2} \nabla \times \mathbf{F}\cdot\mathbf{n}_2\,d\sigma_2$

13. $\nabla \times \mathbf{F} = \begin{vmatrix} \mathbf{i} & \mathbf{j} & \mathbf{k} \\ \frac{\partial}{\partial x} & \frac{\partial}{\partial y} & \frac{\partial}{\partial z} \\ 2z & 3x & 5y \end{vmatrix} = 5\mathbf{i} + 2\mathbf{j} + 3\mathbf{k}$; $\mathbf{r}_r = (\cos\theta)\mathbf{i} + (\sin\theta)\mathbf{j} - 2r\mathbf{k}$ and $\mathbf{r}_\theta = (-r\sin\theta)\mathbf{i} + (r\cos\theta)\mathbf{j}$

$\Rightarrow \mathbf{r}_r \times \mathbf{r}_\theta = \begin{vmatrix} \mathbf{i} & \mathbf{j} & \mathbf{k} \\ \cos\theta & \sin\theta & -2r \\ -r\sin\theta & r\cos\theta & 0 \end{vmatrix} = (2r^2\cos\theta)\mathbf{i} + (2r^2\sin\theta)\mathbf{j} + r\mathbf{k}$; $\mathbf{n} = \frac{\mathbf{r}_r \times \mathbf{r}_\theta}{|\mathbf{r}_r \times \mathbf{r}_\theta|}$ and $d\sigma = |\mathbf{r}_r \times \mathbf{r}_\theta|\,dr\,d\theta$

$\Rightarrow \nabla \times \mathbf{F}\cdot\mathbf{n}\,d\sigma = (\nabla \times \mathbf{F})\cdot(\mathbf{r}_r \times \mathbf{r}_\theta)\,dr\,d\theta = (10r^2\cos\theta + 4r^2\sin\theta + 3r)\,dr\,d\theta \Rightarrow \iint_S \nabla \times \mathbf{F}\cdot\mathbf{n}\,d\sigma$

$= \int_0^{2\pi} \int_0^2 (10r^2\cos\theta + 4r^2\sin\theta + 3r)\,dr\,d\theta = \int_0^{2\pi} \left[\frac{10}{3}r^3\cos\theta + \frac{4}{3}r^3\sin\theta + \frac{3}{2}r^2\right]_0^2 d\theta$

$= \int_0^{2\pi} \left(\frac{80}{3}\cos\theta + \frac{32}{3}\sin\theta + 6\right) d\theta = 6(2\pi) = 12\pi$

15. $\nabla\times\mathbf{F}=\begin{vmatrix}\mathbf{i}&\mathbf{j}&\mathbf{k}\\ \frac{\partial}{\partial x}&\frac{\partial}{\partial y}&\frac{\partial}{\partial z}\\ x^2y&2y^3z&3z\end{vmatrix}=-2y^3\mathbf{i}+0\mathbf{j}-x^2\mathbf{k}$; $\mathbf{r}_r\times\mathbf{r}_\theta=\begin{vmatrix}\mathbf{i}&\mathbf{j}&\mathbf{k}\\ \cos\theta&\sin\theta&1\\ -r\sin\theta&r\cos\theta&0\end{vmatrix}$

$=(-r\cos\theta)\mathbf{i}-(r\sin\theta)\mathbf{j}+r\mathbf{k}$ and $\nabla\times\mathbf{F}\cdot\mathbf{n}\,d\sigma=(\nabla\times\mathbf{F})\cdot(\mathbf{r}_r\times\mathbf{r}_\theta)\,dr\,d\theta$ (see Exercise 13 above)

$\Rightarrow\iint_S\nabla\times\mathbf{F}\cdot\mathbf{n}\,d\sigma=\iint_R(2ry^3\cos\theta-rx^2)\,dr\,d\theta=\int_0^{2\pi}\int_0^1(2r^4\sin^3\theta\cos\theta-r^3\cos^2\theta)\,dr\,d\theta$

$=\int_0^{2\pi}\left(\frac{2}{5}\sin^3\theta\cos\theta-\frac{1}{4}\cos^2\theta\right)d\theta=\left[\frac{1}{10}\sin^4\theta-\frac{1}{4}\left(\frac{\theta}{2}+\frac{\sin 2\theta}{4}\right)\right]_0^{2\pi}=-\frac{\pi}{4}$

17. $\nabla\times\mathbf{F}=\begin{vmatrix}\mathbf{i}&\mathbf{j}&\mathbf{k}\\ \frac{\partial}{\partial x}&\frac{\partial}{\partial y}&\frac{\partial}{\partial z}\\ 3y&5-2x&z^2-2\end{vmatrix}=0\mathbf{i}+0\mathbf{j}-5\mathbf{k}$;

$\mathbf{r}_\phi\times\mathbf{r}_\theta=\begin{vmatrix}\mathbf{i}&\mathbf{j}&\mathbf{k}\\ \sqrt{3}\cos\phi\cos\theta&\sqrt{3}\cos\phi\sin\theta&-\sqrt{3}\sin\phi\\ -\sqrt{3}\sin\phi\sin\theta&\sqrt{3}\sin\phi\cos\theta&0\end{vmatrix}$

$=(3\sin^2\phi\cos\theta)\,\mathbf{i}+(3\sin^2\phi\sin\theta)\,\mathbf{j}+(3\sin\phi\cos\phi)\mathbf{k}$; $\nabla\times\mathbf{F}\cdot\mathbf{n}\,d\sigma=(\nabla\times\mathbf{F})\cdot(\mathbf{r}_\phi\times\mathbf{r}_\theta)\,d\phi\,d\theta$ (see Exercise 13 above) $\Rightarrow\iint_S\nabla\times\mathbf{F}\cdot\mathbf{n}\,d\sigma=\int_0^{2\pi}\int_0^{\pi/2}-15\cos\phi\sin\phi\,d\phi\,d\theta=\int_0^{2\pi}\left[\frac{15}{2}\cos^2\phi\right]_0^{\pi/2}d\theta=\int_0^{2\pi}-\frac{15}{2}\,d\theta=-15\pi$

19. (a) $\mathbf{F}=2x\mathbf{i}+2y\mathbf{j}+2z\mathbf{k}\Rightarrow\text{curl }\mathbf{F}=\mathbf{0}\Rightarrow\oint_C\mathbf{F}\cdot d\mathbf{r}=\iint_S\nabla\times\mathbf{F}\cdot\mathbf{n}\,d\sigma=\iint_S 0\,d\sigma=0$

(b) Let $f(x,y,z)=x^2y^2z^3\Rightarrow\nabla\times\mathbf{F}=\nabla\times\nabla f=\mathbf{0}\Rightarrow\text{curl }\mathbf{F}=\mathbf{0}\Rightarrow\oint_C\mathbf{F}\cdot d\mathbf{r}=\iint_S\nabla\times\mathbf{F}\cdot\mathbf{n}\,d\sigma=\iint_S 0\,d\sigma$ $=0$

(c) $\mathbf{F}=\nabla\times(x\mathbf{i}+y\mathbf{j}+z\mathbf{k})=\mathbf{0}\Rightarrow\nabla\times\mathbf{F}=\mathbf{0}\Rightarrow\oint_C\mathbf{F}\cdot d\mathbf{r}=\iint_S\nabla\times\mathbf{F}\cdot\mathbf{n}\,d\sigma=\iint_S 0\,d\sigma=0$

(d) $\mathbf{F}=\nabla f\Rightarrow\nabla\times\mathbf{F}=\nabla\times\nabla f=\mathbf{0}\Rightarrow\oint_C\mathbf{F}\cdot d\mathbf{r}=\iint_S\nabla\times\mathbf{F}\cdot\mathbf{n}\,d\sigma=\iint_S 0\,d\sigma=0$

21. Let $\mathbf{F}=2y\mathbf{i}+3z\mathbf{j}-x\mathbf{k}\Rightarrow\nabla\times\mathbf{F}=\begin{vmatrix}\mathbf{i}&\mathbf{j}&\mathbf{k}\\ \frac{\partial}{\partial x}&\frac{\partial}{\partial y}&\frac{\partial}{\partial z}\\ 2y&3z&-x\end{vmatrix}=-3\mathbf{i}+\mathbf{j}-2\mathbf{k}$; $\mathbf{n}=\frac{2\mathbf{i}+2\mathbf{j}+\mathbf{k}}{3}\Rightarrow\nabla\times\mathbf{F}\cdot\mathbf{n}=-2$

$\Rightarrow\oint_C 2y\,dx+3z\,dy-x\,dz=\oint_C\mathbf{F}\cdot d\mathbf{r}=\iint_S\nabla\times\mathbf{F}\cdot\mathbf{n}\,d\sigma=\iint_S-2\,d\sigma=-2\iint_S d\sigma$, where $\iint_S d\sigma$ is the area of the region enclosed by C on the plane S: $2x+2y+z=2$

23. Suppose $\mathbf{F}=M\mathbf{i}+N\mathbf{j}+P\mathbf{k}$ exists such that $\nabla\times\mathbf{F}=\left(\frac{\partial P}{\partial y}-\frac{\partial N}{\partial z}\right)\mathbf{i}+\left(\frac{\partial M}{\partial z}-\frac{\partial P}{\partial x}\right)\mathbf{j}+\left(\frac{\partial N}{\partial x}-\frac{\partial M}{\partial y}\right)\mathbf{k}=x\mathbf{i}+y\mathbf{j}+z\mathbf{k}$. Then $\frac{\partial}{\partial x}\left(\frac{\partial P}{\partial y}-\frac{\partial N}{\partial z}\right)=\frac{\partial}{\partial x}(x)\Rightarrow\frac{\partial^2 P}{\partial x\partial y}-\frac{\partial^2 N}{\partial x\partial z}=1$. Likewise, $\frac{\partial}{\partial y}\left(\frac{\partial M}{\partial z}-\frac{\partial P}{\partial x}\right)=\frac{\partial}{\partial y}(y)\Rightarrow\frac{\partial^2 M}{\partial y\partial z}-\frac{\partial^2 P}{\partial y\partial x}=1$ and $\frac{\partial}{\partial z}\left(\frac{\partial N}{\partial x}-\frac{\partial M}{\partial y}\right)=\frac{\partial}{\partial z}(z)\Rightarrow\frac{\partial^2 N}{\partial z\partial x}-\frac{\partial^2 M}{\partial z\partial y}=1$. Summing the calculated equations $\Rightarrow\left(\frac{\partial^2 P}{\partial x\partial y}-\frac{\partial^2 P}{\partial y\partial x}\right)+\left(\frac{\partial^2 N}{\partial z\partial x}-\frac{\partial^2 N}{\partial x\partial z}\right)+\left(\frac{\partial^2 M}{\partial y\partial z}-\frac{\partial^2 M}{\partial z\partial y}\right)=3$ or $0=3$ (assuming the second mixed partials are equal). This result is a contradiction, so there is no field $\mathbf{F}$ such that curl $\mathbf{F}=x\mathbf{i}+y\mathbf{j}+z\mathbf{k}$.

25. $r=\sqrt{x^2+y^2}\Rightarrow r^4=(x^2+y^2)^2\Rightarrow\mathbf{F}=\nabla(r^4)=4x(x^2+y^2)\,\mathbf{i}+4y(x^2+y^2)\,\mathbf{j}=M\mathbf{i}+N\mathbf{j}$

$\Rightarrow\oint_C\nabla(r^4)\cdot\mathbf{n}\,ds=\oint_C\mathbf{F}\cdot\mathbf{n}\,ds=\oint_C M\,dy-N\,dx=\iint_R\left(\frac{\partial M}{\partial x}+\frac{\partial N}{\partial y}\right)dx\,dy$

$=\iint_R\left[4(x^2+y^2)+8x^2+4(x^2+y^2)+8y^2\right]dA=\iint_R 16(x^2+y^2)\,dA=16\iint_R x^2\,dA+16\iint_R y^2\,dA$

$=16I_y+16I_x$.

13.8 THE DIVERGENCE THEOREM AND A UNIFIED THEORY

1. $\mathbf{F} = \frac{-y\mathbf{i}+x\mathbf{j}}{\sqrt{x^2+y^2}} \Rightarrow \text{div } \mathbf{F} = \frac{xy-xy}{(x^2+y^2)^{3/2}} = 0$

3. $\mathbf{F} = -\frac{GM(x\mathbf{i}+y\mathbf{j}+z\mathbf{k})}{(x^2+y^2+z^2)^{3/2}} \Rightarrow \text{div } \mathbf{F} = -GM\left[\frac{(x^2+y^2+z^2)^{3/2}-3x^2(x^2+y^2+z^2)^{1/2}}{(x^2+y^2+z^2)^3}\right]$
$- GM\left[\frac{(x^2+y^2+z^2)^{3/2}-3y^2(x^2+y^2+z^2)^{1/2}}{(x^2+y^2+z^2)^3}\right] - GM\left[\frac{(x^2+y^2+z^2)^{3/2}-3z^2(x^2+y^2+z^2)^{1/2}}{(x^2+y^2+z^2)^3}\right]$
$= -GM\left[\frac{3(x^2+y^2+z^2)^2-3(x^2+y^2+z^2)(x^2+y^2+z^2)}{(x^2+y^2+z^2)^{7/2}}\right] = 0$

5. $\frac{\partial}{\partial x}(y-x) = -1,\ \frac{\partial}{\partial y}(z-y) = -1,\ \frac{\partial}{\partial z}(y-x) = 0 \Rightarrow \nabla \cdot \mathbf{F} = -2 \Rightarrow \text{Flux} = \int_{-1}^{1}\int_{-1}^{1}\int_{-1}^{1} -2\,dx\,dy\,dz = -2(2^3) = -16$

7. $\frac{\partial}{\partial x}(y) = 0,\ \frac{\partial}{\partial y}(xy) = x,\ \frac{\partial}{\partial z}(-z) = -1 \Rightarrow \nabla \cdot \mathbf{F} = x-1;\ z = x^2+y^2 \Rightarrow z = r^2$ in cylindrical coordinates
$\Rightarrow \text{Flux} = \iiint_D (x-1)\,dz\,dy\,dx = \int_0^{2\pi}\int_0^2\int_0^{r^2} (r\cos\theta - 1)\,dz\,r\,dr\,d\theta = \int_0^{2\pi}\int_0^2 (r^3\cos\theta - r^2)\,r\,dr\,d\theta$
$= \int_0^{2\pi}\left[\frac{r^5}{5}\cos\theta - \frac{r^4}{4}\right]_0^2 d\theta = \int_0^{2\pi}\left(\frac{32}{5}\cos\theta - 4\right)d\theta = \left[\frac{32}{5}\sin\theta - 4\theta\right]_0^{2\pi} = -8\pi$

9. $\frac{\partial}{\partial x}(x^2) = 2x,\ \frac{\partial}{\partial y}(-2xy) = -2x,\ \frac{\partial}{\partial z}(3xz) = 3x \Rightarrow \text{Flux} = \iiint_D 3x\,dx\,dy\,dz$
$= \int_0^{\pi/2}\int_0^{\pi/2}\int_0^2 (3\rho\sin\phi\cos\theta)(\rho^2\sin\phi)\,d\rho\,d\phi\,d\theta = \int_0^{\pi/2}\int_0^{\pi/2} 12\sin^2\phi\cos\theta\,d\phi\,d\theta = \int_0^{\pi/2} 3\pi\cos\theta\,d\theta = 3\pi$

11. $\frac{\partial}{\partial x}(2xz) = 2z,\ \frac{\partial}{\partial y}(-xy) = -x,\ \frac{\partial}{\partial z}(-z^2) = -2z \Rightarrow \nabla \cdot \mathbf{F} = -x \Rightarrow \text{Flux} = \iiint_D -x\,dV$
$= \int_0^2\int_0^{\sqrt{16-4x^2}}\int_0^{4-y} -x\,dz\,dy\,dx = \int_0^2\int_0^{\sqrt{16-4x^2}} (xy-4x)\,dy\,dx = \int_0^2\left[\frac{1}{2}x(16-4x^2) - 4x\sqrt{16-4x^2}\right]dx$
$= \left[4x^2 - \frac{1}{2}x^4 + \frac{1}{3}(16-4x^2)^{3/2}\right]_0^2 = -\frac{40}{3}$

13. Let $\rho = \sqrt{x^2+y^2+z^2}$. Then $\frac{\partial\rho}{\partial x} = \frac{x}{\rho},\ \frac{\partial\rho}{\partial y} = \frac{y}{\rho},\ \frac{\partial\rho}{\partial z} = \frac{z}{\rho} \Rightarrow \frac{\partial}{\partial x}(\rho x) = \left(\frac{\partial\rho}{\partial x}\right)x + \rho = \frac{x^2}{\rho} + \rho,\ \frac{\partial}{\partial y}(\rho y) = \left(\frac{\partial\rho}{\partial y}\right)y + \rho$
$= \frac{y^2}{\rho} + \rho,\ \frac{\partial}{\partial z}(\rho z) = \left(\frac{\partial\rho}{\partial z}\right)z + \rho = \frac{z^2}{\rho} + \rho \Rightarrow \nabla \cdot \mathbf{F} = \frac{x^2+y^2+z^2}{\rho} + 3\rho = 4\rho$, since $\rho = \sqrt{x^2+y^2+z^2}$
$\Rightarrow \text{Flux} = \iiint_D 4\rho\,dV = \int_0^{2\pi}\int_0^{\pi}\int_1^{\sqrt{2}} (4\rho)(\rho^2\sin\phi)\,d\rho\,d\phi\,d\theta = \int_0^{2\pi}\int_0^{\pi} 3\sin\phi\,d\phi\,d\theta = \int_0^{2\pi} 6\,d\theta = 12\pi$

15. $\frac{\partial}{\partial x}(5x^3+12xy^2) = 15x^2+12y^2,\ \frac{\partial}{\partial y}(y^3+e^y\sin z) = 3y^2+e^y\sin z,\ \frac{\partial}{\partial z}(5z^3+e^y\cos z) = 15z^2 - e^y\sin z$
$\Rightarrow \nabla \cdot \mathbf{F} = 15x^2+15y^2+15z^2 = 15\rho^2 \Rightarrow \text{Flux} = \iiint_D 15\rho^2\,dV = \int_0^{2\pi}\int_0^{\pi}\int_1^{\sqrt{2}} (15\rho^2)(\rho^2\sin\phi)\,d\rho\,d\phi\,d\theta$
$= \int_0^{2\pi}\int_0^{\pi}\left(12\sqrt{2}-3\right)\sin\phi\,d\phi\,d\theta = \int_0^{2\pi}\left(24\sqrt{2}-6\right)d\theta = \left(48\sqrt{2}-12\right)\pi$

17. (a) $\mathbf{G} = M\mathbf{i}+N\mathbf{j}+P\mathbf{k} \Rightarrow \nabla \times \mathbf{G} = \text{curl } \mathbf{G} = \left(\frac{\partial P}{\partial y} - \frac{\partial N}{\partial z}\right)\mathbf{i} + \left(\frac{\partial M}{\partial z} - \frac{\partial P}{\partial x}\right)\mathbf{k} + \left(\frac{\partial N}{\partial x} - \frac{\partial M}{\partial y}\right)\mathbf{k} \Rightarrow \nabla \cdot \nabla \times \mathbf{G}$
$= \text{div}(\text{curl } \mathbf{G}) = \frac{\partial}{\partial x}\left(\frac{\partial P}{\partial y} - \frac{\partial N}{\partial z}\right) + \frac{\partial}{\partial y}\left(\frac{\partial M}{\partial z} - \frac{\partial P}{\partial x}\right) + \frac{\partial}{\partial z}\left(\frac{\partial N}{\partial x} - \frac{\partial M}{\partial y}\right)$
$= \frac{\partial^2 P}{\partial x\partial y} - \frac{\partial^2 N}{\partial x\partial z} + \frac{\partial^2 M}{\partial y\partial z} - \frac{\partial^2 P}{\partial y\partial x} + \frac{\partial^2 N}{\partial z\partial x} - \frac{\partial^2 M}{\partial z\partial y} = 0$ if all first and second partial derivatives are continuous

(b) By the Divergence Theorem, the outward flux of $\nabla \times \mathbf{G}$ across a closed surface is zero because outward flux of $\nabla \times \mathbf{G} = \iint_S (\nabla \times \mathbf{G}) \cdot \mathbf{n}\, d\sigma$

$= \iiint_D \nabla \cdot \nabla \times \mathbf{G}\, dV$ [Divergence Theorem with $\mathbf{F} = \nabla \times \mathbf{G}$]

$= \iiint_D (0)\, dV = 0$ [by part (a)]

19. (a) $\operatorname{div}(g\mathbf{F}) = \nabla \cdot g\mathbf{F} = \frac{\partial}{\partial x}(gM) + \frac{\partial}{\partial y}(gN) + \frac{\partial}{\partial z}(gP) = \left(g\frac{\partial M}{\partial x} + M\frac{\partial g}{\partial x}\right) + \left(g\frac{\partial N}{\partial y} + N\frac{\partial g}{\partial y}\right) + \left(g\frac{\partial P}{\partial z} + P\frac{\partial g}{\partial z}\right)$

$= \left(M\frac{\partial g}{\partial x} + N\frac{\partial g}{\partial y} + P\frac{\partial g}{\partial z}\right) + g\left(\frac{\partial M}{\partial x} + \frac{\partial N}{\partial y} + \frac{\partial P}{\partial z}\right) = g\nabla \cdot \mathbf{F} + \nabla g \cdot \mathbf{F}$

(b) $\nabla \times (g\mathbf{F}) = \left[\frac{\partial}{\partial y}(gP) - \frac{\partial}{\partial z}(gN)\right]\mathbf{i} + \left[\frac{\partial}{\partial z}(gM) - \frac{\partial}{\partial x}(gP)\right]\mathbf{j} + \left[\frac{\partial}{\partial x}(gN) - \frac{\partial}{\partial y}(gM)\right]\mathbf{k}$

$= \left(P\frac{\partial g}{\partial y} + g\frac{\partial P}{\partial y} - N\frac{\partial g}{\partial z} - g\frac{\partial N}{\partial z}\right)\mathbf{i} + \left(M\frac{\partial g}{\partial z} + g\frac{\partial M}{\partial z} - P\frac{\partial g}{\partial x} - g\frac{\partial P}{\partial x}\right)\mathbf{j} + \left(N\frac{\partial g}{\partial x} + g\frac{\partial N}{\partial x} - M\frac{\partial g}{\partial y} - g\frac{\partial M}{\partial y}\right)\mathbf{k}$

$= \left(P\frac{\partial g}{\partial y} - N\frac{\partial g}{\partial z}\right)\mathbf{i} + \left(g\frac{\partial P}{\partial y} - g\frac{\partial N}{\partial z}\right)\mathbf{i} + \left(M\frac{\partial g}{\partial z} - P\frac{\partial g}{\partial x}\right)\mathbf{j} + \left(g\frac{\partial M}{\partial z} - g\frac{\partial P}{\partial x}\right)\mathbf{j} + \left(N\frac{\partial g}{\partial x} - M\frac{\partial g}{\partial y}\right)\mathbf{k}$

$+ \left(g\frac{\partial N}{\partial x} - g\frac{\partial M}{\partial y}\right)\mathbf{k} = g\nabla \times \mathbf{F} + \nabla g \times \mathbf{F}$

21. The integral's value never exceeds the surface area of S. Since $|\mathbf{F}| \le 1$, we have $|\mathbf{F} \cdot \mathbf{n}| = |\mathbf{F}|\,|\mathbf{n}| \le (1)(1) = 1$ and

$\iiint_D \nabla \cdot \mathbf{F}\, d\sigma = \iint_S \mathbf{F} \cdot \mathbf{n}\, d\sigma$ [Divergence Theorem]

$\le \iint_S |\mathbf{F} \cdot \mathbf{n}|\, d\sigma$ [A property of integrals]

$\le \iint_S (1)\, d\sigma$ [$|\mathbf{F} \cdot \mathbf{n}| \le 1$]

$=$ Area of S.

23. (a) $\frac{\partial}{\partial x}(x) = 1,\ \frac{\partial}{\partial y}(y) = 1,\ \frac{\partial}{\partial z}(z) = 1 \Rightarrow \nabla \cdot \mathbf{F} = 3 \Rightarrow \text{Flux} = \iiint_D 3\, dV = 3\iiint_D dV$

$= 3$(Volume of the solid)

(b) If $\mathbf{F}$ is orthogonal to $\mathbf{n}$ at every point of S, then $\mathbf{F} \cdot \mathbf{n} = 0$ everywhere $\Rightarrow$ Flux $= \iint_S \mathbf{F} \cdot \mathbf{n}\, d\sigma = 0$.

But the flux is 3(Volume of the solid) $\ne 0$, so $\mathbf{F}$ is not orthogonal to $\mathbf{n}$ at every point.

25. $\iint_S \mathbf{F} \cdot \mathbf{n}\, d\sigma = \iiint_D \nabla \cdot \mathbf{F}\, dV = \iiint_D 3\, dV \Rightarrow \frac{1}{3}\iint_S \mathbf{F} \cdot \mathbf{n}\, d\sigma = \iiint_D dV = \text{Volume of D}$

27. (a) From the Divergence Theorem, $\iint_S \nabla f \cdot \mathbf{n}\, d\sigma = \iiint_D \nabla \cdot \nabla f\, dV = \iiint_D \nabla^2 f\, dV = \iiint_D 0\, dV = 0$

(b) From the Divergence Theorem, $\iint_S f\nabla f \cdot \mathbf{n}\, d\sigma = \iiint_D \nabla \cdot f\nabla f\, dV$. Now,

$f\nabla f = \left(f\frac{\partial f}{\partial x}\right)\mathbf{i} + \left(f\frac{\partial f}{\partial y}\right)\mathbf{j} + \left(f\frac{\partial f}{\partial z}\right)\mathbf{k} \Rightarrow \nabla \cdot f\nabla f = \left[f\frac{\partial^2 f}{\partial x^2} + \left(\frac{\partial f}{\partial x}\right)^2\right] + \left[f\frac{\partial^2 f}{\partial y^2} + \left(\frac{\partial f}{\partial y}\right)^2\right] + \left[f\frac{\partial^2 f}{\partial z^2} + \left(\frac{\partial f}{\partial z}\right)^2\right]$

$= f\nabla^2 f + |\nabla f|^2 = 0 + |\nabla f|^2$ since f is harmonic $\Rightarrow \iint_S f\nabla f \cdot \mathbf{n}\, d\sigma = \iiint_D |\nabla f|^2\, dV$, as claimed.

29. $\iint_S f\nabla g \cdot \mathbf{n}\, d\sigma = \iiint_D \nabla \cdot f\nabla g\, dV = \iiint_D \nabla \cdot \left(f\frac{\partial g}{\partial x}\mathbf{i} + f\frac{\partial g}{\partial y}\mathbf{j} + f\frac{\partial g}{\partial z}\mathbf{k}\right) dV$

$= \iiint_D \left(f\frac{\partial^2 g}{\partial x^2} + \frac{\partial f}{\partial x}\frac{\partial g}{\partial x} + f\frac{\partial^2 g}{\partial y^2} + \frac{\partial f}{\partial y}\frac{\partial g}{\partial y} + f\frac{\partial^2 g}{\partial z^2} + \frac{\partial f}{\partial z}\frac{\partial g}{\partial z}\right) dV$

$= \iiint_D \left[f\left(\frac{\partial^2 g}{\partial x^2} + \frac{\partial^2 g}{\partial y^2} + \frac{\partial^2 g}{\partial z^2}\right) + \left(\frac{\partial f}{\partial x}\frac{\partial g}{\partial x} + \frac{\partial f}{\partial y}\frac{\partial g}{\partial y} + \frac{\partial f}{\partial z}\frac{\partial g}{\partial z}\right)\right] dV = \iiint_D (f\nabla^2 g + \nabla f \cdot \nabla g)\, dV$

CHAPTER 13 PRACTICE AND ADDITIONAL EXERCISES

1. Path 1: $\mathbf{r} = t\mathbf{i} + t\mathbf{j} + t\mathbf{k} \Rightarrow x = t, y = t, z = t, 0 \le t \le 1 \Rightarrow f(g(t), h(t), k(t)) = 3 - 3t^2$ and $\frac{dx}{dt} = 1, \frac{dy}{dt} = 1,$
$\frac{dz}{dt} = 1 \Rightarrow \sqrt{\left(\frac{dx}{dt}\right)^2 + \left(\frac{dy}{dt}\right)^2 + \left(\frac{dz}{dt}\right)^2}\, dt = \sqrt{3}\, dt \Rightarrow \int_C f(x, y, z)\, ds = \int_0^1 \sqrt{3}\,(3 - 3t^2)\, dt = 2\sqrt{3}$
Path 2: $\mathbf{r}_1 = t\mathbf{i} + t\mathbf{j}, 0 \le t \le 1 \Rightarrow x = t, y = t, z = 0 \Rightarrow f(g(t), h(t), k(t)) = 2t - 3t^2 + 3$ and $\frac{dx}{dt} = 1, \frac{dy}{dt} = 1,$
$\frac{dz}{dt} = 0 \Rightarrow \sqrt{\left(\frac{dx}{dt}\right)^2 + \left(\frac{dy}{dt}\right)^2 + \left(\frac{dz}{dt}\right)^2}\, dt = \sqrt{2}\, dt \Rightarrow \int_{C_1} f(x, y, z)\, ds = \int_0^1 \sqrt{2}\,(2t - 3t^2 + 3)\, dt = 3\sqrt{2};$
$\mathbf{r}_2 = \mathbf{i} + \mathbf{j} + t\mathbf{k} \Rightarrow x = 1, y = 1, z = t \Rightarrow f(g(t), h(t), k(t)) = 2 - 2t$ and $\frac{dx}{dt} = 0, \frac{dy}{dt} = 0, \frac{dz}{dt} = 1$
$\Rightarrow \sqrt{\left(\frac{dx}{dt}\right)^2 + \left(\frac{dy}{dt}\right)^2 + \left(\frac{dz}{dt}\right)^2}\, dt = dt \Rightarrow \int_{C_2} f(x, y, z)\, ds = \int_0^1 (2 - 2t)\, dt = 1$
$\Rightarrow \int_C f(x, y, z)\, ds = \int_{C_1} f(x, y, z)\, ds + \int_{C_2} f(x, y, z) = 3\sqrt{2} + 1$

3. $\mathbf{r} = (a \cos t)\mathbf{j} + (a \sin t)\mathbf{k} \Rightarrow x = 0, y = a \cos t, z = a \sin t \Rightarrow f(g(t), h(t), k(t)) = \sqrt{a^2 \sin^2 t} = a\,|\sin t|$ and
$\frac{dx}{dt} = 0, \frac{dy}{dt} = -a \sin t, \frac{dz}{dt} = a \cos t \Rightarrow \sqrt{\left(\frac{dx}{dt}\right)^2 + \left(\frac{dy}{dt}\right)^2 + \left(\frac{dz}{dt}\right)^2}\, dt = a\, dt$
$\Rightarrow \int_C f(x, y, z)\, ds = \int_0^{2\pi} a^2\, |\sin t|\, dt = \int_0^{\pi} a^2 \sin t\, dt + \int_{\pi}^{2\pi} -a^2 \sin t\, dt = 4a^2$

5. $\frac{\partial P}{\partial y} = -\frac{1}{2}(x + y + z)^{-3/2} = \frac{\partial N}{\partial z}, \frac{\partial M}{\partial z} = -\frac{1}{2}(x + y + z)^{-3/2} = \frac{\partial P}{\partial x}, \frac{\partial N}{\partial x} = -\frac{1}{2}(x + y + z)^{-3/2} = \frac{\partial M}{\partial y}$
$\Rightarrow M\, dx + N\, dy + P\, dz$ is exact; $\frac{\partial f}{\partial x} = \frac{1}{\sqrt{x+y+z}} \Rightarrow f(x, y, z) = 2\sqrt{x + y + z} + g(y, z) \Rightarrow \frac{\partial f}{\partial y} = \frac{1}{\sqrt{x+y+z}} + \frac{\partial g}{\partial y}$
$= \frac{1}{\sqrt{x+y+z}} \Rightarrow \frac{\partial g}{\partial y} = 0 \Rightarrow g(y, z) = h(z) \Rightarrow f(x, y, z) = 2\sqrt{x + y + z} + h(z) \Rightarrow \frac{\partial f}{\partial z} = \frac{1}{\sqrt{x+y+z}} + h'(z)$
$= \frac{1}{\sqrt{x+y+z}} \Rightarrow h'(x) = 0 \Rightarrow h(z) = C \Rightarrow f(x, y, z) = 2\sqrt{x + y + z} + C \Rightarrow \int_{(-1,1,1)}^{(4,-3,0)} \frac{dx + dy + dz}{\sqrt{x+y+z}}$
$= f(4, -3, 0) - f(-1, 1, 1) = 2\sqrt{1} - 2\sqrt{1} = 0$

7. $\frac{\partial M}{\partial z} = -y \cos z \ne y \cos z = \frac{\partial P}{\partial x} \Rightarrow \mathbf{F}$ is not conservative; $\mathbf{r} = (2 \cos t)\mathbf{i} + (2 \sin t)\mathbf{j} - \mathbf{k}, 0 \le t \le 2\pi$
$\Rightarrow d\mathbf{r} = (-2 \sin t)\mathbf{i} - (2 \cos t)\mathbf{j} \Rightarrow \int_C \mathbf{F} \cdot d\mathbf{r} = \int_0^{2\pi} [-(-2 \sin t)(\sin(-1))(-2 \sin t) + (2 \cos t)(\sin(-1))(-2 \cos t)]\, dt$
$= 4 \sin(1) \int_0^{2\pi} (\sin^2 t + \cos^2 t) dt = 8\pi \sin(1)$

9. Let $M = 8x \sin y$ and $N = -8y \cos x \Rightarrow \frac{\partial M}{\partial y} = 8x \cos y$ and $\frac{\partial N}{\partial x} = 8y \sin x \Rightarrow \int_C 8x \sin y\, dx - 8y \cos x\, dy$
$= \iint_R (8y \sin x - 8x \cos y)\, dy\, dx = \int_0^{\pi/2} \int_0^{\pi/2} (8y \sin x - 8x \cos y)\, dy\, dx = \int_0^{\pi/2} (\pi^2 \sin x - 8x)\, dx = -\pi^2 + \pi^2 = 0$

11. Let $z = 1 - x - y \Rightarrow f_x(x, y) = -1$ and $f_y(x, y) = -1 \Rightarrow \sqrt{f_x^2 + f_y^2 + 1} = \sqrt{3} \Rightarrow$ Surface Area $= \iint_R \sqrt{3}\, dx\, dy$
$= \sqrt{3}$(Area of the circular region in the xy-plane) $= \pi\sqrt{3}$

13. $\nabla f = 2x\mathbf{i} + 2y\mathbf{j} + 2z\mathbf{k}, \mathbf{p} = \mathbf{k} \Rightarrow |\nabla f| = \sqrt{4x^2 + 4y^2 + 4z^2} = 2\sqrt{x^2 + y^2 + z^2} = 2$ and $|\nabla f \cdot \mathbf{p}| = |2z| = 2z$ since
$z \ge 0 \Rightarrow$ Surface Area $= \iint_R \frac{2}{2z}\, dA = \iint_R \frac{1}{z}\, dA = \iint_R \frac{1}{\sqrt{1 - x^2 - y^2}}\, dx\, dy = \int_0^{2\pi} \int_0^{1/\sqrt{2}} \frac{1}{\sqrt{1 - r^2}}\, r\, dr\, d\theta$
$\int_0^{2\pi} \left[-\sqrt{1 - r^2}\right]_0^{1/\sqrt{2}} d\theta = \int_0^{2\pi} \left(1 - \frac{1}{\sqrt{2}}\right) d\theta = 2\pi\left(1 - \frac{1}{\sqrt{2}}\right)$

15. $f(x,y,z)=\frac{x}{a}+\frac{y}{b}+\frac{z}{c}=1 \Rightarrow \nabla f=\left(\frac{1}{a}\right)\mathbf{i}+\left(\frac{1}{b}\right)\mathbf{j}+\left(\frac{1}{c}\right)\mathbf{k} \Rightarrow |\nabla f|=\sqrt{\frac{1}{a^2}+\frac{1}{b^2}+\frac{1}{c^2}}$ and $\mathbf{p}=\mathbf{k} \Rightarrow |\nabla f\cdot\mathbf{p}|=\frac{1}{c}$ since $c>0 \Rightarrow$ Surface Area $=\iint\limits_R \frac{\sqrt{\frac{1}{a^2}+\frac{1}{b^2}+\frac{1}{c^2}}}{\left(\frac{1}{c}\right)}\,dA=c\sqrt{\frac{1}{a^2}+\frac{1}{b^2}+\frac{1}{c^2}}\iint\limits_R dA=\frac{1}{2}abc\sqrt{\frac{1}{a^2}+\frac{1}{b^2}+\frac{1}{c^2}}$, since the area of the triangular region R is $\frac{1}{2}ab$. To check this result, let $\mathbf{v}=a\mathbf{i}+c\mathbf{k}$ and $\mathbf{w}=-a\mathbf{i}+b\mathbf{j}$; the area can be found by computing $\frac{1}{2}|\mathbf{v}\times\mathbf{w}|$.

17. $\nabla f=2y\mathbf{j}+2z\mathbf{k},\ \mathbf{p}=\mathbf{k} \Rightarrow |\nabla f|=\sqrt{4y^2+4z^2}=2\sqrt{y^2+z^2}=10$ and $|\nabla f\cdot\mathbf{p}|=2z$ since $z\ge 0$ $\Rightarrow d\sigma=\frac{10}{2z}\,dx\,dy=\frac{5}{z}\,dx\,dy=\iint\limits_S g(x,y,z)\,d\sigma=\iint\limits_R (x^4y)(y^2+z^2)\left(\frac{5}{z}\right)dx\,dy$ $=\iint\limits_R (x^4y)(25)\left(\frac{5}{\sqrt{25-y^2}}\right)dx\,dy=\int_0^4\int_0^1 \frac{125y}{\sqrt{25-y^2}}\,x^4\,dx\,dy=\int_0^4 \frac{25y}{\sqrt{25-y^2}}\,dy=50$

19. A possible parametrization is $\mathbf{r}(\phi,\theta)=(6\sin\phi\cos\theta)\mathbf{i}+(6\sin\phi\sin\theta)\mathbf{j}+(6\cos\phi)\mathbf{k}$ (spherical coordinates); now $\rho=6$ and $z=-3 \Rightarrow -3=6\cos\phi \Rightarrow \cos\phi=-\frac{1}{2} \Rightarrow \phi=\frac{2\pi}{3}$ and $z=3\sqrt{3} \Rightarrow 3\sqrt{3}=6\cos\phi$ $\Rightarrow \cos\phi=\frac{\sqrt{3}}{2} \Rightarrow \phi=\frac{\pi}{6} \Rightarrow \frac{\pi}{6}\le\phi\le\frac{2\pi}{3}$; also $0\le\theta\le 2\pi$

21. A possible parametrization is $\mathbf{r}(r,\theta)=(r\cos\theta)\mathbf{i}+(r\sin\theta)\mathbf{j}+(1+r)\mathbf{k}$ (cylindrical coordinates); now $r=\sqrt{x^2+y^2} \Rightarrow z=1+r$ and $1\le z\le 3 \Rightarrow 1\le 1+r\le 3 \Rightarrow 0\le r\le 2$; also $0\le\theta\le 2\pi$

23. Let $x=u\cos v$ and $z=u\sin v$, where $u=\sqrt{x^2+z^2}$ and v is the angle in the xz-plane with the x-axis $\Rightarrow \mathbf{r}(u,v)=(u\cos v)\mathbf{i}+2u^2\mathbf{j}+(u\sin v)\mathbf{k}$ is a possible parametrization; $0\le y\le 2 \Rightarrow 2u^2\le 2 \Rightarrow u^2\le 1$ $\Rightarrow 0\le u\le 1$ since $u\ge 0$; also, for just the upper half of the paraboloid, $0\le v\le\pi$

25. $\mathbf{r}_u=\mathbf{i}+\mathbf{j},\ \mathbf{r}_v=\mathbf{i}-\mathbf{j}+\mathbf{k} \Rightarrow \mathbf{r}_u\times\mathbf{r}_v=\begin{vmatrix}\mathbf{i}&\mathbf{j}&\mathbf{k}\\1&1&0\\1&-1&1\end{vmatrix}=\mathbf{i}-\mathbf{j}-2\mathbf{k} \Rightarrow |\mathbf{r}_u\times\mathbf{r}_v|=\sqrt{6}$

$\Rightarrow$ Surface Area $=\iint\limits_{R_{uv}} |\mathbf{r}_u\times\mathbf{r}_v|\,du\,dv=\int_0^1\int_0^1 \sqrt{6}\,du\,dv=\sqrt{6}$

27. $\mathbf{r}_r=(\cos\theta)\mathbf{i}+(\sin\theta)\mathbf{j},\ \mathbf{r}_\theta=(-r\sin\theta)\mathbf{i}+(r\cos\theta)\mathbf{j}+\mathbf{k} \Rightarrow \mathbf{r}_r\times\mathbf{r}_\theta=\begin{vmatrix}\mathbf{i}&\mathbf{j}&\mathbf{k}\\\cos\theta&\sin\theta&0\\-r\sin\theta&r\cos\theta&1\end{vmatrix}$

$=(\sin\theta)\mathbf{i}-(\cos\theta)\mathbf{j}+r\mathbf{k} \Rightarrow |\mathbf{r}_r\times\mathbf{r}_\theta|=\sqrt{\sin^2\theta+\cos^2\theta+r^2}=\sqrt{1+r^2} \Rightarrow$ Surface Area $=\iint\limits_{R_{r\theta}} |\mathbf{r}_r\times\mathbf{r}_\theta|\,dr\,d\theta$

$=\int_0^{2\pi}\int_0^1 \sqrt{1+r^2}\,dr\,d\theta=\int_0^{2\pi}\left[\frac{r}{2}\sqrt{1+r^2}+\frac{1}{2}\ln\left(r+\sqrt{1+r^2}\right)\right]_0^1 d\theta=\int_0^{2\pi}\left[\frac{1}{2}\sqrt{2}+\frac{1}{2}\ln\left(1+\sqrt{2}\right)\right]d\theta$

$=\pi\left[\sqrt{2}+\ln\left(1+\sqrt{2}\right)\right]$

29. $\frac{\partial P}{\partial y}=0=\frac{\partial N}{\partial z},\ \frac{\partial M}{\partial z}=0=\frac{\partial P}{\partial x},\ \frac{\partial N}{\partial x}=0=\frac{\partial M}{\partial y} \Rightarrow$ Conservative

31. $\frac{\partial P}{\partial y}=0\ne ye^z=\frac{\partial N}{\partial z} \Rightarrow$ Not Conservative

33. $\frac{\partial f}{\partial x}=2 \Rightarrow f(x,y,z)=2x+g(y,z) \Rightarrow \frac{\partial f}{\partial y}=\frac{\partial g}{\partial y}=2y+z \Rightarrow g(y,z)=y^2+zy+h(z)$ $\Rightarrow f(x,y,z)=2x+y^2+zy+h(z) \Rightarrow \frac{\partial f}{\partial z}=y+h'(z)=y+1 \Rightarrow h'(z)=1 \Rightarrow h(z)=z+C$ $\Rightarrow f(x,y,z)=2x+y^2+zy+z+C$

35. Over Path 1: $\mathbf{r} = t\mathbf{i} + t\mathbf{j} + t\mathbf{k},\ 0 \le t \le 1 \Rightarrow x = t,\ y = t,\ z = t$ and $d\mathbf{r} = (\mathbf{i} + \mathbf{j} + \mathbf{k})\,dt \Rightarrow \mathbf{F} = 2t^2\mathbf{i} + \mathbf{j} + t^2\mathbf{k}$
$\Rightarrow \mathbf{F} \cdot d\mathbf{r} = (3t^2 + 1)\,dt \Rightarrow \text{Work} = \int_0^1 (3t^2 + 1)\,dt = 2;$
Over Path 2: $\mathbf{r}_1 = t\mathbf{i} + t\mathbf{j},\ 0 \le t \le 1 \Rightarrow x = t,\ y = t,\ z = 0$ and $d\mathbf{r}_1 = (\mathbf{i} + \mathbf{j})\,dt \Rightarrow \mathbf{F}_1 = 2t^2\mathbf{i} + \mathbf{j} + t^2\mathbf{k}$
$\Rightarrow \mathbf{F}_1 \cdot d\mathbf{r}_1 = (2t^2 + 1)\,dt \Rightarrow \text{Work}_1 = \int_0^1 (2t^2 + 1)\,dt = \frac{5}{3};\ \mathbf{r}_2 = \mathbf{i} + \mathbf{j} + t\mathbf{k},\ 0 \le t \le 1 \Rightarrow x = 1,\ y = 1,\ z = t$ and
$d\mathbf{r}_2 = \mathbf{k}\,dt \Rightarrow \mathbf{F}_2 = 2\mathbf{i} + \mathbf{j} + \mathbf{k} \Rightarrow \mathbf{F}_2 \cdot d\mathbf{r}_2 = dt \Rightarrow \text{Work}_2 = \int_0^1 dt = 1 \Rightarrow \text{Work} = \text{Work}_1 + \text{Work}_2 = \frac{5}{3} + 1 = \frac{8}{3}$

37. (a) $\mathbf{r} = (e^t \cos t)\mathbf{i} + (e^t \sin t)\mathbf{j} \Rightarrow x = e^t \cos t,\ y = e^t \sin t$ from $(1, 0)$ to $(e^{2\pi}, 0) \Rightarrow 0 \le t \le 2\pi$
$\Rightarrow \frac{d\mathbf{r}}{dt} = (e^t \cos t - e^t \sin t)\,\mathbf{i} + (e^t \sin t + e^t \cos t)\,\mathbf{j}$ and $\mathbf{F} = \frac{x\mathbf{i} + y\mathbf{j}}{(x^2 + y^2)^{3/2}} = \frac{(e^t \cos t)\mathbf{i} + (e^t \sin t)\mathbf{j}}{(e^{2t}\cos^2 t + e^{2t}\sin^2 t)^{3/2}}$
$= \left(\frac{\cos t}{e^{2t}}\right)\mathbf{i} + \left(\frac{\sin t}{e^{2t}}\right)\mathbf{j} \Rightarrow \mathbf{F} \cdot \frac{d\mathbf{r}}{dt} = \left(\frac{\cos^2 t}{e^t} - \frac{\sin t \cos t}{e^t} + \frac{\sin^2 t}{e^t} + \frac{\sin t \cos t}{e^t}\right) = e^{-t}$
$\Rightarrow \text{Work} = \int_0^{2\pi} e^{-t}\,dt = 1 - e^{-2\pi}$
(b) $\mathbf{F} = \frac{x\mathbf{i} + y\mathbf{j}}{(x^2 + y^2)^{3/2}} \Rightarrow \frac{\partial f}{\partial x} = \frac{x}{(x^2 + y^2)^{3/2}} \Rightarrow f(x, y, z) = -(x^2 + y^2)^{-1/2} + g(y, z) \Rightarrow \frac{\partial f}{\partial y} = \frac{y}{(x^2 + y^2)^{3/2}} + \frac{\partial g}{\partial y}$
$= \frac{y}{(x^2 + y^2)^{3/2}} \Rightarrow g(y, z) = C \Rightarrow f(x, y, z) = -(x^2 + y^2)^{-1/2}$ is a potential function for $\mathbf{F} \Rightarrow \int_C \mathbf{F} \cdot d\mathbf{r}$
$= f(e^{2\pi}, 0) - f(1, 0) = 1 - e^{-2\pi}$

39. $\nabla \times \mathbf{F} = \begin{vmatrix} \mathbf{i} & \mathbf{j} & \mathbf{k} \\ \frac{\partial}{\partial x} & \frac{\partial}{\partial y} & \frac{\partial}{\partial z} \\ y^2 & -y & 3z^2 \end{vmatrix} = -2y\mathbf{k}$; unit normal to the plane is $\mathbf{n} = \frac{2\mathbf{i} + 6\mathbf{j} - 3\mathbf{k}}{\sqrt{4 + 36 + 9}} = \frac{2}{7}\mathbf{i} + \frac{6}{7}\mathbf{j} - \frac{3}{7}\mathbf{k}$
$\Rightarrow \nabla \times \mathbf{F} \cdot \mathbf{n} = \frac{6}{7}y;\ \mathbf{p} = \mathbf{k}$ and $f(x, y, z) = 2x + 6y - 3z \Rightarrow |\nabla f \cdot \mathbf{p}| = 3 \Rightarrow d\sigma = \frac{|\nabla f|}{|\nabla f \cdot \mathbf{p}|}\,dA = \frac{7}{3}\,dA$
$\Rightarrow \oint_C \mathbf{F} \cdot d\mathbf{r} = \iint_R \frac{6}{7}y\,d\sigma = \iint_R \left(\frac{6}{7}y\right)\left(\frac{7}{3}\,dA\right) = \iint_R 2y\,dA = \int_0^{2\pi}\int_0^1 2r \sin\theta\,r\,dr\,d\theta = \int_0^{2\pi} \frac{2}{3}\sin\theta\,d\theta = 0$

41. $\mathbf{r} = t\mathbf{i} + \left(\frac{2\sqrt{2}}{3}t^{3/2}\right)\mathbf{j} + \left(\frac{t^2}{2}\right)\mathbf{k},\ 0 \le t \le 2 \Rightarrow x = t,\ y = \frac{2\sqrt{2}}{3}t^{3/2},\ z = \frac{t^2}{2} \Rightarrow \frac{dx}{dt} = 1,\ \frac{dy}{dt} = \sqrt{2}\,t^{1/2},\ \frac{dz}{dt} = t$
$\Rightarrow \sqrt{\left(\frac{dx}{dt}\right)^2 + \left(\frac{dy}{dt}\right)^2 + \left(\frac{dz}{dt}\right)^2}\,dt = \sqrt{1 + 2t + t^2}\,dt = \sqrt{(t + 1)^2}\,dt = |t + 1|\,dt = (t + 1)\,dt$ on the domain given.
Then $M = \int_C \delta\,ds = \int_0^2 \left(\frac{1}{t+1}\right)(t + 1)\,dt = \int_0^2 dt = 2;\ M_{yz} = \int_C x\delta\,ds = \int_0^2 t\left(\frac{1}{t+1}\right)(t + 1)\,dt = \int_0^2 t\,dt = 2;$
$M_{xz} = \int_C y\delta\,ds = \int_0^2 \left(\frac{2\sqrt{2}}{3}t^{3/2}\right)\left(\frac{1}{t+1}\right)(t + 1)\,dt = \int_0^2 \frac{2\sqrt{2}}{3}t^{3/2}\,dt = \frac{32}{15};\ M_{xy} = \int_C z\delta\,ds$
$= \int_0^2 \left(\frac{t^2}{2}\right)\left(\frac{1}{t+1}\right)(t + 1)\,dt = \int_0^2 \frac{t^2}{2}\,dt = \frac{4}{3} \Rightarrow \bar{x} = \frac{M_{yz}}{M} = \frac{2}{2} = 1;\ \bar{y} = \frac{M_{xz}}{M} = \frac{\left(\frac{32}{15}\right)}{2} = \frac{16}{15};\ \bar{z} = \frac{M_{xy}}{M}$
$= \frac{\left(\frac{4}{3}\right)}{2} = \frac{2}{3};\ I_x = \int_C (y^2 + z^2)\,\delta\,ds = \int_0^2 \left(\frac{8}{9}t^3 + \frac{t^4}{4}\right)dt = \frac{232}{45};\ I_y = \int_C (x^2 + z^2)\,\delta\,ds = \int_0^2 \left(t^2 + \frac{t^4}{4}\right)dt = \frac{64}{15};$
$I_z = \int_C (y^2 + x^2)\,\delta\,ds = \int_0^2 \left(t^2 + \frac{8}{9}t^3\right)dt = \frac{56}{9}$

43. $\mathbf{r}(t) = (e^t \cos t)\mathbf{i} + (e^t \sin t)\mathbf{j} + e^t\mathbf{k},\ 0 \le t \le \ln 2 \Rightarrow x = e^t \cos t,\ y = e^t \sin t,\ z = e^t \Rightarrow \frac{dx}{dt} = (e^t \cos t - e^t \sin t),$
$\frac{dy}{dt} = (e^t \sin t + e^t \cos t),\ \frac{dz}{dt} = e^t \Rightarrow \sqrt{\left(\frac{dx}{dt}\right)^2 + \left(\frac{dy}{dt}\right)^2 + \left(\frac{dz}{dt}\right)^2}\,dt$
$= \sqrt{(e^t \cos t - e^t \sin t)^2 + (e^t \sin t + e^t \cos t)^2 + (e^t)^2}\,dt = \sqrt{3e^{2t}}\,dt = \sqrt{3}\,e^t\,dt;\ M = \int_C \delta\,ds = \int_0^{\ln 2} \sqrt{3}\,e^t\,dt$
$= \sqrt{3};\ M_{xy} = \int_C z\delta\,ds = \int_0^{\ln 2} \left(\sqrt{3}\,e^t\right)(e^t)\,dt = \int_0^{\ln 2} \sqrt{3}\,e^{2t}\,dt = \frac{3\sqrt{3}}{2} \Rightarrow \bar{z} = \frac{M_{xy}}{M} = \frac{\left(\frac{3\sqrt{3}}{2}\right)}{\sqrt{3}} = \frac{3}{2};$
$I_z = \int_C (x^2 + y^2)\,\delta\,ds = \int_0^{\ln 2} (e^{2t}\cos^2 t + e^{2t}\sin^2 t)\left(\sqrt{3}\,e^t\right)dt = \int_0^{\ln 2} \sqrt{3}\,e^{3t}\,dt = \frac{7\sqrt{3}}{3}$

45. $M = 2xy + x$ and $N = xy - y \Rightarrow \frac{\partial M}{\partial x} = 2y + 1,\ \frac{\partial M}{\partial y} = 2x,\ \frac{\partial N}{\partial x} = y,\ \frac{\partial N}{\partial y} = x - 1 \Rightarrow \text{Flux} = \iint_R \left(\frac{\partial M}{\partial x} + \frac{\partial N}{\partial y}\right) dx\,dy$
$= \iint_R (2y + 1 + x - 1)\,dy\,dx = \int_0^1 \int_0^1 (2y + x)\,dy\,dx = \frac{3}{2}$; $\text{Circ} = \iint_R \left(\frac{\partial N}{\partial x} - \frac{\partial M}{\partial y}\right) dx\,dy$
$= \iint_R (y - 2x)\,dy\,dx = \int_0^1 \int_0^1 (y - 2x)\,dy\,dx = -\frac{1}{2}$

47. $\frac{\partial}{\partial x}(2xy) = 2y,\ \frac{\partial}{\partial y}(2yz) = 2z,\ \frac{\partial}{\partial z}(2xz) = 2x \Rightarrow \nabla \cdot \mathbf{F} = 2y + 2z + 2x \Rightarrow \text{Flux} = \iiint_D (2x + 2y + 2z)\,dV$
$= \int_0^1 \int_0^1 \int_0^1 (2x + 2y + 2z)\,dx\,dy\,dz = \int_0^1 \int_0^1 (1 + 2y + 2z)\,dy\,dz = \int_0^1 (2 + 2z)\,dz = 3$

49. $\mathbf{F} = y\mathbf{i} + z\mathbf{j} + x\mathbf{k} \Rightarrow \nabla \cdot \mathbf{F} = 0 \Rightarrow \text{Flux} = \iint_S \mathbf{F} \cdot \mathbf{n}\,d\sigma = \iiint_D \nabla \cdot \mathbf{F}\,dV = 0$